Electrodynamics and Special Theory of Relativity

Reiner M. Dreizler • Cora S. Lüdde

Electrodynamics and Special Theory of Relativity

Reiner M. Dreizler
Institut f. Theoretische Physik
Goethe Universität Frankfurt
Frankfurt/Main, Hessen, Germany

Cora S. Lüdde
Institut f. Theoretische Physik
Goethe Universität Frankfurt
Frankfurt/Main, Hessen, Germany

ISBN 978-3-662-69941-6   ISBN 978-3-662-69942-3 (eBook)
https://doi.org/10.1007/978-3-662-69942-3

Translation of the German language edition: "Theoretische Physik 2" by Reiner M. Dreizler and Cora S. Lüdde, © Springer-Verlag Berlin Heidelberg 2005. Published by Springer Berlin Heidelberg. All Rights Reserved.

© The Editor(s) (if applicable) and The Author(s), under exclusive license to Springer-Verlag GmbH, DE, part of Springer Nature 2025

This work is subject to copyright. All rights are solely and exclusively licensed by the Publisher, whether the whole or part of the material is concerned, specifically the rights of translation, reprinting, reuse of illustrations, recitation, broadcasting, reproduction on microfilms or in any other physical way, and transmission or information storage and retrieval, electronic adaptation, computer software, or by similar or dissimilar methodology now known or hereafter developed.
The use of general descriptive names, registered names, trademarks, service marks, etc. in this publication does not imply, even in the absence of a specific statement, that such names are exempt from the relevant protective laws and regulations and therefore free for general use.
The publisher, the authors and the editors are safe to assume that the advice and information in this book are believed to be true and accurate at the date of publication. Neither the publisher nor the authors or the editors give a warranty, expressed or implied, with respect to the material contained herein or for any errors or omissions that may have been made. The publisher remains neutral with regard to jurisdictional claims in published maps and institutional affiliations.

This Springer imprint is published by the registered company Springer-Verlag GmbH, DE, part of Springer Nature.
The registered company address is: Heidelberger Platz 3, 14197 Berlin, Germany

If disposing of this product, please recycle the paper.

Preface

The historical development of electrodynamics and of the special theory of relativity is interconnected. The formulation of the theory of relativity is a consequence of the fact that electromagnetism is not compatible with the principle of relativity formulated on the basis of classical mechanics. The central equations of electrodynamics, the Maxwell equations, are not form invariant under Galilei transformations. This would imply that different results should be observed for electromagnetic phenomena in different inertial systems. Experiments, in particular those by Michelson and Morley, show that this is not the case.

The material presented here follows the historical development of the theory of electromagnetism from its beginnings until its completion with the theory of relativity. Each of the chapters contains two parts. The first part retraces the development of the theory with the topics:

- Basic concepts and equations of electrostatics (Dreizler and Lüdde, 2024, Electrostatics and Magnetostatics (Springer Berlin Heidelberg), Chaps. 1 and 2)
- Calculation of stationary electric fields for given distributions of charges and materials. Practical aspects: electrical circuits and response of materials to stationary electric fields (Dreizler and Lüdde, 2024, Electrostatics and Magnetostatics (Springer Berlin Heidelberg), Chaps. 3 and 4)
- Generation of stationary magnetic fields by stationary electric currents, response of materials to magnetic fields: para- and ferro-magnetism—on the basis of classical models (Dreizler and Lüdde, 2024, Electrostatics and Magnetostatics (Springer Berlin Heidelberg), Chap. 5)
- Law of induction, Maxwell equations, the wave solutions of the inhomogeneous equations (transmitters), the wave solutions of the homogeneous equations (propagation), energy and momentum of electromagnetic fields (Chap. 1)
- Additional applications: transformers, crystal optics, metal optics, wave guides, diffraction, electromagnetic radiation by antennas, bremsstrahlung, Čerenkov radiation (Chap. 2)
- Michelson-Morley experiments, simple form of Lorentz transformation, consequences of the Lorentz transformation: addition theorems (velocity and accelerations), length contraction, time dilatation, equal time problem, linear spaces with non-Euclidean metric, Minkowski space, relativistic formulation of mechanics and electrodynamics, historical remarks (Chap. 3)

The second part tries to expand and amplify the coverage of this material by additional questions and suggestions under the label *Details*. In order to cope with the length of the material, we have split this book into two separate parts:

- Part 1: Stationary Electric and Magnetic Problems (in: Dreizler and Lüdde, 2024, Electrostatics and Magnetostatics (Springer Berlin Heidelberg), ISBN: 978-3-662-69932-4)
- Part 2: Dynamic Problems and the Special Theory of Relativity (in: Dreizler and Lüdde, 2024, Electrodynamics and Special Theory of Relativity (Springer Berlin Heidelberg), ISBN: 978-3-662-69941-6)

It is necessary to add some comments concerning the different systems of units, which are used for electric and magnetic quantities. The two commonly used units in electrodynamics are extensions of two systems of mechanics, namely the SI-system and the CGS (centimeter gram second)-system. These extensions are:

- The rationalised MKSA (meter, kilogram, second, Ampère)-system, also called the SI-system. The letter A stands for the unit of electrical currents 'Ampère'. This system is usually preferred for technical applications.
- The Gauss CGS-system, which is usually just called the Gauss system. The unit of the electric current in this system, the 'statamp', can be directly expressed in terms the mechanical CGS units. This system is the usual choice for theoretical considerations.
- Other systems of units, which are used, are:
  The electrostatic system of units (esu)
  The electromagnetic system of units (emu)
  The Heaviside-Lorentz system of units

Five constant factors are needed to handle the equations involved in electrodynamics and to accommodate the different systems of units. These constants,

$$k_\mu \longrightarrow k_e, k_d, k_m, k_f \,,$$

are associated with specific physical quantities. They are discussed and justified in Appendix B. Additional lists of the units of various physical quantities are reproduced in Appendix B.2. In the main text all basic equations are given in a form employing these constants, so that they are valid for all systems of units. The exception is Chap. 2, where only the units of the Gauss system are used.

Mathematical topics as, e.g., special functions of mathematical physics, the theory of distributions, Green's functions, etc. are interspersed in the main text. A short compilation of mathematical material for the subjects,

- Equations of vector analysis,
- Angular functions,

- Radial functions,
- Linear spaces with non-Euclidean metric,

is also presented in the appendix of both parts as a support for working with the relations of electrodynamics.

Frankfurt/Main, Germany　　　　　　　　　　　　　　　　　　　Reiner M. Dreizler
November 2023　　　　　　　　　　　　　　　　　　　　　　　　　Cora S. Lüdde

It is with a heavy heart that I must add an addendum to this foreword. Unfortunately, Prof. Dr Dreizler passed away shortly before this work was completed. He was a very patient teacher who always endeavoured to impart knowledge without placing himself at the centre. The collaboration with him was always peaceful, but also characterised by many fruitful discussions. It was an honour to work with this generous and tolerant person and I thank him posthumously for this.

I would also like to thank my family and friends for their patience, understanding and support during this project. Last but not least, I would also like to express my thanks to the team of Springer Verlag, Heidelberg, for the friendly cooperation and technical support.

Frankfurt/Main, Germany　　　　　　　　　　　　　　　　　　　　　　　Cora S. Lüdde
April 2024

Contents

1 Electrodynamics: Foundations 1
  1.1 The Law of Induction 2
      1.1.1 The Law of Faraday 2
      1.1.2 Self- and Mutual Induction 6
  1.2 The Maxwell Equations 10
  1.3 Electromagnetic Waves 15
      1.3.1 Wave Equations 15
      1.3.2 Wave Solutions of the Maxwell Equations 24
      1.3.3 Properties of Electromagnetic Waves 27
  1.4 Energy and Momentum of the Electromagnetic Field 35
      1.4.1 Energy Transport by Electromagnetic Waves 35
      1.4.2 Momentum Conservation 39
  1.5 Electromagnetic Potentials 42
  1.6 The Solution of the Inhomogeneous Wave Equation 45
  1.7 Details 50
      1.7.1 The Transformation of Faraday's Law 50
      1.7.2 Some Properties of Plane Electromagnetic Waves 52
      1.7.3 Illustration of the Dispersion of Wave Packets 54
      1.7.4 The Temporal Mean Value of the Poynting Vector 60
      1.7.5 The Elements of the Maxwell Stress Tensor 62

2 Electrodynamics: Applications 65
  2.1 Technical Implementation of Induction 65
      2.1.1 The AC Generator 66
      2.1.2 The Transformer 66
  2.2 Wave Propagation 68
      2.2.1 Reflection and Refraction in Crystal Optics 69
      2.2.2 Wave Propagation in Metals 79
      2.2.3 Wave Guides and Wire Waves 83
      2.2.4 Diffraction 91
  2.3 Generation of Waves: Transmitters 101
      2.3.1 Specification of Transmitters 101
      2.3.2 The Hertz Dipole 104
      2.3.3 Contributions of Higher Multipoles 108

		2.3.4	The Complete Multipole Expansion	112
		2.3.5	An Exactly Solvable Transmitter Problem	115
	2.4	Generation of Waves: The Radiation of Moving Point Charges		119
		2.4.1	The Liénard-Wiechert Potentials	120
		2.4.2	The Classical Bremsstrahlung	125
		2.4.3	Remarks on the Čerenkov Radiation	130
	2.5	Details		133
		2.5.1	The Real and the Ideal Transformer	133
		2.5.2	Refraction: Fresnel Formulae and Energy Flow	140
		2.5.3	The Telegraph Equation	146
		2.5.4	Diffraction	147
		2.5.5	The Hertz Dipole	157
		2.5.6	Multipole Radiation (Higher Multipole Moments)	160
		2.5.7	The Complete Multipole Expansion	164
		2.5.8	The Centrally Fed, Thin Antenna	168
		2.5.9	Calculation of the Liénard-Wiechert Fields	176
		2.5.10	The Bremsstrahlung	187
3	**Theory of Relativity and Electromagnetism**			**195**
	3.1	The Lorentz Transformation		197
		3.1.1	The Michelson-Morley Experiment	198
		3.1.2	A Simple Derivation of Alternative Transformation Equations	201
	3.2	Implications of the Lorentz Transformation		205
		3.2.1	The Addition Theorem for Velocities	205
		3.2.2	The Lorentz Contraction	210
		3.2.3	The Time Dilatation	213
	3.3	The Minkowski Space		216
		3.3.1	Definition	216
		3.3.2	The Lorentz Transformation in the $\mathcal{M}(1, 1)$-world	219
		3.3.3	Formal Aspects 1: Co- and Contravariant Coordinates	222
		3.3.4	Formal Variant 2: Imaginary Time Coordinate	229
	3.4	Relativistic Mechanics		233
		3.4.1	The Four-Velocity	234
		3.4.2	The Four-Momentum and the Relativistic Energy	236
		3.4.3	The Equations of Motion in Relativistic Mechanics	241
	3.5	Electrodynamics and the Theory of Relativity		245
		3.5.1	The Potential Equations	245
		3.5.2	The Field Equations	247
		3.5.3	The Relativistic Forces on Charges	253
		3.5.4	The Lagrange Equations	256
	3.6	A Short History of the Aether and Other Historical Remarks		260
	3.7	Details		263
		3.7.1	The Aberration	263
		3.7.2	The Simple Lorentz Transformation	265

Contents   xi

	3.7.3	A More General Form of the Lorentz Transformation	267
	3.7.4	The Maxwell Equations in Relativistic Notation	277
	3.7.5	The Energy-Momentum Tensor	281
	3.7.6	The Relativistic Lagrange/Hamilton Formalism	288

A Literature 293
   A.1 Books and Literature Quoted in the Text 293
   A.2 Introductory Texts 294
   A.3 Electrodynamics 294
   A.4 Special Relativity 295
   A.5 Mathematics 295
   A.6 Special Functions and Handbooks 295

B Systems of Units in Electrodynamics 297
   B.1 The Systems 297
   B.2 Tables 304

C Additional Mathematical Topics 309
   C.1 Equations of Vector Analysis 309
   C.2 Multiple Products of Vectors 309
   C.3 Product Rules for the Application of the ∇-Operator 310
   C.4 Double Application of ∇ 310
   C.5 Differential Operators in Spherical and Cylinder Coordinates 311
        C.5.1 Spherical Coordinates 311
        C.5.2 Cylinder Coordinates 311
   C.6 Angular Functions 312
        C.6.1 Legendre Polynomials $P_l(x)$ 312
        C.6.2 The Functions $Q_l(x)$ 314
        C.6.3 Associated Legendre Functions 315
        C.6.4 Spherical Harmonics 316
   C.7 Radial Functions 318
        C.7.1 The Hypergeometric Functions $F(a,b,c;x)$ 318
        C.7.2 The Confluent Hypergeometric Functions $F(a,c;x)$ 320
   C.8 Linear Spaces with Non-Euclidean Metric 322

Index 327

Electrodynamics: Foundations

Electric and magnetic fields are coupled in dynamical (time dependent) situations. The fact, that a magnetic field, which changes in time, produces a time dependent electric field, which induces a magnetic field changing with time, is incorporated in Maxwell's equations. These equations separate in a stationary situation, which is characterised by

$$\frac{\partial \boldsymbol{B}(\boldsymbol{r},t)}{\partial t} = \boldsymbol{0} \quad \text{and} \quad \frac{\partial \boldsymbol{E}(\boldsymbol{r},t)}{\partial t} = \boldsymbol{0},$$

into the set of differential equations for stationary fields, which have been discussed in the first chapters. The treatment of the Maxwell equations starts with the discussion of the experimental basis of electrodynamics, the law of induction (Chap. 1.1). One of the variants of this law corresponds to one of the four Maxwell equations, which will be assembled in Chap. 1.2, on the basis of the original argumentation of Maxwell. The physical content of these equations is presented in two sections. The homogeneous or free Maxwell equations describe the situation in domains of space without any sources as charges or currents changing with time. They are wave equations with solutions, which describe the propagation of electromagnetic waves (Chap. 1.3). The solution of the inhomogeneous or complete Maxwell equations deals with questions concerning the generation of electromagnetic waves by charges and currents changing with time, the topic of senders (transmitters). An introduction to this topic is discussed in Chap. 1.4, in the form of the energy and momentum aspects of electrodynamics: Electrodynamic fields transport energy and momentum. An economical formulation of the transmission problem calls for the introduction of time dependent electromagnetic potentials (Chap. 1.5), which are a way to access the solution of the inhomogeneous wave equation (Chap. 1.6). Applications of electrodynamics as optical problems, the mode of operation of transformers and others are found in Chap. 2.

1.1 The Law of Induction

The law of induction, which have been investigated by Michael Faraday starting in 1836, have contributed considerably to the foundation of electrodynamics. Electrodynamics proper has been developed by James Clark Maxwell after some work on the results of Faraday.[1]

The law of induction is explained in this section, beginning with the experiments on the generation of a time dependent magnetic field by a time dependent electric field. An essential role towards an abstract interpretation of the experiments was the concept of fields. One particular aspect, which accompanied the induction, was the feedback on the electric field by the generated magnetic field, which can be observed by the self-induction in current loops or in coils.

1.1.1 The Law of Faraday

The basic experiments on induction can be summarised in the following fashion: A current loop is placed into a magnetic field. One observes that a time dependent current is induced in the loop, if the magnetic field changes in time

$$\boldsymbol{B}(t) \quad \longrightarrow \quad i_{\text{loop}}(t) \,.$$

The change of the magnetic field can be effected by either moving a permanent magnet or by changing the current producing the magnetic field. Possible variants are a motion of the loop in an inhomogeneous magnetic field or a rotation of the loop in a homogeneous magnetic field or a change of the form of the loop in time. One can observe, that all variants produce an electric current in the loop.

The variety of possibilities can be combined by considering the magnetic flux through the loop

$$\phi_B(t) = \iint_F \boldsymbol{B}(\boldsymbol{r}, t) \cdot \mathbf{d} \boldsymbol{f} \,.$$

The magnetic flux is changed in each of the cases, either by varying the field $\boldsymbol{B}$, the surface of the loop F or the angle between the field $\boldsymbol{B}$ and the infinitesimal surface element. The result of the experiments can therefore be written in the form

$$i_{\text{ind}}(t) = -k' \frac{\mathrm{d}}{\mathrm{d}t} \phi_B(t) \qquad (k' > 0) \,. \tag{1.1}$$

---

[1] 1865 with the contribution entitled 'Dynamical Theory of the Electromagnetic Field' and as a summary of the completed theory in 1873 with 'Treatise on Electricity and Magnetism'.

1.1 The Law of Induction

Fig. 1.1 Basic experiment

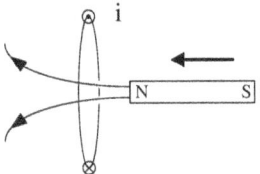

The sign on the right hand side corresponds to the rule of Lenz, which says: The currents generated in the process of induction are directed in such a way, that their magnetic field would counteract the change of the inducing field. This rule is a consequence of energy conservation. For example: If the north pole of a permanent magnet is moved in the direction of the loop, the change of the magnetic field points (as in Fig. 1.1) to the left. The magnetic field of the induced current $\boldsymbol{B}_{\text{ind}}$ should point to the right, so that the current flows in the direction indicated in Fig. 1.1.

The magnet would be pulled into the loop at the slightest nudge, if the induced field $\boldsymbol{B}_{\text{ind}}$ would be a vector pointing in the opposite direction. It would be possible to gain mechanical energy without any cost.

For the discussion of the constant k' one can use Ohm's law (Dreizler and Lüdde, 2024, Electrostatics and Magnetostatics (Springer Berlin Heidelberg), Chap. 5.1)

$$U = R \cdot i$$

in order to make the transition from the induced current to the induced voltage

$$U_{\text{ind}}(t) = R\, i_{\text{ind}}(t) = -(k'R)\frac{d}{dt}\phi_B(t)$$

or

$$U_{\text{ind}}(t) = -k_{\text{ind}}\frac{d}{dt}\phi_B(t) \qquad \text{(Faraday's law)}. \tag{1.2}$$

This is the form of Faraday's law, which is used for practical applications. The induced voltage could be measured by opening the loop and applying a voltage meter. The constant k_{ind} can be identified after a brief consideration of the dimensions involved. In the case of the SI-system one finds because of

$$[U] = \left[\frac{\text{kg} \cdot \text{m}^2}{\text{C} \cdot \text{s}^2}\right] \quad \text{and} \quad [\phi_B] = \left[\frac{\text{kg} \cdot \text{m}^2}{\text{C} \cdot \text{s}}\right]$$

the constant

$$k_{\text{ind, SI}} = 1 \,.$$

For the CGS-system one obtains with

$$[U] = \left[\frac{g^{1/2} \cdot cm^{1/2}}{s}\right] \quad \text{and} \quad [\phi_B] = \left[\frac{g^{1/2} \cdot cm^{3/2}}{s}\right]$$

for the units of the constant

$$[k_{\text{ind, CGS}}] = s/cm \ .$$

The constant has the value $k_{\text{ind, CGS}} = 1/c$, which agrees with the constant k_{ind}, which has been introduced in Dreizler and Lüdde, 2024, Electrostatics and Magnetostatics (Springer Berlin Heidelberg), Chap. 5.2.3

$$k_{\text{ind}} \equiv k_f \ .$$

The practical form of the law of Faraday is the basis of electrical engineering. Some applications will be addressed in the following chapters (Chaps. 2.1.1 and 2.1.2).

A form of the Faraday law, which is more adapted to the theoretical aspects can be found, if the induced voltage is represented by a line integral with the electric field

$$U_{\text{ind}}(t) = \oint_K \boldsymbol{E}(\boldsymbol{r}, t) \cdot \mathbf{dr} \ .$$

This equation expresses the fact, that the electric field is not vortex free. It generates the transport of charges, which leads to the induced current. The theoretical variant of the law of induction (1.2) is

$$\oint_K \boldsymbol{E}(\boldsymbol{r}, t) \cdot \mathbf{dr} = -k_f \frac{d}{dt} \iint_{F(K)} \boldsymbol{B}(\boldsymbol{r}, t) \cdot \mathbf{df} \ . \tag{1.3}$$

The line integral should be evaluated along the border K of the surface $F(K)$, over which the flux is changed. The orientation of the vectors $\mathbf{dr}$ and $\mathbf{df}$ are matched in the usual fashion (right hand rule, Fig. 1.2). In order to discuss the variant (1.3) of the Faraday law further, it is useful to abstract from this still practically oriented form the following idea: One argues, that the induced vortex field is created even without the wires, in which the flow of the current can be observed. The integral has to be evaluated over an arbitrary surface with a **fixed** border K. The right hand

Fig. 1.2 The surface $F(K)$

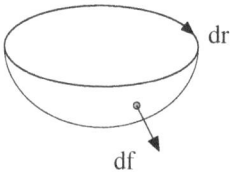

1.1 The Law of Induction

side of this equation can be reformulated, so that the time derivative acts only on the B-field

$$\oint_K E(r,t) \cdot dr = -k_f \iint_{F(K)} \left(\frac{d}{dt} B(r,t)\right) \cdot df .$$

The flux can be changed by changing the magnetic induction (e.g. moving magnet) or by changing the position of the loop.

It is easy to show (see Detail 1.7.1) that the result of the total time derivative of flux through the moving circuit is

$$\iint_{F(K)} \left(\frac{d}{dt} B(r,t)\right) \cdot df = \iint_{F(K)} \frac{\partial}{\partial t} B(r,t) \cdot df$$

$$+ \iint_{F(K)} \nabla \times (B \times v) \cdot df .$$

The curve integral can be transformed with the theorem of Stokes

$$\oint_K E(r,t) \cdot dr = \iint_{F(K)} \nabla \times E(r,t) \cdot df$$

and one finally arrives with

$$\iint_{F(K)} \nabla \times E(r,t) \cdot df = -k_f \iint_{F(K)} \frac{\partial}{\partial t} B(r,t) \cdot df$$

$$+ k_f \iint_{F(K)} \nabla \times (v \times B) \cdot df$$

at the differential form of Faraday's law

$$\nabla \times E(r,t) - k_f (\nabla \times (v \times B)) = -k_f \frac{\partial}{\partial t} B(r,t) . \tag{1.4}$$

If the loop is at rest and the flux is changed because of a changing magnetic induction the equation can be reduced to

$$\nabla \times E(r,t) = -k_f \frac{\partial}{\partial t} B(r,t)$$

This equation stands for the statement, that an electric vortex field is generated by a time dependent B-field. The stationary limit of this equation is

$$\frac{\partial B}{\partial t} = 0 \quad \longrightarrow \quad \nabla \times E(r) = 0 .$$

Fig. 1.3 Self-induction

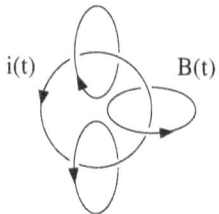

The crucial question, whether the abstraction is correct, has yet to be tested. The simple induction experiments with current loops are consistent with this interpretation, but do not prove the abstraction fully. The correctness of the abstraction is finally demonstrated by the success of the Maxwell equations. Before the detailed discussion of the complete set of the basic equations of electrodynamics, one should, on the basis of the practical form of the Faraday law, look more closely at some remarks concerning the concept of self-induction.

1.1.2 Self- and Mutual Induction

A circular current i produces a magnetic field $\boldsymbol{B}$. The flux of this field is present over the surface of the ring (Fig. 1.3). If the current changes with time, so does the magnetic field and the magnetic field will, due to induction, act back on the current in the ring. This is a verbal description of the effect, which is referred to as **induction**. For a quantitative description, one calculates the magnetic flux through the area of the ring by using the stationary Biot-Savart formula (Dreizler and Lüdde, 2024, Electrostatics and Magnetostatics (Springer Berlin Heidelberg), Eq. (5.21))

$$\boldsymbol{B}(\boldsymbol{r}) = \operatorname{rot} \boldsymbol{A}(\boldsymbol{r}) = \nabla \times \boldsymbol{A}(\boldsymbol{r})$$

for the magnetic field of the ring and integrate over the (thin) ring or, more generally, over a loop K

$$\begin{aligned}\phi_B(t) &= \iint_F \boldsymbol{B}(\boldsymbol{r}) \cdot \mathbf{d}\boldsymbol{f} \\ &= k_m \iint_F \oint_K \frac{\mathbf{d}\boldsymbol{r}' \times (\boldsymbol{r}-\boldsymbol{r}')}{|\boldsymbol{r}-\boldsymbol{r}'|^3} \cdot \mathbf{d}\boldsymbol{f}\, i(t) = \gamma\, i(t)\,.\end{aligned}$$

The constant γ could be calculated, it depends only on the geometry of the loop. If the loop is embedded in a magnetisable material, one has to replace the flux ϕ_B by $(\mu\, \phi_B)$, in case of the simplest material equation.

1.1 The Law of Induction

A simpler representation of the situation can be obtained with the representation of the magnetic field by the vector potential.[2] Application of the theorem of Stokes allows then to write

$$\phi_B(t) = \iint_F \boldsymbol{B}(\boldsymbol{r},t) \cdot \mathbf{d}\boldsymbol{f} = \iint_F (\nabla \times \boldsymbol{A}(\boldsymbol{r},t)) \cdot \mathbf{d}\boldsymbol{f} = \oint_K \boldsymbol{A}(\boldsymbol{r},t) \cdot \mathbf{d}\boldsymbol{r}$$

and with the Biot-Savart formula (Dreizler and Lüdde, 2024, Electrostatics and Magnetostatics (Springer Berlin Heidelberg), Eq. (5.29))

$$\boldsymbol{A}(\boldsymbol{r}) = \left[\frac{i\,k_m}{2} \oint (\boldsymbol{r}' \times \mathbf{d}\boldsymbol{r}')\right] \times \frac{\boldsymbol{r}}{r^3} + \dots$$

for the vector potential

$$\phi_B(t) = k_m \oint_K \oint_{K'} \frac{\mathbf{d}\boldsymbol{r} \cdot \mathbf{d}\boldsymbol{r}'}{|\boldsymbol{r}-\boldsymbol{r}'|}\, i(t) = \gamma\, i(t)\,.$$

If this relation is inserted into Faraday law (1.2), one finds

$$U_{\text{ind}}(t) = -k_f \frac{d\phi_B(t)}{dt} = -\gamma k_f \frac{d}{dt} i(t) = -L \frac{d}{dt} i(t)\,. \tag{1.5}$$

The constant quantity L, which is given by

$$L = k_f k_m \iint_F \oint_K \frac{\mathbf{d}\boldsymbol{r}' \times (\boldsymbol{r}-\boldsymbol{r}')}{|\boldsymbol{r}-\boldsymbol{r}'|^3} \cdot \mathbf{d}\boldsymbol{f} = k_f k_m \oint_K \oint_{K'} \frac{\mathbf{d}\boldsymbol{r} \cdot \mathbf{d}\boldsymbol{r}'}{|\boldsymbol{r}-\boldsymbol{r}'|} \tag{1.6}$$

is the **coefficient of self-induction**, or shortened as the self-induction. The unit in the CGS-system is

$$[L]_{\text{CGS}} = \frac{\text{statvolt}}{\text{statamp/s}} = \frac{\text{s}^2}{\text{cm}}\,.$$

A corresponding statement in the SI-system is

$$[L]_{\text{SI}} = \frac{\text{V}}{\text{A/s}} = 1\ \text{Henry}\,.$$

The factor for the conversion (in good approximation) is

$$[L]_{\text{SI}} = \frac{1}{9} \cdot 10^{11}\, [L]_{\text{CGS}}\,.$$

---

[2] Chapter 1.4 demonstrates that the dynamical extension of the stationary relation $\boldsymbol{B}(\boldsymbol{r}) = \nabla \times \boldsymbol{A}(\boldsymbol{r})$ is the equation $\boldsymbol{B}(\boldsymbol{r},t) = \nabla \times \boldsymbol{A}(\boldsymbol{r},t)$.

The currents $i_1(t), i_2(t), \ldots$ in a system of n current loops generate in the k-th loop a flux

$$\phi_k(t) = \iint_{F(k)} \sum_{l=1}^{n} \left[\boldsymbol{B}_l(\boldsymbol{r}_l, t) \cdot \mathbf{d}\boldsymbol{f}_k\right] = \oint_{K(k)} \sum_{l=1}^{n} \left[\boldsymbol{A}_l(\boldsymbol{r}_l, t) \cdot \mathbf{d}\boldsymbol{r}_k\right].$$

Each current loop is submitted to the flux of the other current loops. If the Biot-Savart form for thin loops is used again, one has for the flux through the k-th loop

$$\phi_k(t) = \sum_{l=1}^{n} k_m \oint_{K(l)} \oint_{K(k)} \frac{\mathbf{d}\boldsymbol{r}_l \cdot \mathbf{d}\boldsymbol{r}_k}{|\boldsymbol{r}_l - \boldsymbol{r}_k|} \, i_l(t),$$

so that the voltage induced in the k-th loop is

$$U_k(t) = -\sum_{l=1}^{n} L_{kl} \frac{di_l(t)}{dt}. \tag{1.7}$$

The quantities L_{kk} are the extensions of the self-induction discussed above. The quantities L_{kl} ($l \neq k$) are the **mutual or counter induction coefficients**. They are given by (under the assumptions of simple material equations in the total space)

$$L_{kl} = \mu k_f k_m \oint_{K(l)} \oint_{K(k)} \frac{\mathbf{d}\boldsymbol{r}_l \cdot \mathbf{d}\boldsymbol{r}_k}{|\boldsymbol{r}_l - \boldsymbol{r}_k|} = L_{lk}. \tag{1.8}$$

The calculation of the mutual induction does not provide any difficulties, as the denominators in the formula are not singular, if the loops are well separated. The calculation of the self-induction needs some care due to the singularity present. If e.g. the approximation for the magnitude of a uniform field inside of a solenoid (see Dreizler and Lüdde, 2024, Electrostatics and Magnetostatics (Springer Berlin Heidelberg), Eq. (5.20)) is used (with N windings, length l, cross section F and a magnetised core)

$$\boldsymbol{B} = \frac{4\pi \mu k_m N}{l} i(t),$$

one obtains for the self-induction with

$$\phi(t) = B(t) F \quad \text{and} \quad U_{\text{ind}}(t) = -N k_f \dot{\phi}(t)$$

the result

$$L = \frac{4\pi \mu F}{l} k_f k_m N^2. \tag{1.9}$$

1.1 The Law of Induction

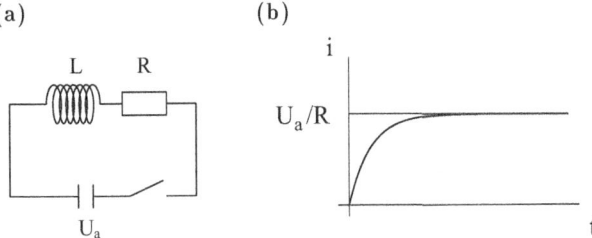

Fig. 1.4 Current at the start of a $L-R$ circuit. (**a**) The circuit. (**b**) The time dependence of the current

The self-induction leads to time lag effects, if the current circuits are switched on or off. For a circuit with a coil (L) and a resistance (R) (Fig. 1.4a) the statement is

$$U_W = R\,i = U_a + U_{\text{ind}}\,.$$

The voltage U_W across the resistance is given by the voltage applied and the voltage across the coil. This leads to the differential equation for this simple circuit

$$L\frac{di(t)}{dt} + R i(t) = U_a(t)\,. \tag{1.10}$$

The solution of this linear differential equation with constant coefficients, $U_a = \text{const.}$ (for a well charged battery) and the initial condition $i(t = 0) = 0$ is

$$i(t) = \frac{U_a}{R}\left(1 - e^{-(R/L)t}\right)\,. \tag{1.11}$$

The time lag effect, which is governed by the ratio U_a/R, is apparent. The current U_a/R is only reached some time after the circuit is switched on (Fig. 1.4b). The limit $R \to 0$ is also of interest, as it implies

$$i(t) = \frac{U_a}{R}\left(1 - 1 + \frac{R}{L}t + O(R^2)\right) \xrightarrow{R \to 0} \frac{U_a}{L}t\,.$$

The current grows linearly (until the battery is exhausted or the gadget is destroyed), stated differently: Never connect an electric gadget without internal resistance to a voltage source.

1.2 The Maxwell Equations

It is useful to look again at the assembly of the field equations for the time independent case. One can then think about the modifications for the time dependent case. The following equations have been discussed for the stationary aspects of the electromagnetic world:

(1) The Coulomb law in Gauss form

$$\text{div}\,\mathbf{D}(\mathbf{r}) = 4\pi k_d \, \rho_{tr}(\mathbf{r}) \qquad \nabla \cdot \mathbf{D}(\mathbf{r}) = 4\pi k_d \, \rho_{tr}(\mathbf{r}). \qquad (1.12)$$

The dielectric displacement $\mathbf{D}$ is created by the true charges.

(2) The Ampère law

$$\text{rot}\,\mathbf{H}(\mathbf{r}) = 4\pi k_h \, \mathbf{j}_{tr}(\mathbf{r}) \qquad \nabla \times \mathbf{H}(\mathbf{r}) = 4\pi k_h \, \mathbf{j}_{tr}(\mathbf{r}). \qquad (1.13)$$

The magnetic field strength $\mathbf{H}$ is created by the true (stationary) currents.

(3)

$$\text{rot}\,\mathbf{E}(\mathbf{r}) = \mathbf{0} \qquad \nabla \times \mathbf{E}(\mathbf{r}) = \mathbf{0}. \qquad (1.14)$$

The stationary electric field $\mathbf{E}$ is vortex free. It can be represented by a scalar potential.

(4)

$$\text{div}\,\mathbf{B}(\mathbf{r}) = \mathbf{0} \qquad \nabla \cdot \mathbf{B}(\mathbf{r}) = \mathbf{0}. \qquad (1.15)$$

The magnetic induction $\mathbf{B}$ does not have any sources: Magnetic charges do not exist.

Besides the equations for the fields an additional set of equations needs to be discussed, which connects the true macroscopic fields $\mathbf{E}$ and $\mathbf{B}$ with the subsidiary fields $\mathbf{D}$ and $\mathbf{H}$

$$\mathbf{D}(\mathbf{r}) = \frac{k_d}{k_e} \mathbf{E}(\mathbf{r}) + 4\pi k_d \, \mathbf{P}(\mathbf{r})$$

$$\mathbf{B}(\mathbf{r}) = \frac{k_m}{k_h} \mathbf{H}(\mathbf{r}) + 4\pi \frac{k_m}{k_f} \mathbf{M}(\mathbf{r}) \,.$$

These relations can be used, if the polarisation $\mathbf{P}$ and the magnetisation $\mathbf{M}$ (that is the response of the material) has been calculated with a simple or with a realistic model. Usually these microscopic relations are replaced by empirical material equations (in a simple form)

$$\mathbf{D}(\mathbf{r}) = \varepsilon \frac{k_d}{k_e} \mathbf{E}(\mathbf{r}) \qquad \mathbf{B}(\mathbf{r}) = \mu \frac{k_m}{k_h} \mathbf{H}(\mathbf{r}) \,.$$

1.2 The Maxwell Equations

Additional statements are e.g. the differential form of Ohm's law, that an electric field can initiate a current flow (in conductors or in a plasma)

$$j_{tr}(r) = \sigma E(r) \, .$$

In stationary situations electric and magnetic effects are only coupled via the current flow

$$\nabla \times H(r) = 4\pi k_h \sigma E(r) \, . \tag{1.16}$$

This equation is obtained, if Ohm's law is inserted into the law of Ampère. An additional coupling of magnetic and electric fields occurs in dynamical situations: A time dependent magnetic field creates an electric field. This is expressed by the law of induction

$$\nabla \times E(r,t) = -k_f \frac{\partial}{\partial t} B(r,t) \, .$$

The law of induction replaces the statement concerning the vortex free character of the stationary electric field (see (1.14) above).

The question, that needs to be answered, is: How should the relations (1.12), (1.13) and (1.15) be modified in order to deal correctly with time dependent phenomena? The simplest modification of the law of Coulomb would be

$$\nabla \cdot D(r,t) = 4\pi k_d \rho_{tr}(r,t) \, . \tag{1.17}$$

Instead of a stationary charge distribution one is faced with a charge distribution, which changes in time. As an example, one might think of a uniformly charged sphere, which is moved through space. Equation (1.17) would than predict the production of a time dependent D-field. The moving charge represents on the other side a current $i(t)$, which produces a magnetic field. This production of magnetic fields is governed by Ampère's law. The next step would therefore involve a modification of this law. The simplest modification could be

$$\nabla \times H(r,t) = 4\pi k_h j_{tr}(r,t)$$

which is, however, not satisfactory. The divergence of this equation

$$\nabla \cdot (\nabla \times H(r,t)) = 4\pi k_h \nabla \cdot j_{tr}(r,t)$$

shows, that the left hand side gives the value zero. This result is true for every differentiable function

$$\nabla \cdot (\nabla \times H(r,t)) = 0 \, .$$

On the other hand the right hand side is not equal to zero, as a consequence of the equation of continuity

$$\nabla \cdot \boldsymbol{j}_{tr}(\boldsymbol{r},t) = -\frac{\partial \rho_{tr}(\boldsymbol{r},t)}{\partial t} \neq 0.$$

Ampère's law (2) is only consistent for stationary situations ($\partial \rho_{tr}/\partial t = 0$). The simplest modification is not compatible with the demand of charge conservation. A simple way out has been suggested in 1865 by Maxwell. Helmholtz was able to demonstrate a few weeks later (in 1870) by experiment, that Maxwell's proposition is correct. Even today there is no reason to doubt its correctness.

The argument of Maxwell can be summed up in the following manner: The equation of continuity can be combined with the directly extended Coulomb law (1.17) to give

$$0 = \nabla \cdot \boldsymbol{j}_{tr}(\boldsymbol{r},t) + \frac{\partial \rho_{tr}(\boldsymbol{r},t)}{\partial t} = \nabla \cdot \boldsymbol{j}_{tr}(\boldsymbol{r},t) + \frac{1}{4\pi k_d}\frac{\partial}{\partial t}(\nabla \cdot \boldsymbol{D}(\boldsymbol{r},t))$$

$$= \nabla \cdot \left(\boldsymbol{j}_{tr}(\boldsymbol{r},t) + \frac{1}{4\pi k_d}\frac{\partial}{\partial t}\boldsymbol{D}(\boldsymbol{r},t)\right) = 0.$$

One can find a consistent equation, if the current density in the law of Ampère is replaced by the expression in brackets

$$\nabla \times \boldsymbol{H}(\boldsymbol{r},t) = 4\pi k_h \left(\boldsymbol{j}_{tr}(\boldsymbol{r},t) + \frac{1}{4\pi k_d}\frac{\partial}{\partial t}\boldsymbol{D}(\boldsymbol{r},t)\right). \quad (1.18)$$

The left as well as the right side yield the value zero after application of the divergence operator. In the stationary limit ($\boldsymbol{D}(\boldsymbol{r},t) \to \boldsymbol{D}(\boldsymbol{r})$) the original law of Ampère is recovered. The additional term expresses the fact, that not only the current density but also an electric field, which varies in free space with the time, can produce a magnetic vortex field. The definition of the **displacement current density** $\boldsymbol{j}_v$ yields

$$\boldsymbol{j}_v(\boldsymbol{r},t) = \frac{1}{4\pi k_d}\frac{\partial}{\partial t}\boldsymbol{D}(\boldsymbol{r},t), \quad (1.19)$$

so that one has

$$\nabla \times \boldsymbol{H}(\boldsymbol{r},t) = 4\pi k_h \left(\boldsymbol{j}_{tr}(\boldsymbol{r},t) + \boldsymbol{j}_v(\boldsymbol{r},t)\right). \quad (1.20)$$

It is useful (as it constitutes the core of the argument), to execute the transition from the stationary to the dynamical Ampère equation in a more descriptive (but equivalent) manner. For this purpose one can look at a current loop with a source of an alternating current and a capacity C (Fig. 1.5a). The alternating current in the

1.2 The Maxwell Equations

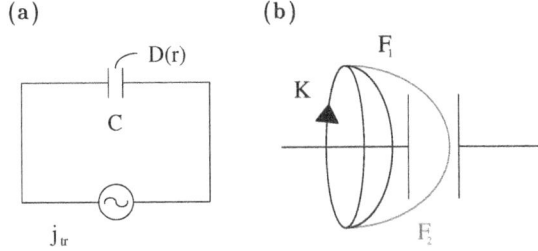

Fig. 1.5 Maxwell's displacement current. (**a**) Circuit. (**b**) Argument

wire can be represented by a current density $j_{tr}(r,t)$. The plates of the capacitor are reloaded in a periodic manner. In the space between the plates exists a **D**-field, which changes in time. Look now at the integral form of the simply extended Ampère law

$$\oint_K H(r,t) \cdot dr = 4\pi k_h \iint_{F(K)} j_{tr}(r,t) \cdot df ,$$

for a situation, in which the curve K is supposed to enclose the feed cable (Fig. 1.5b). The choice of the surface $F(K)$ is, according the theorem of Stokes, arbitrary as long as K is its border. Everything is right, if the surface F_1 is chosen, so that it is pierced by the feed cable

$$\iint_{F_1} j_{tr}(r,t) \cdot df \neq 0 .$$

If, the mathematically equivalent surface F_2, which traverses the space between the plates of the condenser, is used, the result is

$$\iint_{F_2} j_{tr}(r,t) \cdot df = 0 .$$

The displacement current suggested by Maxwell sorts the situation. It is possible to regard, in the vein of a practical variant, Maxwell's displacement current as an abstraction of the actual current, which is produced by the electric field between the plates.

The consequences of this modification are far reaching. An alternating current in a conductor, which is produced by an alternating electric field, generates, according to the (extended) Ampère law a magnetic field changing with time. This produces, according to the Faraday law a time dependent electric vortex field. This produces, in turn, according to the extended Ampère law another **B**-field, and so on. This chain of time dependent **E**- and **B**-fields, propagating in space and time is an **electromagnetic wave** (Fig. 1.6).

The existence of electromagnetic waves is ensured. The generation and propagation is described correctly by the simple extension of the Coulomb law and the

Fig. 1.6 Indication of an electromagnetic wave

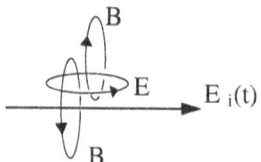

extension of the Ampère law suggested by Maxwell. The absence of sources for the **B**-field (4) is still valid in the dynamical case. The additional statements concerning the response of materials to the fields might have to be modified in comparison with the provisional remarks offered here. It is possible, that the polarisation of a material (Fig. 1.7) is not able to follow the variation of the external field or that it reacts more easily to certain frequencies. This point will be addressed in Chap. 1.3.3.

The basic equations of electrodynamics, the **Maxwell equations** are

$$
\begin{aligned}
&(1)\ \text{Coulomb's Law}\\
&\qquad \nabla \cdot \boldsymbol{D}(\boldsymbol{r},t) = 4\pi k_d\, \rho_{tr}(\boldsymbol{r},t)\\[4pt]
&(2)\ \text{Ampère's Law}\\
&\qquad \nabla \times \boldsymbol{H}(\boldsymbol{r},t) = 4\pi k_h\, \boldsymbol{j}_{tr}(\boldsymbol{r},t) + \frac{k_h}{k_d}\frac{\partial \boldsymbol{D}(\boldsymbol{r},t)}{\partial t}\\[4pt]
&(3)\ \text{Faraday's Law}\\
&\qquad \nabla \times \boldsymbol{E}(\boldsymbol{r},t) = -k_f\, \frac{\partial \boldsymbol{B}(\boldsymbol{r},t)}{\partial t}\\[4pt]
&(4)\ \text{Magnetic Sources}\\
&\qquad \nabla \cdot \boldsymbol{B}(\boldsymbol{r},t) = 0\ .
\end{aligned}
\qquad (1.21)
$$

This set of eight differential equations reduce to the equations, which have been discussed under the heading electro- and magnetostatics in the stationary limit (all quantities involved are time independent).

Fig. 1.7 Experiment: Frequency dependent material response

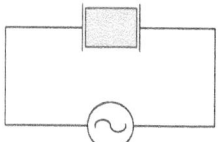

1.3 Electromagnetic Waves

They take the form

$$\nabla \cdot \boldsymbol{D}(\boldsymbol{r},t) = 4\pi \rho_{tr}(\boldsymbol{r},t) \qquad \nabla \times \boldsymbol{E}(\boldsymbol{r},t) = -\frac{1}{c}\frac{\partial}{\partial t}\boldsymbol{B}(\boldsymbol{r},t)$$

$$\nabla \cdot \boldsymbol{B}(\boldsymbol{r},t) = 0 \qquad \nabla \times \boldsymbol{H}(\boldsymbol{r},t) = \frac{1}{c}\frac{\partial \boldsymbol{D}(\boldsymbol{r},t)}{\partial t} + \frac{4\pi}{c}\boldsymbol{j}_{tr}(\boldsymbol{r},t), \tag{1.22}$$

in the CGS-system. In the SI-system[3] they are

$$\nabla \cdot \boldsymbol{D}(\boldsymbol{r},t) = \rho_{tr}(\boldsymbol{r},t) \qquad \nabla \times \boldsymbol{E}(\boldsymbol{r},t) = -\frac{\partial}{\partial t}\boldsymbol{B}(\boldsymbol{r},t)$$

$$\nabla \cdot \boldsymbol{B}(\boldsymbol{r},t) = 0 \qquad \nabla \times \boldsymbol{H}(\boldsymbol{r},t) = \frac{\partial \boldsymbol{D}(\boldsymbol{r},t)}{\partial t} + \boldsymbol{j}_{tr}(\boldsymbol{r},t). \tag{1.23}$$

A confirmation of the correctness, based on the (special) theory of relativity, will be presented in Chap. 3. A selection of applications, which illustrates, that experimental statements on classical electromagnetic phenomena can be understood in terms of Maxwell's equations, is found in Chap. 2. The discussions in the next sections deals with the basic solution of the *free* Maxwell equations (there exist no true charges and no true currents in the section of space, which is of interest).

1.3 Electromagnetic Waves

The following topics will be addressed in this section:

(1) How can waves be characterised?
(2) In how far are wave phenomena described by Maxwell's equation?
(3) What kind of waves are electromagnetic waves?

1.3.1 Wave Equations

As a useful path to the answer of the first of these questions one can look at the differential equation of an elastic string, which is a topic from mechanics.

---

[3] In some of the text books the quantities $\boldsymbol{D}_n = \varepsilon_0 \boldsymbol{D}$ and $\boldsymbol{H}_n = \boldsymbol{H}/\mu_0$ are introduced, so that the Maxwell equations are all together free of any constants.

1.3.1.1 The Wave Equation for One Degree of Freedom

Imagine, that a string under tension is in the rest position arranged along the x-axis.[4] It is then plugged by applying a displacement of a section of the string in the y-direction. The displacement of each section of a string from the rest position $y = 0$ changes with the position x along the string and the time t. The function $y(x, t)$ satisfies the partial differential equation

$$\frac{\partial^2 y(x,t)}{\partial x^2} - \frac{1}{v^2}\frac{\partial^2 y(x,t)}{\partial t^2} = 0, \qquad (1.24)$$

or in short hand

$$y'' - \frac{1}{v^2}\ddot{y} = 0.$$

The quantity v is defined as

$$v = \sqrt{\frac{\tau}{\rho}},$$

where τ is the coefficient of the harmonic restoring force and ρ the linear density of the material of the string. This differential equation is a **one dimensional wave equation**. The process of the solution of this differential equation for given initial conditions is quite similar to the process of the solution of the Poisson or Laplace equations. One obtains particular solutions with the ansatz

$$y(x,t) = G(x)H(t)$$

for the separation of variables. One inserts the ansatz into the differential equation and finds

$$v^2 \frac{G''(x)}{G(x)} = \frac{\ddot{H}(t)}{H(t)} = -\omega^2$$

for each possible value of ω^2 an ordinary differential equation in the form of oscillator equations

$$\ddot{H}(t) + \omega^2 H(t) = 0$$

$$G''(x) + k^2 G(x) = 0 \qquad \left(k^2 = \frac{\omega^2}{v^2}\right).$$

---

[4] The terminology 'one degree of freedom' refers to the direction of the propagation.

1.3 Electromagnetic Waves

The separation constant of the equation for the time dependence is traditionally denoted by ω^2, the separation constant[5] for the space part by k^2. The fundamental solutions of the oscillator equations for each value of ω are

$$H(t) = \left\{ e^{i\omega t}, e^{-i\omega t} \right\} \qquad G(x) = \left\{ e^{ikx}, e^{-ikx} \right\} .$$

The basis functions $y(x, t)$ are therefore

$$y(x, t) = \{\exp[\pm ikx \pm i\omega t]\} .$$

They have to be real, if they represent a variable as the displacement, so that one could use the alternative functions

$$y(x, t) = \{\cos k\,(x \pm vt), \sin k\,(x \pm vt)\} .$$

The complex form (with the understanding, that only the real part is physically relevant) is usually preferred, as it can be handled more easily. In order to interpret the basic solutions, one can take a closer look at the real function

$$y = a \sin(kx + \omega t) \qquad (a \text{ real}) .$$

A snapshot (t_{fixed}) corresponds to a sine function with the amplitude a, the phase $\omega\, t_{\text{fixed}}$ and the wavelength $\lambda = 2\pi/k$ (Fig. 1.8a). The quantity k is called the **wave number**

$$k = \frac{2\pi}{\lambda} \qquad [k] = \text{length}^{-1} . \qquad (1.25)$$

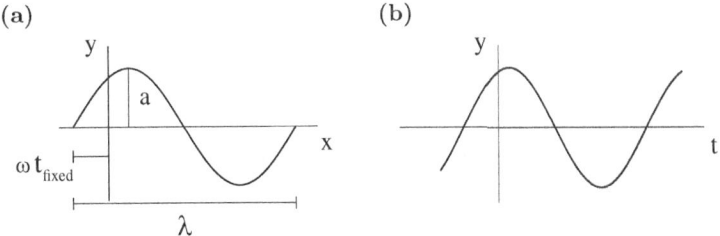

Fig. 1.8 One dimensional plane waves. (a) Snapshot for $t = t_{\text{fixed}}$. (b) Variation of $y(t)$ at the location x_{fixed}

---

[5] The traditionally used quantity k^2 is not identical with the constants, which define the system of units.

Fig. 1.9 The definition of the speed of propagation

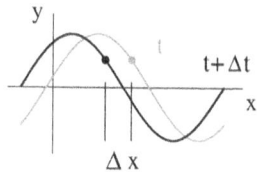

It indicates how many waves with the wave length λ can fit into the standard interval 2π. The displacement as a function of time at a fixed location varies also with the sine function (Fig. 1.8b). The oscillation in time is characterised by the frequency

$$f = \frac{\omega}{2\pi}. \qquad (1.26)$$

The definition $\omega = kv$ states

$$2\pi f = \frac{2\pi}{\lambda} v \quad \text{or} \quad f\lambda = v. \qquad (1.27)$$

The constant v represents the speed of propagation of the harmonic wave. This can be justified in the following fashion: Look at the deflection at the time t and the time $t + \Delta t$ (Fig. 1.9)

$$y(x, t) = a \sin(kx \pm \omega t)$$

$$y(x + \Delta x, t + \Delta t) = a \sin(k(x + \Delta x) \pm \omega(t + \Delta t)).$$

The wave functions have the same value for $k\Delta x \pm \omega \Delta t = 0$. The whole wave shifts by the amount Δx in the time $\Delta t > 0$, so that

$$\frac{\Delta x}{\Delta t} = v = \mp \frac{\omega}{k}$$

follows. A wave, which is moving to the left, is characterised by a negative velocity of propagation. The four basic solutions describe **monochromatic waves**, that means waves, which have a definite wave length and frequency. The four basic types differ in the form and the direction of motion

$\sin(kx + \omega t)$ sine wave with velocity v towards the left

$\cos(kx + \omega t)$ cosine wave with velocity v towards the left,

$\sin(kx - \omega t)$ sine wave with velocity v towards the right

$\cos(kx - \omega t)$ cosine wave with velocity v towards the right.

1.3 Electromagnetic Waves

The wave equation is linear. The general solution (with selected speed of propagation, as v features as a parameter in the differential equation) has to be represented by a superposition of the particular solutions, in real form

$$y(x,t) = \int_0^\infty \Big\{ a(k) \sin[k(x+vt)] + b(k) \cos[k(x+vt)]$$

$$+ c(k) \sin[k(x-vt)] + d(k) \cos[k(x-vt)] \Big\} dk \ .$$

It is also possible to use a range of wave numbers with $[-\infty, \infty]$, e.g. for a wave moving to the left

$$\int_{-\infty}^\infty dk \{ a_1(k) \sin[k(x+vt)] + b_1(k) \cos[k(x+vt)] \}$$

or

$$= \int_0^\infty dk \{ (a_1(k) - a_1(-k)) \sin[k(x+vt)]$$

$$+ (b_1(k) + b_1(-k)) \cos[k(x+vt)] \} \ .$$

The second form is often more practical.

The complex expression can usually be written in a more compact form

$$y(x,t) = \int_{-\infty}^\infty \left\{ A(k) e^{ik(x+vt)} + B(k) e^{ik(x-vt)} \right\} dk \ . \tag{1.28}$$

The two complex coefficients A, B, for each value of k correspond to four real coefficients. This superposition of monochromatic waves are in mathematical language Fourier integrals. Fourier integration allows to represent (practically) every wave form, as e.g.

- $A = 0$ and $B \neq 0$ wave form moving towards the right as moving wave pulses (Fig. 1.10a), moving wave packets (Fig. 1.10b) or moving saw teeth (Fig. 1.10c). Some additional examples are

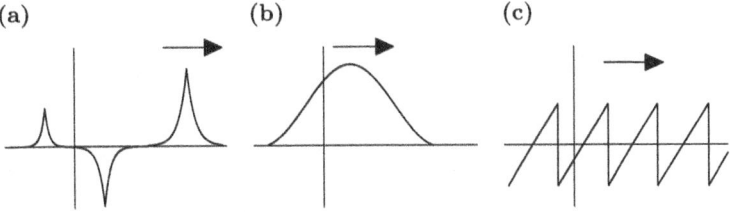

Fig. 1.10 Diverse wave forms

- For $A = 0$ and $B(k) = B\,\delta(k - k_0)$ with $y = B\,e^{ik_0(x-vt)}$, one has a monochromatic cosine wave moving to the right, if B is real. A representation of the δ-function

$$\delta(b) = \frac{1}{2\pi} \int_{-\infty}^{\infty} e^{k\,b}\, dk$$

is used in this case.
- For $A = 0$ and $B(k) = B_1\,\delta(k - k_1) + B_2\,\delta(k - k_2)$ is a superposition of two monochromatic waves.
- The specification $B(k) = 1/(2\pi)$ corresponds to $y = \delta(x - vt)$, a wave pulse, with infinite height and extremely localised features.
- For $A \neq 0$ and $B = 0$, one finds with the corresponding functions in the Fourier integral a wave form moving to the left.
- For $A \neq 0$ and $B \neq 0$ one can obtain standing waves, as e.g. for $A = B = \delta(k - k_0)$

$$\mathrm{Re}\,(y(x,t)) = \mathrm{Re}\left[e^{ik_0 x}\left(e^{i\omega_0 t} + e^{-i\omega_0 t}\right)\right]$$

$$= (2\cos\omega_0 t)\cos k_0 x \qquad (\omega_0 = k_0 v)\,.$$

The wave is a standing cosine curve with an amplitude, which varies with time (Fig. 1.11a).

A special solution can be selected for physical problems by stating initial and boundary conditions. The following types of problems can be distinguished:

(a) **boundary-initial value problems** with the specification

$$y(a,t) = y_a \qquad y(b,t) = y_b$$

for all times t. A typical problem of this type is the fixed spring with $y_a = y_b = 0$. The wave form has to fit into the specified interval $[a, b]$ (Fig. 1.11b).

Fig. 1.11 Examples of wave mechanics. (**a**) Standing wave. (**b**) Solution of a boundary value problem

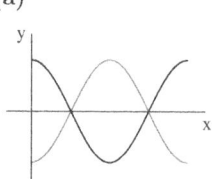
(a)

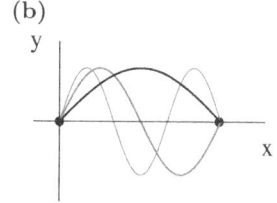
(b)

1.3 Electromagnetic Waves

This is only possible for certain wave numbers k_n ($n = 0, 1, \ldots$). These numbers are the eigenvalues of the boundary value problem

$$y(x, t) = \sum_n \left[A_n\, e^{ik_n(x+vt)} + B_n\, e^{ik_n(x-vt)} \right]. \tag{1.29}$$

The Fourier integral is replaced by a Fourier series. The final wave form is uniquely selected by the additional condition

$$y(x, 0) = f(x) \qquad \left[\frac{\partial}{\partial t} y(x, t)\right]_{t=0} = g(x)$$

(for all x, in $[a, b]$).

(ii) **Initial value problems**. The wave is not restricted to a finite interval. It can move in the complete (one-dimensional) space. The specification

$$y(x, 0) = f(x) \qquad \left[\frac{\partial}{\partial t} y(x, t)\right]_{t=0} = g(x)$$

(for all x) selects the coefficients of the Fourier integral. This means: The wave equation regulates the (one dimensional) development of the wave pattern, if the initial form of the wave and the starting time is prescribed.

The discussion of the wave equation can be extended to higher dimensions of space.

1.3.1.2 The Wave Equation in Two Space Dimensions

The two-dimensional wave equation has the form

$$\left(\frac{\partial^2}{\partial x^2} + \frac{\partial^2}{\partial y^2}\right)\psi(x, y, t) - \frac{1}{v^2}\frac{\partial^2 \psi(x, y, t)}{\partial t^2} = 0. \tag{1.30}$$

The function ψ represents a scalar quantity, which changes with time in the x-y plane. It could e.g. be the coordinate z (for a vibrating membrane or surface water waves), so that the differential equation describes the displacements with respect to that plane. The quantity ψ could be a more abstract quantity as a surface density.

The general solution of the two-dimensional wave equation is also determined by finding particular solutions and their superposition. The details follow the pattern of the one-dimensional case. As there are now three independent variables, there exist two separation constants. With the variables $\boldsymbol{r} = (x, y)$ and a wave vector $\boldsymbol{k} = (k_x, k_y)$, as a summary of the separation constants, a solution of the form

$$\psi_{\text{part.}}(\boldsymbol{r}, t) = e^{i(\boldsymbol{k}\cdot\boldsymbol{r} \pm \omega t)}$$

Fig. 1.12 Two dimensional plane wave. (**a**) View. (**b**) Top view

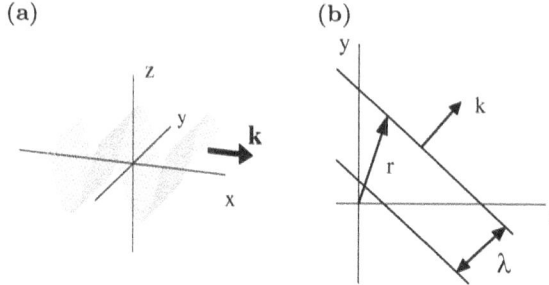

with

$$\omega = vk = v\sqrt{k_x^2 + k_y^2} \qquad (1.31)$$

would represent a two-dimensional plane wave. The real part of such a solution could describe physically measurable quantities as

$$\psi_{\text{phys.}}(\boldsymbol{r}, t) = \text{Re}\left[A\, e^{i(\boldsymbol{k}\cdot\boldsymbol{r}\pm\omega t)}\right]$$

$$= (\text{Re } A)\cos(\boldsymbol{k}\cdot\boldsymbol{r}\pm\omega t) - (\text{Im } A)\sin(\boldsymbol{k}\cdot\boldsymbol{r}\pm\omega t)\ .$$

All points, for which the scalar product of $\boldsymbol{k}\cdot\boldsymbol{r}$ has the same value at a time t, have the same value for the quantity ψ. The scalar product indicates, that all points on a straight line perpendicular to the vector $\boldsymbol{k}$ form a wave front. The distance between two neighbouring peaks or valleys is the wave length λ (see Fig. 1.12a,b).

One can picture a two-dimensional harmonic wave as a giant corrugated sine- or cosine shaped sheet, which moves with the velocity $v = \omega/k$ in the direction $\boldsymbol{k}$. The magnitude of this vector is connected with the wave length

$$\lambda = \frac{2\pi}{k} = \frac{2\pi}{\left[k_x^2 + k_y^2\right]^{1/2}}\ . \qquad (1.32)$$

For a two-dimensional plane wave the relation (as in the one-dimensional case)

$$\omega = 2\pi f \qquad \lambda = \frac{2\pi}{k} \quad \longrightarrow \quad v = \frac{\omega}{k} = \lambda \cdot f$$

is valid. The general solution of the two-dimensional wave equation is given by the double Fourier integral

$$\psi(\boldsymbol{r}, t) = \int_{-\infty}^{\infty} dk_x \int_{-\infty}^{\infty} dk_y \left[A(k_x, k_y)\, e^{i(\boldsymbol{k}\cdot\boldsymbol{r}+\omega(k)t)}\right] \qquad (1.33)$$

1.3 Electromagnetic Waves

$$+ B(k_x, k_y) \, e^{i(k \cdot r - \omega(k)t)} \Big] \qquad \left(\omega(k) = v \sqrt{k_x^2 + k_y^2} \right) ,$$

which can also represent a multitude of shapes for the waves. One could, e.g., consider a tip, which dips once or periodically into a surface of water, and produces a two dimensional wave pulse or wave rings, which spread out in a circular fashion. The amplitudes are fixed either by initial-boundary conditions (e.g. for a drum) or only by initial conditions. For the initial value problem one has to specify

$$\psi(r, 0) = f(r) , \quad \dot{\psi}(r, 0) = g(r) \quad \text{for all} \quad r .$$

1.3.1.3 The Wave Equation in Three Space Dimensions

The discussion of the three-dimensional wave equation

$$\Delta \psi(x, y, z, t) - \frac{1}{v^2} \frac{\partial^2}{\partial t^2} \psi(x, y, z, t) = 0 \qquad (1.34)$$

follows the same pattern. The particular solutions, the plane wave solutions, do not differ much from the two dimensional case

$$\psi_{\text{part.}}(r, t) = e^{i(k \cdot r \pm \omega t)} .$$

One needs three separation constants, which can be represented by a wave number vector

$$k = (k_x, k_y, k_z) ,$$

with the relations

$$\lambda = \frac{2\pi}{k} , \quad k = \sqrt{k_x^2 + k_y^2 + k_z^2} , \quad \omega = 2\pi f \quad \longrightarrow \quad v = \frac{\omega}{k} = \lambda \cdot f .$$

In order to visualise plane waves in a space with three dimensions one can consider the following: All vectors r, for which the end points lie in a plane perpendicular to the wave vector, describe the same state of oscillation if they have the same length. An example could be a wave function ψ, which describes the density of a medium. All end points of vectors in a plane perpendicular to the vector k at a given time, which have the same length, represent points with the same density. The points with the same density move with the same velocity in or against the direction of k.

The general solution of the wave equation is given by a three-dimensional Fourier integral

$$\psi(r, t) = \iiint d^3 k \left[A(k) \, e^{i(k \cdot r + \omega(k)t)} + B(k) \, e^{i(k \cdot r - \omega(k)t)} \right] . \qquad (1.35)$$

Special solutions for initial value problems are, as before, selected by demanding

$$\psi(r,0) = f(r) \qquad \frac{\partial \psi(r,t)}{\partial t}\bigg|_{t=0} = g(r) \qquad \text{for all } r.$$

As an example one can consider the three-dimensional Fourier integral for the δ-function

$$\delta(k) = \frac{1}{(2\pi)^3} \iiint e^{ik\cdot r} d^3r.$$

If

$$\psi(r,0) = f(r) = \iiint d^3k \left[A(k) e^{ik\cdot r} + B(k) e^{ik\cdot r} \right]$$

is multiplied with $e^{-ik'\cdot r}$ one obtains after integration over the complete space

$$\iiint f(r) e^{-ik'\cdot r} d^3r = \iiint d^3r \iiint d^3k\, e^{i(k-k')\cdot r} (A(k) + B(k)).$$

The integration on the right hand side yields (up to a factor) the δ-function. Consecutive integration over the wave number leads then to

$$A(k') + B(k') = \frac{1}{(2\pi)^3} \iiint f(r) e^{-ik'\cdot r} d^3r.$$

The specification of the derivative yields in the same vein

$$i\omega(k')(A(k') - B(k')) = \frac{1}{(2\pi)^3} \iiint g(r) e^{-ik'\cdot r} d^3r.$$

The two sets of equations would allow (if all integrals involved have been evaluated) to find the functions $A(k)$ and $B(k)$.

1.3.2 Wave Solutions of the Maxwell Equations

The questions at this point are: What are electromagnetic waves? How are they produced? How do they propagate after they have been produced? The first step to answer these questions, is the demonstration, that wave equations can be extracted from Maxwell's equations. Leaving aside the question of the generation for the moment, which will be discussed in Chap. 2.3, one can assume, the electromagnetic

1.3 Electromagnetic Waves

wave enters a region of space, in which the simple material equations

$$D(r,t) = \varepsilon \frac{k_d}{k_e} E(r,t) \qquad B(r,t) = \mu \frac{k_m}{k_h} H(r,t)$$

hold and in which there exist no true charges and no true currents

$$\rho_{tr}(r,t) = 0 \qquad j_{tr}(r,t) = 0.$$

The Maxwell equations, which contain in this case only the measurable fields B and E, are referred to as the free Maxwell equations. They are

$$\nabla B \cdot (r,t) = \nabla \cdot E(r,t) = 0 \tag{1.36}$$

$$\nabla \times E(r,t) = -k_f \frac{\partial B}{\partial t}(r,t) \tag{1.37}$$

$$\nabla \times B(r,t) = \varepsilon \mu \frac{k_m}{k_e} \frac{\partial E(r,t)}{\partial t}. \tag{1.38}$$

Some steps are necessary in order to get from these equations to the associated wave equations:

Step 1: Calculate the rotation of Eq. (1.38)

$$\nabla \times (\nabla \times B(r,t)) = \varepsilon \mu \frac{k_m}{k_e} \nabla \times \frac{\partial}{\partial t} E(r,t)$$

and use the alternative form of the double rotation in order to write the relation

$$\nabla \times (\nabla \times B(r,t)) = \nabla (\nabla \cdot B(r,t)) - \Delta B(r,t) = -\Delta B(r,t)$$

or in explicit form

$$\text{rot}(\text{rot } B(r,t)) = \text{grad}(\text{div } B(r,t)) - \Delta B(r,t) = -\Delta B(r,t).$$

After changing the order of the differentiation with respect to time and to space one finds on the right hand side

$$\Delta B(r,t) + \varepsilon \mu \frac{k_m}{k_e} \frac{\partial}{\partial t} (\nabla \times E(r,t)) = 0.$$

Step 2: Insert this equation into the law of induction for $\nabla \times E$ and find

$$\Delta B(r,t) - \varepsilon \mu \frac{k_m k_f}{k_e} \frac{\partial^2}{\partial t^2} B(r,t) = 0. \tag{1.39}$$

The result is a set of wave equations for the three components of the B-field, e.g. for the x-component

$$\Delta B_x(r,t) - \varepsilon\mu \frac{k_m k_f}{k_e} \frac{\partial^2}{\partial t^2} B_x(r,t) = 0 \,.$$

The same steps for $\nabla \times E(r,t)$, beginning with (1.37), yields the corresponding wave equation for the components of the E-field

$$\Delta E(r,t) - \varepsilon\mu \frac{k_m k_f}{k_e} \frac{\partial^2}{\partial t^2} E(r,t) = 0 \,. \tag{1.40}$$

The two sets of (vector) wave equations govern the propagation of the components of the E and B-fields in a material, for which the simple material equation is acceptable. The form of the wave, which propagates, depends on the mechanism for its production, which will have to be discussed later. The wave equations show, however, directly that the basic plane E-vector waves and the basic plane B-vector waves propagate with the same velocity

$$v = \left[\frac{k_e}{\varepsilon\mu k_m k_f} \right]^{1/2} . \tag{1.41}$$

Experiment shows, that the velocity of propagation for electromagnetic waves in vacuum with $\varepsilon = \mu = 1$ is equal to the velocity of light c. The consequence is: the combination $[k_e/(k_m k_f)]^{1/2}$ of the three constants has to be equal to the value c

$$\left[\frac{k_e}{k_m k_f} \right]^{1/2}_{\text{CGS}} = \left[\frac{k_e}{k_m k_f} \right]^{1/2}_{\text{SI}} = c \,. \tag{1.42}$$

In CGS-system with $k_e = 1$ and $k_m = k_f = 1/c$ this is directly apparent. In the SI-system one has, due to $k_e = 1/(4\pi\varepsilon_0)$, $k_m = \mu_0/(4\pi)$ and $k_f = 1$, the relation

$$v_{\text{SI}}(\text{vacuum}) = \frac{1}{\sqrt{\varepsilon_0 \mu_0}} \,. \tag{1.43}$$

The values for these quantities

$$\varepsilon_0 = 8.8542 \cdot 10^{-12} \left[\frac{\text{C}^2 \cdot \text{s}^2}{\text{kg} \cdot \text{m}^3} \right] \qquad \mu_0 = 1.2566 \cdot 10^{-6} \left[\frac{\text{kg} \cdot \text{m}}{\text{C}^2} \right]$$

1.3 Electromagnetic Waves

lead indeed to $v_{SI}(\text{vacuum}) = c = 2.99\ldots \cdot 10^8$ m/s. For all materials one finds $\varepsilon\mu > 1$, so that (independent of the system of units) the statement holds

$$v(\text{material}) = \frac{c}{\sqrt{\varepsilon\mu}} < v(\text{vacuum}) = c.$$

In Maxwell's equations the two vector fields are coupled. This coupling is not evident in the wave equations. It is a fact that the wave equations can be derived from the Maxwell equations, but the Maxwell equations can not be reconstructed from the wave equations. This implies, that the Maxwell equations contain additional information. It is this additional information which fixes the actual properties of the electromagnetic waves.

1.3.3 Properties of Electromagnetic Waves

The simplest wave form, that can be expected as a solution of the wave equations are monochromatic plane waves. One can summarise the fundamental solutions of the six wave equations (1.39) and (1.40), if the transmitter produces such a wave form[6]

$$E(r,t) = E_0\, e^{i(k\cdot r - \omega t)} \qquad B(r,t) = B_0\, e^{i(k\cdot r - \omega t)} \qquad (1.44)$$

$$\omega = vk = \sqrt{\frac{k_e}{k_f k_m}}\frac{k}{\sqrt{\varepsilon\mu}} = \frac{ck}{\sqrt{\varepsilon\mu}}. \qquad (1.45)$$

These equations describe a situation, in which the waves propagate only in the direction of the vector of the wave number. The discussion in the opposite direction would be similar. The amplitudes of the plane waves require in the general case six complex numbers E_x, E_y, E_z, B_x, B_y, B_z. It is possible though to restrict oneself to the choice

$$\mathbf{E}_0 = E_0\, \mathbf{e}_1 \qquad \mathbf{B}_0 = B_0\, \mathbf{e}_2$$

(a complex number multiplied by an arbitrary real unit vector) as only the real part of the wave function has any physical significance. In order to extract the additional information inherent in Maxwell's equations, one inserts the solution (1.44) into these equations. The following rules for the handling of the action of the nabla

---

[6] Monochromatic waves can only be produced by a monochromatic sender. A sender which oscillates with the frequency f produces a wave with the wave length $\lambda = v/f$ or c/f in the vacuum. The emission is, however, not restricted to a plane, it has a different geometry (see Chap. 2.3). One can obtain a good approximation of a plane wave by gating a beam from the sender. Only one direction for the propagation is selected.

operator and the time derivative on the plane waves are quite useful for this task

$$\nabla e^{i(k \cdot r \pm \omega t)} = ik\, e^{i(k \cdot r \pm \omega t)}$$

$$\frac{\partial}{\partial t} e^{i(k \cdot r \pm \omega t)} = \pm i\omega\, e^{i(k \cdot r \pm \omega t)}.$$

One obtains in this fashion:

(a) From the equations with the operator div (1.36)

$$\nabla \cdot E(r,t) = 0 \quad \longrightarrow \quad i(e_1 \cdot k) E_0 e^{i(k \cdot r - \omega t)} = 0 \quad \longrightarrow \quad (e_1 \cdot k) = 0$$

and correspondingly

$$\nabla \cdot B(r,t) = 0 \quad \longrightarrow \quad (e_2 \cdot k) = 0.$$

The field vectors are perpendicular to the direction of propagation. The plane electromagnetic wave is a **transverse wave**.

(b) From the equations with the operator rot ((1.37), (1.38))

$$\nabla \times E(r,t) + k_f \frac{\partial}{\partial t} B(r,t) = 0$$

$$\longrightarrow \quad i\left[(k \times e_1) E_0 - \omega k_f e_2 B_0\right] e^{i(k \cdot r - \omega t)} = 0$$

$$\nabla \times B(r,t) - \varepsilon\mu \frac{k_m}{k_e} \frac{\partial}{\partial t} E(r,t) = 0$$

$$\longrightarrow \quad i\left[(k \times e_2) B_0 + \varepsilon\mu \frac{\omega k_m}{k_e} e_1 E_0\right] e^{i(k \cdot r - \omega t)} = 0.$$

There result two vector equations (the expressions in the bracket results in a null vector), which lead to a relation between the three vectors e_1, e_2 and k

$$e_2 = \frac{1}{k}(k \times e_1),$$

as well as a relation between the amplitudes

$$B_0 = \frac{\sqrt{\varepsilon\mu}}{ck_f} E_0.$$

1.3 Electromagnetic Waves

(These statements and additional details concerning electromagnetic plane waves are explained in more detail in (Detail 1.7.2)).

The additional information contained in Maxwell's equations can be summed up in the following form

$$k \cdot E(r,t) = k \cdot B(r,t) = 0 \tag{1.46}$$

$$B(r,t) = \frac{\sqrt{\varepsilon\mu}}{c k k_f}(k \times E(r,t)) = \frac{1}{\omega k_f}(k \times E(r,t)). \tag{1.47}$$

The three vectors B, k, and E form (in each point of space and for each time) a trihedron, so that they constitute (in the order given) a right hand system. The amplitudes of the fields are coupled. They are equal in the vacuum, in a medium one has $|B| > |E|$.

In order to discuss further details of electromagnetic waves, it is necessary to go over to the real form. It is also convenient to choose the z-direction as the direction of propagation

$$k = (0, 0, k).$$

The complex form of the electric field is then

$$E(r,t) = \left(E_x\, e^{i(kz-\omega t)},\ E_y\, e^{i(kz-\omega t)},\ 0 \right).$$

Using the structure 'real magnitude times phase' for the complex amplitudes

$$E_x = \tilde{E}_x\, e^{i\alpha_x} \qquad E_y = \tilde{E}_y\, e^{i\alpha_y} \qquad (\tilde{E}, \alpha\ \text{real})$$

leads to

$$E_{\text{real}}(r,t) = \left\{ \tilde{E}_x \cos(kz - \omega t + \alpha_x),\ \tilde{E}_y \cos(kz - \omega t + \alpha_y),\ 0 \right\}. \tag{1.48}$$

The real part of the B-field is then, according to the relation (1.47),

$$B_{\text{real}}(r,t) = \frac{1}{\omega k_f} \left\{ -\tilde{E}_y \cos(kz - \omega t + \alpha_y),\right.$$

$$\left. \tilde{E}_x \cos(kz - \omega t + \alpha_x),\ 0 \right\}. \tag{1.49}$$

Fig. 1.13 Linearly polarised plane wave (moving in the z-direction). (**a**) Field vectors in the coordinate planes. (**b**) Field vectors in arbitrary planes

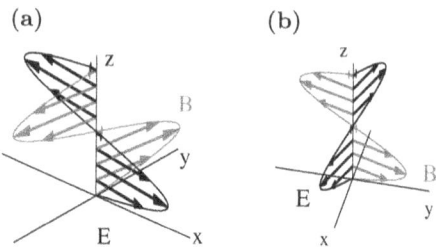

Some special cases can be used to obtain an idea of the nature of electromagnetic waves:

(i) The case with $\tilde{E}_y = 0$ represents a **linearly polarised** plane wave. The E-vector oscillates in the x-z plane. The corresponding B-vector oscillates perpendicular to the E-vector in the y-z plane (Fig. 1.13a). One should then imagine, that this picture is repeated for each straight line parallel to the z-axis and that the complete arrangement moves with the velocity v in the direction of the z-axis.

(ii) The case with $\tilde{E}_x \neq \tilde{E}_y \neq 0$ and $\alpha_x = \alpha_y$ represents also a linearly polarised plane wave. The difference is an oscillation of the E-vector (and the B-vector) in parallel planes with an inclination given by $\tilde{E}_x/\tilde{E}_y$ (Fig. 1.13b).

(iii) In the case with $\alpha_y = \alpha_x \pm \pi/2$ and $\tilde{E}_x = \tilde{E}_y = \tilde{E}$ one has for the E-vector

$$E_{\text{real}}(r, t) = \left\{ \tilde{E} \cos(kz - \omega t + \alpha_x), \pm \tilde{E} \sin(kz - \omega t + \alpha_x), 0 \right\}.$$

This is the parametric representation of a circle for a given value of z. The end point of the E-vector rotates in each point of a plane with $z = const.$ on a circle. This is a **circularly polarised plane wave** (Fig. 1.14a).

The positive/negative sign of $\pi/2$ distinguishes the sense of the circulation (viewed against the z-direction), as indicated in Fig. 1.14a in the counter-clockwise/clock-wise direction. One discerns left and right circularly polarised plane waves. A snap shot of the E-vectors along the z-axis would show a helix. This

Fig. 1.14 Circularly polarised electromagnetic wave. (**a**) Field vectors along the z-axis. (**b**) Snapshot of the E-vectors

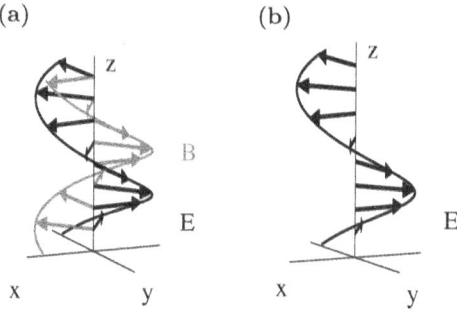

1.3 Electromagnetic Waves

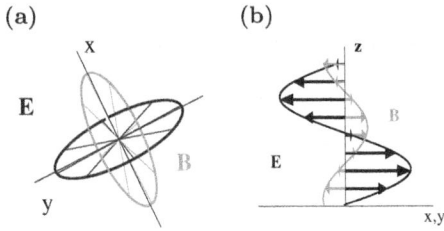

Fig. 1.15 Elliptically polarised electromagnetic wave. (a) Planes of the oscillations of the fields. (b) Snapshot

helix winds with a velocity v in the z-direction (Fig. 1.14b). The $\boldsymbol{B}$-field should be represented by the appropriate, orthogonal helix.

(iv) For $\tilde{E}_x \neq \tilde{E}_y$ and $\alpha_x \neq \alpha_y$, the plane wave is **elliptically polarised**. An end point of the $\boldsymbol{E}$-vectors moves on an elliptical helix. In the general case the momentary ellipses of the $\boldsymbol{E}$- and the $\boldsymbol{B}$-direction differ by an angle of $(\alpha_x - \alpha_y)/2$ (Fig. 1.15a). It is a rather difficult exercise in projective geometry to produce snapshots of these waves (Fig. 1.15b).

There exists a broad spectrum of monochromatic electromagnetic waves, which are classified by the frequency or wavelength. The wavelength in the vacuum (equivalent to the one in air as $\varepsilon \mu_{air} \approx 1$) are

$\lambda = 10^6 - 10^{-1}$ cm   radio frequencies (AM, FM), micro waves

$\Longrightarrow$ macroscopic sender

$\lambda = 10^{-2} - 10^{-4}$ cm   infrared (IR) light (heat radiation)

$\lambda \approx 10^{-4}$ cm   visible light

$\lambda = 10^{-5} - 10^{-6}$ cm   ultraviolet (UV) light

$\lambda = 10^{-6} - 10^{-10}$ cm   x-rays

$\Longrightarrow$ atomic sender

$\lambda = 10^{-9} - 10^{-11}$ cm γ-rays

$\Longrightarrow$ nuclear sender, elementary processes.

In optics the ratio of velocity of light (in vacuum) to the velocity of propagation in the medium

$$\frac{c}{v} = \sqrt{\varepsilon \mu} = n \qquad (1.50)$$

is referred to as the **index of refraction** of the medium, or in this form also as **Maxwell's relation**. It relates the optical properties with the general properties of a material. Some caution is necessary in applying this equation, as it does not yield the dependence of n on the frequency. An example is distilled water, for which one finds the values

$$\mu \approx 1 \quad \varepsilon \approx 80$$

and with (1.50) an index of refraction $n(H_2O) \approx 8.9$. This value is indeed correct for radio waves. The value measured in the frequency range of observable light is however $n(H_2O, \lambda \approx 10^{-4}\,\text{cm}) \approx 1.33$. The dependence on the frequency of the material constants is considerable

$$\varepsilon = \varepsilon(f) \quad \mu = \mu(f) \quad \longrightarrow n = n(f).$$

A qualitative understanding of the frequency dependence is not difficult. The polarisation is due to the displacement of the constituents of the atoms. If the charges have to respond to a higher frequency, one can expect differences in comparison with the quasi-stationary situation. The calculation of the frequency dependence of the material constants is not an easy matter.[7] The term **dispersion** is used for different phenomena associated with the frequency dependence of physical properties. A direct one is the spreading of wave packets in a medium. This is actually a topic in an introductory course on quantum mechanics, but should be touched upon briefly here.

The velocity of propagation in vacuum is the same for every monochromatic wave

$$v = c \quad \text{resp.} \quad \omega(k) = ck.$$

A wave packet, that is constructed by a superposition of plane waves with different wave numbers, moves in vacuum with the same velocity as each of its Fourier components, so that it does not change its form. In a dispersive medium the velocity is a function of the wave number

$$v = v(k) = \frac{c}{n(k)} \quad \text{resp.} \quad \omega(k) = v(k)k,$$

which, in general, is not a linear function of k. The Fourier components propagate differently in the moving wave packet. The consequence is:

- The wave packet spreads.
- The centre of the wave packet moves with a velocity, which is different from the velocity of the mean frequency of the components of the packet.

---

[7] Interested readers find some details in N. Ashcroft and D. Mermin, 'Solid State Physics' (Saunders Publications, Philadelphia, 1976).

1.3 Electromagnetic Waves

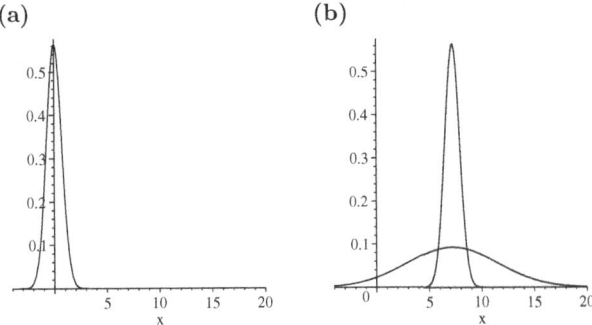

Fig. 1.16 Dispersion of wave packets (Gauss packet). (**a**) Situation at start: $t = 0$. (**b**) Situation at a time $t > 0$, without and with dispersion

This spreading is illustrated for a special one-dimensional wave packet, a Gauss packet, in Fig. 1.16a, b (see Detail 1.7.3 for a more detailed discussion). This wave packet is represented by the Fourier expansion

$$\Psi(x, t) = \frac{1}{2\pi} \int_{-\infty}^{\infty} dk\, e^{-k^2/4a^2} e^{i(kx - \omega(k)t)},$$

where the angular frequency $\omega(k)$ is either equal to $\omega_1(k) = ct$ (non-dispersive packet) or to $\omega_2(k) = ct + ht^2$ (dispersive packet). The result of the integration for the dispersive packet is a more involved complex function. For this reason the figure shows the square of the absolute value of the wave function rather than the real part

$$|\Psi(x, t)|^2 = \Psi^*(x, t)\, \Psi(x, t).$$

The integration yields for the non-dispersive Gauss packet

$$|\Psi_1(x, t)|^2 = \frac{a^2}{\pi} e^{-2a^2(x-ct)^2}.$$

The parameter a controls the width of the packet. The velocity of propagation of the maximum is $v_1 = c$.

The result for the dispersive Gauss packet is

$$|\Psi_2(x, t)|^2 = \frac{a^2}{\pi \left[1 + 16a^4 h^2 t^2\right]^{1/2}} \exp\left[-2a^2 \frac{(x - ct)^2}{\left[1 + 16a^4 h^2 t^2\right]}\right].$$

The maximum of the packet still moves with the velocity c, but the factor in front and the additional time dependence of the exponent are the reason for the spreading (by keeping the area under the curves constant).

A short summary of the concepts, which are used in the discussion of the present aspect of the term 'dispersion', can be best explained for a scalar wave packet in one space dimension

$$\Psi(x,t) = \int_{-\infty}^{\infty} dk\, A(k)\, e^{i(kx-\omega(k)t)} .$$

The medium value of the wave number of this packet is

$$k_0 = \frac{\int_{-\infty}^{\infty} k A(k)\, dk}{\int_{-\infty}^{\infty} A(k)\, dk} .$$

The velocity of the propagation of the component of the wave packet with k_0 is called the **phase velocity**

$$v_{ph}(k_0) = \frac{\omega(k_0)}{k_0} = \frac{c}{n(k_0)} , \qquad (1.51)$$

which is, in general, different from the velocity of the barycentre of the packet, the **group velocity** $v_{gr}(k_0)$.

The expansion of the argument of the exponential function about k_0

$$kx - \omega(k)t = k_0 x + (k-k_0)x + \omega(k_0)t + \left.\frac{d\omega(k)}{dk}\right|_{k=k_0} (k-k_0)t + \ldots$$

factorises the wave function into a plane wave with the wave number k_0 and a part, which describes the actual wave group, which in turn is determined by the weight function $A(k)$

$$\Psi(x,t) \approx e^{i(k_0 x - \omega(k_0)t)} \int_{-\infty}^{\infty} dq\, A(q+k_0)$$

$$\cdot \exp\left[iq\left(x - \left.\frac{d\omega(k)}{dk}\right|_{k=k_0} t\right)\right] ,$$

where the variable q is given by $q = k - k_0$. The whole packet moves (if this approximation is adequate) with the velocity

$$v_{gr}(k_0) = \left.\frac{d\omega(k)}{dk}\right|_{k=k_0} \qquad (1.52)$$

(compare the argument on p. 18). The two velocities are equal in a vacuum with a linear relation between ω and the wave number

$$v_{ph}(\text{vacuum}) = \frac{ck_0}{k_0} = c \qquad v_{gr}(\text{vacuum}) = \left.\frac{d(ck)}{dk}\right|_{k=k_0} = c.$$

1.4 Energy and Momentum of the Electromagnetic Field

Mechanical waves (e.g. water waves) transport energy and momentum. This could be checked by a simple experiment (e.g. at the south-east coast of Australia or in Hawaii). Electromagnetic waves do also transport energy and momentum. This statement can be verified with the aid of Maxwell's equations.

1.4.1 Energy Transport by Electromagnetic Waves

Starting point for the discussion of the energy law of electrodynamics are the equations

$$\nabla \times \boldsymbol{E}(\boldsymbol{r},t) = -k_f \frac{\partial}{\partial t} \boldsymbol{B}(\boldsymbol{r},t)$$

$$\nabla \times \boldsymbol{H}(\boldsymbol{r},t) = 4\pi k_h \boldsymbol{j}_{tr}(\boldsymbol{r},t) + \frac{k_h}{k_d} \frac{\partial}{\partial t} \boldsymbol{D}(\boldsymbol{r},t).$$

The first one is used to form the scalar product with $\boldsymbol{H}$, the second with $\boldsymbol{E}$. Subtraction of the resulting equation gives

$$\boldsymbol{H} \cdot (\nabla \times \boldsymbol{E}) - \boldsymbol{E} \cdot (\nabla \times \boldsymbol{H})$$
$$= -\left(k_f \boldsymbol{H} \cdot \frac{\partial}{\partial t} \boldsymbol{B} + \frac{k_h}{k_d} \boldsymbol{E} \cdot \frac{\partial}{\partial t} \boldsymbol{D}\right) - 4\pi k_h \boldsymbol{j}_{tr} \cdot \boldsymbol{E}.$$

In order to interpret this result, the individual terms are written differently. For the left hand side one writes

$$\nabla \cdot (\boldsymbol{E} \times \boldsymbol{H}) = \boldsymbol{H} \cdot (\nabla \times \boldsymbol{E}) - \boldsymbol{E} \cdot (\nabla \times \boldsymbol{H}).$$

The term with the derivative with respect to time is written in the form

$$k_f \boldsymbol{H} \frac{\partial}{\partial t} \boldsymbol{B} + \frac{k_m}{k_e} \boldsymbol{E} \frac{\partial}{\partial t} \boldsymbol{D} = \frac{1}{2} \frac{\partial}{\partial t} \left(k_f \boldsymbol{H} \cdot \boldsymbol{B} + \frac{k_h}{k_d} \boldsymbol{E} \cdot \boldsymbol{D} \right)$$

so that direct sorting yields the relation

$$j_{tr}(r,t) \cdot E(r,t) + \frac{1}{4\pi k_h} \nabla \cdot [E(r,t) \times H(r,t)]$$
$$+ \frac{1}{8\pi} \frac{\partial}{\partial t} \left(\frac{k_f}{k_h} B(r,t) \cdot H(r,t) + \frac{1}{k_d} D(r,t) \cdot E(r,t) \right) = 0 . \quad (1.53)$$

The term with $(D \cdot E)/(8\pi k_d)$ has already appeared in the discussion of the stationary energy. This scalar product describes the energy density w_{el}, which is stored in the electric fields in the stationary limit. This energy density depends on time in electrodynamics. The term in B and H represents the energy density, which is stored in the magnetic fields

$$w_{\mathrm{mag}}(r,t) = \frac{k_f}{8\pi k_h} B(r,t) \cdot H(r,t) . \quad (1.54)$$

The interpretation of the term with the scalar product $j_{tr} \cdot E$ can be obtained with the following argument: The current density represents charges moving in the fields and produce the fields. The time dependent work of a homogeneous field on a point charge q, which moves in the combined E-, B-fields, can be calculated by

$$\frac{dA}{dt} = F \cdot \frac{ds}{dt} = q E \cdot v .$$

The magnetic field does not do any work on the charge, because of the form of the Lorentz force. The explicit result for the work can be changed into a form with the current density, if the relation $qv = j_{tr} dV$ is used. One then finds, that the work per unit time, which is expended by an electric field on moving charges, is generally

$$\frac{dA(r,t)}{dt} = \left[E(r,t) \cdot j_{tr}(r,t) \right] dV .$$

This term describes, according to this interpretation, transfer of electromagnetic energy into the mechanical energy of charged particles. If the charges move in a medium, as e.g. a conductor, the mechanical energy is transferred into thermal energy by collision processes with the metal lattice. For this reason this term is referred to as **Joule's heat term.**

The total electromagnetic energy density can be abbreviated in the form

$$w_{\mathrm{em}}(r,t) = \frac{1}{8\pi} \left(\frac{1}{k_d} D(r,t) \cdot E(r,t) + \frac{k_f}{k_h} B(r,t) \cdot H(r,t) \right) , \quad (1.55)$$

1.4 Energy and Momentum of the Electromagnetic Field

so that the **energy conservation law of electrodynamics** can be stated in differential form as

$$\boldsymbol{j}_{tr}(\boldsymbol{r},t) \cdot \boldsymbol{E}(\boldsymbol{r},t) + \frac{\partial}{\partial t} w_{em}(\boldsymbol{r},t) + \frac{1}{4\pi k_m} \nabla \cdot [\boldsymbol{E}(\boldsymbol{r},t) \times \boldsymbol{H}(\boldsymbol{r},t)] = 0. \quad (1.56)$$

The term $\nabla \cdot (\boldsymbol{E} \times \boldsymbol{H})/(4\pi k_m)$ can be interpreted if one considers energies rather than the energy densities. Integration of the differential form (1.56) over a volume of space with the fields

$$\iiint_V \boldsymbol{j}_{tr} \cdot \boldsymbol{E}\, dV + \frac{\partial}{\partial t} \iiint_V w_{em}\, dV + \frac{1}{4\pi k_h} \iiint_V \nabla \cdot [\boldsymbol{E} \times \boldsymbol{H}]\, dV = 0$$

and use of the divergence theorem for the term in question leads to

$$\iiint_V \nabla \cdot [\boldsymbol{E} \times \boldsymbol{H}]\, dV = \oiint_{O(V)} [\boldsymbol{E} \times \boldsymbol{H}] \cdot d\boldsymbol{f}\,.$$

The surface integral can (as usual) be interpreted as a flux, in this case according to the dimension of the integrand a flux of electromagnetic energy through the surface of the given volume. This flux of energy is given by the vector

$$\boldsymbol{S}(\boldsymbol{r},t) = \frac{1}{4\pi k_h} [\boldsymbol{E}(\boldsymbol{r},t) \times \boldsymbol{H}(\boldsymbol{r},t)]\,. \quad (1.57)$$

This vector is referred to as the **Poynting vector**. The unit of this vector is

$$[S] = \left[\frac{\text{energy}}{\text{surface} \cdot \text{time}}\right].$$

An alternative interpretation on the basis of the dimensions involved, which does more justice to the vector character of this quantity, can be obtained by considering

$$\left[\frac{S}{c^2}\right] = \left[\frac{\text{momentum} \cdot \text{length}}{\text{time}} \frac{1}{\text{surface} \cdot \text{time}} \frac{\text{time}^2}{\text{length}^2}\right]$$

$$= \left[\frac{\text{momentum}}{\text{volume}}\right].$$

The vector $\boldsymbol{S}/c^2$ thus represents the momentum density of the electromagnetic field.

The energy conservation law of electrodynamics in integral form can be expressed with this definition as

$$\frac{d}{dt}\iiint_V w_{em}(r,t)\,dV = -\iiint_V (j_{tr}(r,t)\cdot E(r,t))\,dV$$

$$-\oiint_{O(V)} S(r,t)\cdot df\,. \quad (1.58)$$

In words: The variation of electromagnetic field energy with time in a volume manifests itself in mechanical energy (motion of charges and generation of heat) and the energy, which is introduced or taken out through the surface of the volume (by electromagnetic waves). The Poynting vector is the proper quantity for the discussion of the energy flow of an electromagnetic field. It describes the direction and the strength. Under the assumption of a simple material equation as (1.47)

$$H(r,t) = \sqrt{\frac{\varepsilon}{\mu}}\frac{k_h}{ckk_f k_m}(k\times E(r,t))$$

the expression for the Poynting vector of a plane wave S_{pw} is

$$S_{pw}(r,t) = \frac{c}{4\pi k_e}\sqrt{\frac{\varepsilon}{\mu}}\,[E(r,t)\times(k\times E(r,t))]\,\frac{1}{k}\,.$$

An alternative form of this expression can be given, if a standard formula for the vector product (and the relation $E\cdot k = 0$) is used

$$S_{pw}(r) = \frac{c}{4\pi k_e}\sqrt{\frac{\varepsilon}{\mu}}\,(E(r,t)\cdot E(r,t))\,\frac{k}{k}\,. \quad (1.59)$$

If one works with the complex representation of the fields, only the real part should be inserted here

$$\mathrm{Re}\,(E(r,t)) = \tilde{E}\,\cos(kr - \omega t + \varphi)\,.$$

The mean value of the Poynting vector in time is of interest for the discussion of the situation of intensities

$$\bar{S}_{pw}(r) = \frac{1}{T}\int_0^T S_{pw}(r,t)\,dt \qquad \left(T = \frac{2\pi}{\omega}\right)\,.$$

The result of a direct calculation (see Detail 1.7.4) is

$$\bar{S}_{\text{pw}}(r) = \frac{c}{8\pi k_e} \sqrt{\frac{\varepsilon}{\mu}} |\tilde{E}|^2 \frac{k}{k} \equiv \bar{S}_{\text{pw}} . \quad (1.60)$$

The mean value of the Poynting vector of a plane wave is the same for every point of space. It can also be given in complex form

$$\bar{S}_{\text{pw}}(r,t) = \frac{1}{8\pi k_h} \left[E(r,t) \times H^*(r,t) \right] . \quad (1.61)$$

It is sufficient for a proof of this statement to insert the expression for H into the expression above. The mean value of the Poynting vector in time is proportional to the intensity $|\tilde{E}|^2$ of the electromagnetic radiation. It also contains the information concerning the direction of the flow of energy. As expected this is the direction of the propagation of the plane wave.

1.4.2 Momentum Conservation

Momentum conservation in electrodynamics follows from the discussion of a distribution of charges in an electromagnetic field. The charges q_i and the velocities v_i in the equation of motion for the total momentum of a system of charges $(q_1, q_2, \ldots)$[8]

$$\frac{d p_{\text{mech}}(t)}{dt} = \sum_i q_i \left\{ E(r,t) + k_f [v_i \times B(r,t)] \right\}$$

are first replaced

$$\sum_i q_i \longrightarrow \iiint \rho_{tr}(r,t) \, dV \quad \text{and} \quad \sum_i q_i v_i \longrightarrow \iiint j_{tr}(r,t) \, dV .$$

The result is

$$\frac{d p_{\text{mech}}(t)}{dt} = \iiint \left\{ \rho_{tr}(r,t) E(r,t) + k_f [j_{tr}(r,t) \times B(r,t)] \right\} dV . \quad (1.62)$$

Note, that the magnetic forces contribute to the balance of the momentum.

---

[8] Compare Dreizler and Lüdde, 2024, Electrostatics and Magnetostatics (Springer Berlin Heidelberg), Eq. (5.48) $m\dot{v} = qE + q k_f [v \times B]$.

The charge and current distributions are replaced in the next step with the aid of the Maxwell equations (1.21). The resulting expression

$$\frac{d\boldsymbol{p}_{\text{mech}}}{dt} = \iiint \left\{ \frac{(\nabla \cdot \boldsymbol{D})\boldsymbol{E}}{4\pi k_d} - \frac{k_f}{4\pi k_d} \frac{\partial \boldsymbol{D}}{\partial t} \times \boldsymbol{B} \right.$$

$$\left. + \frac{k_f}{4\pi k_h} (\nabla \times \boldsymbol{H}) \times \boldsymbol{B} \right\} dV$$

is then sorted in the following fashion: The term with the partial derivatives is replaced by

$$\frac{\partial(\boldsymbol{D} \times \boldsymbol{B})}{\partial t} = \frac{\partial \boldsymbol{D}}{\partial t} \times \boldsymbol{B} + \boldsymbol{D} \times \frac{\partial \boldsymbol{B}}{\partial t}$$

and elimination of the time derivative of the $\boldsymbol{B}$-field with the law of induction (1.4) gives

$$= \frac{\partial \boldsymbol{D}}{\partial t} \times \boldsymbol{B} - \frac{1}{k_f} \boldsymbol{D} \times (\nabla \times \boldsymbol{E}) \,.$$

The derivative of the cross product of the $\boldsymbol{D}$- and the $\boldsymbol{B}$-field corresponds, up to a factor to the derivative of the Poynting vector (1.57) (provided one can assume simple material equations),

$$\frac{k_f}{4\pi k_d} \frac{\partial}{\partial t} (\boldsymbol{D}(\boldsymbol{r},t) \times \boldsymbol{B}(\boldsymbol{r},t)) = \frac{\varepsilon \mu}{c^2} \frac{\partial \boldsymbol{S}(\boldsymbol{r},t)}{\partial t} \,.$$

The quantity $\boldsymbol{S}/c^2$ has been identified in Chap. 1.4.1 as a momentum density. This suggests that the volume integral of this quantity can be used to define the momentum of the electromagnetic field

$$\boldsymbol{p}_{\text{field}}(t) = \frac{\varepsilon \mu}{c^2} \iiint \frac{\partial \boldsymbol{S}(\boldsymbol{r},t)}{\partial t} dV \,. \tag{1.63}$$

The reorganisation of these equations up to this point can be summed up with

$$\frac{d}{dt}(\boldsymbol{p}_{\text{mech}} + \boldsymbol{p}_{\text{field}}) = \iiint \left\{ \frac{1}{4\pi k_d} \left[(\nabla \cdot \boldsymbol{D})\boldsymbol{E} - \boldsymbol{D} \times (\nabla \times \boldsymbol{E}) \right] dV \right.$$

$$\left. - \frac{k_f}{4\pi k_h} \boldsymbol{B} \times (\nabla \times \boldsymbol{H}) \right\} \,. \tag{1.64}$$

The left hand side represents the time derivative of the total momentum of the system charges and electromagnetic field. It is necessary to convert the contributions on the right hand side into a surface integral, which represents the flow of momentum out of

1.4 Energy and Momentum of the Electromagnetic Field

the volume under consideration, in order to find a conversation law. The integrand of the volume integral contains an electrical and a magnetic contribution. For the vector of the electric fields one finds, again with the simple material equation, the components ($i = 1, 2, 3$) (see Detail 1.7.5)

$$\left[(\nabla \cdot \boldsymbol{D})\boldsymbol{E} - \boldsymbol{D} \times (\nabla \times \boldsymbol{E})\right]_i = \varepsilon \frac{k_d}{k_e} \sum_{k=1}^{3} \frac{\partial}{\partial x_k} \left(E_i E_k - \frac{1}{2} E^2 \delta_{i,k} \right) .$$

The magnetic contribution has a similar structure, as it can be complemented with a term proportional to $(\nabla \cdot \boldsymbol{B})\boldsymbol{H} = \boldsymbol{0}$ in order to obtain a corresponding set of components of a magnetic vector.

One can now define the elements of a symmetric tensor (of second rank)

$$T_{ik} = \frac{\varepsilon}{4\pi k_e} \left(E_i E_k - \frac{1}{2} E^2 \delta_{i,k} \right) + \frac{k_f}{4\pi \mu k_m} \left(B_i B_k - \frac{1}{2} B^2 \delta_{i,k} \right) , \qquad (1.65)$$

which is called **Maxwell's stress tensor**. The intermediate result (1.64) can be written with the elements of the stress tensor in a component form[9]

$$\frac{\mathrm{d}}{\mathrm{d}t} \left[\boldsymbol{p}_{\text{mech}}(t) + \boldsymbol{p}_{\text{field}}(t) \right]_i = \iiint \sum_{k=1}^{3} \frac{\partial}{\partial x_k} T_{ik}(\boldsymbol{r}, t) \, \mathrm{d}V .$$

The integrand can be interpreted as the divergence of a vector

$$\boldsymbol{T}_i = (T_{i1}, T_{i2}, T_{i3}) ,$$

so that one writes with the theorem of Gauss

$$\frac{\mathrm{d}}{\mathrm{d}t} \left[\boldsymbol{p}_{\text{mech}}(t) + \boldsymbol{p}_{\text{field}}(t) \right]_i = \iiint_V \nabla \cdot \boldsymbol{T}_i \, \mathrm{d}V = \oiint_{S(V)} \mathrm{d}\boldsymbol{f} \cdot \boldsymbol{T}_i \qquad (1.66)$$

or explicitly

$$\frac{\mathrm{d}}{\mathrm{d}t} \left[\boldsymbol{p}_{\text{mech}}(t) + \boldsymbol{p}_{\text{field}}(t) \right]_i = \oiint_{S(V)} \sum_n e_n T_{in} \, \mathrm{d}f .$$

The quantity $\boldsymbol{e}_n \cdot \boldsymbol{T}_i$, which involves a unit vector representing the normal with respect to the surface, can be interpreted as the components of a force per unit area, which acts on the surface $S(V)$. An alternative interpretation would be: This

---

[9] A relativistic variant is discussed in Chap. 3.4.3.

quantity is the flow of momentum per unit area out of the volume through the surface in the direction of the corresponding coordinate. Equation (1.66) can be applied for the calculation of the force on an object in the electromagnetic field.

The conservation law of momentum in electrodynamics states: The change in time of the total momentum of a system of charges and fields in a volume is equal to the flow of the momentum through the surface of the volume. If the flow vanishes, the total momentum $\boldsymbol{p}_{\text{mech}}(t) + \boldsymbol{p}(t)$ is conserved.

This is the statement, which was indicated in Dreizler and Lüdde, 2024, Electrostatics and Magnetostatics (Springer Berlin Heidelberg), Chap. 5.5 concerning the fact, that inclusion of the momentum of the fields lead to a law of momentum conservation, even though the magnetic forces between charges do not satisfy the third axiom of mechanics.

1.5 Electromagnetic Potentials

An alternative version of the complete Maxwell equations (with a time dependent distribution of charges and currents) is useful for their solution. These are

$$\nabla \cdot \boldsymbol{B}(\boldsymbol{r},t) = 0 \qquad \nabla \times \boldsymbol{B}(\boldsymbol{r},t) = \frac{\varepsilon \mu k_m}{k_e} \frac{\partial \boldsymbol{E}(\boldsymbol{r},t)}{\partial t} + 4\pi \mu k_m \boldsymbol{j}_{tr}(\boldsymbol{r},t)$$

$$\nabla \cdot \boldsymbol{E}(\boldsymbol{r},t) = \frac{4\pi k_e}{\varepsilon} \rho_{tr}(\boldsymbol{r},t) \qquad \nabla \times \boldsymbol{E}(\boldsymbol{r},t) = -k_f \frac{\partial}{\partial t} \boldsymbol{B}(\boldsymbol{r},t)$$

for the case that a simple material equation can be used. The task is the calculation of $\boldsymbol{E}(\boldsymbol{r},t)$ and $\boldsymbol{B}(\boldsymbol{r},t)$ for a given set of functions for the sources $\rho_{tr}(\boldsymbol{r},t)$ and $\boldsymbol{j}_{tr}(\boldsymbol{r},t)$. A simplification of the task might be possible, if one attempts to find a way to the solution with, as in the stationary limit, the aid of a formulation of the problem at hand in terms of time dependent potentials.

The relation $\nabla \cdot \boldsymbol{B} = 0$ allows, as in the stationary case, the introduction of a magnetic vector potential

$$\boldsymbol{B}(\boldsymbol{r},t) = \nabla \times \boldsymbol{A}(\boldsymbol{r},t) \ . \tag{1.67}$$

The electric field can, however, not be represented simply by a scalar potential, as the relation $\nabla \times \boldsymbol{E} = 0$ is not valid. It is possible, on the other hand, to insert the representation (1.67) of the $\boldsymbol{B}$-field into the law of induction, so that

$$\nabla \times \boldsymbol{E}(\boldsymbol{r},t) = -k_f \frac{\partial}{\partial t} \left(\nabla \times \boldsymbol{A}(\boldsymbol{r},t) \right)$$

follows. After an interchange of the sequence of differentiation

$$\nabla \times \left(\boldsymbol{E}(\boldsymbol{r},t) + k_f \frac{\partial \boldsymbol{A}(\boldsymbol{r},t)}{\partial t} \right) = 0$$

1.5 Electromagnetic Potentials

one realises, that the vector function in brackets can be represented as the gradient of a scalar function

$$E(r,t) + k_f \frac{\partial A(r,t)}{\partial t} = -\nabla V(r,t)$$

or

$$E(r,t) = -\left(\nabla V(r,t) + k_f \frac{\partial A(r,t)}{\partial t}\right). \tag{1.68}$$

The representation of the two fields with six components by four functions has to be introduced into the remaining two Maxwell equations. The resulting equations for the potential functions are

$$\Delta V(r,t) + k_f \frac{\partial}{\partial t} \nabla A(r,t) = -\frac{4\pi k_e}{\varepsilon} \rho_{tr}(r,t)$$

$$\tag{1.69}$$

$$\Delta A(r,t) - \frac{\varepsilon\mu}{c^2} \frac{\partial^2 A(r,t)}{\partial t^2} - \nabla\left(\nabla \cdot A(r,t) + \frac{\varepsilon\mu k_m}{k_e} \frac{\partial V(r,t)}{\partial t}\right)$$

$$= -4\pi \mu k_m j_{tr}(r,t).$$

These differential equations do not look very promising, as the potentials are coupled. A possible way out of this complication can be found by recalling, that the stationary limit of the potential equations could be handled more easily if one relied on the gauge freedom. The gauge introduced in this limit was the Coulomb gauge, which can be extended to

$$\nabla \cdot A(r,t) = 0. \tag{1.70}$$

This gauge can be used in the time dependent situation in order to treat the case of wave propagation with ($\rho_{tr} = 0$, $j_{tr} = 0$). The equation, which is obtained for the potential $V(r,t)$, is then

$$\Delta V(r,t) = 0$$

with the possible solution $V(r,t) = 0$. The equation for the vector potential $A(r,t)$ is

$$\Delta A(r,t) - \frac{\varepsilon\mu}{c^2} \frac{\partial^2 A(r,t)}{\partial t^2} = 0.$$

The solution of this (vector) differential equation, a wave equation, can be applied to calculate the measurable fields

$$E(r,t) = -k_f \frac{\partial A(r,t)}{\partial t} \qquad B(r,t) = \nabla \times A(r,t) \,.$$

The condition $\nabla \cdot A = 0$ is equivalent to the statement: The vector A is (as are the fields) perpendicular to the direction of propagation of the wave. The extension of the Coulomb gauge in the dynamical case is therefore also called the **transverse gauge**.

A different gauge is required to deal with the sender problem. The gauge, which is preferred for this problem, is the **Lorentz gauge**, which is defined by the transformation

$$A'(r,t) = A(r,t) + \nabla f(r,t)$$

$$V'(r,t) = V(r,t) - k_f \frac{\partial}{\partial t} f(r,t) \,.$$

The function f is an arbitrary function, which is a solution of the wave equation

$$\Delta f(r,t) - \frac{\varepsilon \mu}{c^2} \frac{\partial^2 f(r,t)}{\partial t^2} = 0 \,.$$

The four potential functions have to satisfy the gauge conditions

$$\nabla \cdot A(r,t) + \frac{\varepsilon \mu k_m}{k_e} \frac{\partial V(r,t)}{\partial t} = \nabla \cdot A'(r,t) + \frac{\varepsilon \mu k_m}{k_e} \frac{\partial V'(r,t)}{\partial t} = 0 \,,$$

because of the relations

$$\nabla \cdot A' + \frac{\varepsilon \mu k_m}{k_e} \frac{\partial V'}{\partial t} = \nabla \cdot A + \Delta f + \frac{\varepsilon \mu k_m}{k_e} \frac{\partial V}{\partial t} - \frac{\varepsilon \mu}{c^2} \frac{\partial^2 f}{\partial t^2}$$

$$= \nabla \cdot A + \frac{\varepsilon \mu k_m}{k_e} \frac{\partial V}{\partial t} = 0 \,.$$

The physical content is not changed by this transformation

$$B' = \nabla \times A' = \nabla \times A = B$$

$$E' = -\nabla V' - k_f \frac{\partial A'}{\partial t} = -\nabla V + k_f \nabla \frac{\partial}{\partial t} f - k_f \frac{\partial A}{\partial t} - k_f \frac{\partial}{\partial t} \nabla f$$

$$= -\nabla V - k_f \frac{\partial A}{\partial t} = E \,.$$

In the Lorentz gauge

$$\nabla \cdot \mathbf{A}(\mathbf{r},t) + \frac{\varepsilon \mu k_m}{k_e}\frac{\partial V(\mathbf{r},t)}{\partial t} = 0 \qquad (1.71)$$

the differential equations (1.69) for the potential functions are

$$\Delta \mathbf{A}(\mathbf{r},t) - \frac{\varepsilon \mu}{c^2}\frac{\partial^2}{\partial t^2}\mathbf{A}(\mathbf{r},t) = -4\pi \mu k_m \, \mathbf{j}_{tr}(\mathbf{r},t) \qquad (1.72)$$

$$\Delta V(\mathbf{r},t) - \frac{\varepsilon \mu}{c^2}\frac{\partial^2}{\partial t^2}V(\mathbf{r},t) = -\frac{4\pi k_e}{\varepsilon}\rho_{tr}(\mathbf{r},t) \ . \qquad (1.73)$$

One finds four inhomogeneous wave equation. The same number of inhomogeneous equations is also encountered in the discussion of original Maxwell theory in terms of fields. The formulation with electromagnetic potentials is completely equivalent to the original theory with fields. The homogeneous Maxwell equations are taken care of by the definition of the potentials.

The differential operator on the left hand side of (1.72) and (1.73) is usually abbreviated in the form ($\varepsilon = \mu = 1$)

$$\Box \mathbf{A}(\mathbf{r},t) \qquad \text{resp.} \qquad \Box V(\mathbf{r},t)$$

for the vacuum theory. The operator $\Box$

$$\Box = \Delta - \frac{1}{c^2}\frac{\partial^2}{\partial t^2} \qquad (1.74)$$

is the **d'Alembert operator**. The differential equations (1.72) and (1.73) are also referred to as the inhomogeneous d'Alembert equations.

The solution of the inhomogeneous wave equation will be discussed after this formal but useful transcription of the theory.

1.6 The Solution of the Inhomogeneous Wave Equation

The four inhomogeneous differential equations, which are discussed in this section, have the same structure. They are linear, inhomogeneous partial differential equations of second order in four variables, as e.g.

$$\Delta V(\mathbf{r},t) - \frac{\varepsilon \mu}{c^2}\frac{\partial^2 V(\mathbf{r},t)}{\partial t^2} = -\frac{4\pi k_e}{\varepsilon}\rho_{tr}(\mathbf{r},t) \ .$$

The structure of the general solution is the same as for ordinary differential equations, namely the sum of the general solution of the homogeneous differential equation and a special solution of the inhomogeneous differential equation

$$V(\mathbf{r}, t) = V_{\text{hom}}(\mathbf{r}, t) + V_{\text{part}}(\mathbf{r}, t) \,.$$

The general solution of the homogeneous differential equation can be given in the form of a Fourier integral

$$V_{\text{hom}}(\mathbf{r}, t) = \iiint d^3k \left\{ c_+(\mathbf{k}) \, e^{i(\mathbf{k} \cdot \mathbf{r} + \omega t)} + c_-(\mathbf{k}) \, e^{i(\mathbf{k} \cdot \mathbf{r} - \omega t)} \right\} \,. \tag{1.75}$$

The discussion of an appropriate particular solution of the inhomogeneous differential equations is sufficient for the treatment of the sender problem. Such solutions can, as in the stationary limit, be obtained with the method of Green's functions

$$V_{\text{part}}(\mathbf{r}, t) = \frac{k_e}{\varepsilon} \iiint d^3r' \int dt' \, G(\mathbf{r}, t, \mathbf{r}', t') \, \rho_{tr}(\mathbf{r}', t') \,. \tag{1.76}$$

The Green's function is now a function of eight variables. Insertion of the ansatz (1.76) into the differential equation for the potential

$$\iiint d^3r' \int dt' \left\{ \Delta G(\mathbf{r}, t, \mathbf{r}', t') - \frac{\varepsilon\mu}{c^2} \frac{\partial^2}{\partial t^2} G(\mathbf{r}, t, \mathbf{r}', t') \right\} \rho_{tr}(\mathbf{r}', t')$$

$$= -4\pi \, \rho_{tr}(\mathbf{r}, t)$$

shows, that the Green's function has to satisfy the differential equation

$$\left\{ \Delta - \frac{\varepsilon\mu}{c^2} \frac{\partial^2}{\partial t^2} \right\} G(\mathbf{r}, t, \mathbf{r}', t') = -4\pi \, \delta(\mathbf{r} - \mathbf{r}') \delta(t - t') \tag{1.77}$$

in a medium or

$$\Box G(\mathbf{r}, t, \mathbf{r}', t') = -4\pi \, \delta(\mathbf{r} - \mathbf{r}') \delta(t - t')$$

in a vacuum. The derivatives act on the unprimed coordinates. In extension of the remarks for the stationary Green's function one can offer the following comments on the dynamic Green's function: The differential equation (1.77) is an inhomogeneous wave equation for a point source with the strength $q = 1$, which can be found at the position $\mathbf{r}'$ at the time t'. This moving charge produces a wave, which propagates from the space-time point $\mathbf{r}', t'$ to the space-time point $\mathbf{r}, t$.

In order to find a solution one can, as in the case of an inhomogeneous, linear, ordinary differential equation, rely on the **method of indefinite coefficients**. For a partial differential equation it is necessary to have an ansatz with an arbitrary

1.6 The Solution of the Inhomogeneous Wave Equation

number of coefficients. This requirement can be satisfied with an expansion of the particular solution in terms of 'plane waves' with the coefficients $g(\mathbf{k}, \omega)$

$$V_{\text{part}}(\mathbf{r}, t) = \frac{k_e}{\varepsilon} \int_{-\infty}^{\infty} d\omega \iiint d^3k \, g(\mathbf{k}, \omega) \, e^{i(\mathbf{k}\cdot\mathbf{r} - \omega t)}. \tag{1.78}$$

This is a Fourier representation in a four-dimensional space (position and time). The four parameters ω and $\mathbf{k}$ are independent quantities. The usual relation $k^2 = \omega^2/c^2$ is not valid in this case. The orthogonality relation of the four-dimensional plane waves is

$$\int dt \iiint d^3r \, e^{-i(\mathbf{k}\cdot\mathbf{r} - \omega t)} e^{i(\mathbf{k}'\cdot\mathbf{r} - \omega' t)} = (2\pi)^4 \, \delta^{(3)}(\mathbf{k} - \mathbf{k}')\delta(\omega - \omega').$$

One obtains four δ-functions as the exponential functions factorise. Insertion of (1.78) into the differential equation (1.73) yields

$$\int d\omega \iiint d^3k \left(-k^2 + \frac{\varepsilon\mu\omega^2}{c^2}\right) g(\mathbf{k}, \omega) \, e^{i(\mathbf{k}\cdot\mathbf{r} - \omega t)} = -4\pi \, \rho_{tr}(\mathbf{r}, t).$$

This equation can be resolved with respect to the function g. One multiplies the equation with $e^{-i(\mathbf{k}'\cdot\mathbf{r} - \omega' t)}$, integrates over the primed space and time coordinates to find

$$\int d\omega \iiint d^3k \left(-k^2 + \frac{\varepsilon\mu\omega^2}{c^2}\right) g(\mathbf{k}, \omega) \delta^{(3)}(\mathbf{k} - \mathbf{k}')\delta(\omega - \omega')$$

$$= -\frac{1}{4\pi^3} \int dt' \iiint d^3r' \, e^{-i(\mathbf{k}'\cdot\mathbf{r}' - \omega' t')} \rho_{tr}(\mathbf{r}', t').$$

This intermediate result can be resolved directly

$$g(\mathbf{k}, \omega) = \frac{1}{4\pi^3} \frac{1}{\left(k^2 - \frac{\varepsilon\mu\omega^2}{c^2}\right)} \int dt' \iiint d^3r' \, e^{-i(\mathbf{k}\cdot\mathbf{r} - \omega t)} \rho_{tr}(\mathbf{r}', t').$$

(1.79)

One can finally obtain the Green's function, if this result is inserted into the ansatz (1.78) for the particular solution and compared with (1.76)

$$G(\boldsymbol{r}, t, \boldsymbol{r}', t') = \frac{1}{4\pi^3} \int_{-\infty}^{\infty} d\omega \iiint d^3k \left[\right.$$

$$\left. \frac{1}{\left(k^2 - \frac{\varepsilon\mu\omega^2}{c^2}\right)} e^{i(\boldsymbol{k}\cdot(\boldsymbol{r}-\boldsymbol{r}'))} e^{-i\omega(t-t')} \right]. \quad (1.80)$$

Two points can be observed directly:

1. The Green's function depends only on the time difference $t - t'$ and the position difference $\boldsymbol{r} - \boldsymbol{r}'$

$$G(\boldsymbol{r}, t, \boldsymbol{r}', t') = G(\boldsymbol{r} - \boldsymbol{r}', t - t'). \quad (1.81)$$

2. The expression for the Green's function is singular. The integrand diverges for $k = \pm\omega\sqrt{\varepsilon\mu}/c$. The reason for this complication is found in the fact, that the differential equation for the Green's function has been solved correctly (check this statement by inserting the result into (1.73)), but that some additional physical conditions have not been respected.

One of the conditions in a dynamical theory is **causality**. The solution should incorporate the fact, that the motion of a charge starting at position $\boldsymbol{r}'$ at the time t' should arrive at a later time t at the position $\boldsymbol{r}$. One expects, that a Green's function, which respects causality, has the property

$$G(\boldsymbol{r} - \boldsymbol{r}', t - t') = 0 \quad \text{for} \quad t - t' \le 0. \quad (1.82)$$

The observation, that the wave has arrived at the position $\boldsymbol{r}$, cannot take place before the charge has been moved. A Green's function, which satisfies the condition (1.82) is called a **retarded Green's function** $G^{(+)}$. These quantities are also needed in other branches of physics.

In order to obtain a retarded Green's function, one has to modify the frequency integration.[10] For this reason one has to replace the integration along the real axis

$$I_{\text{real}} = \int_{-\infty}^{\infty} \frac{e^{-i\omega(t-t')}}{\left(k^2 - (\omega^2/c_{\text{med}}^2)\right)} d\omega \quad (c_{\text{med}} = c/\sqrt{\varepsilon\mu}),$$

which is not well defined, by a complex contour integration

$$I_{\text{complex}} = \oint_{C_i} \frac{e^{-i\omega(t-t')}}{\left(k^2 - (\omega^2/c_{\text{med}}^2)\right)} d\omega \quad C_i \to C_1, C_2.$$

---

[10] The background is the analysis of functions of complex variables.

1.6 The Solution of the Inhomogeneous Wave Equation

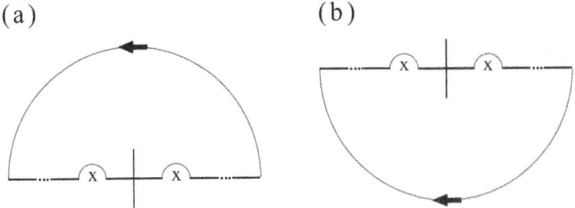

Fig. 1.17 Contour integration for the calculation of the retarded Green's function. (a) For $t - t' < 0$. (b) For $t - t' > 0$

The contour C_1, which is to be used for $(t - t') < 0$ (Fig. 1.17a), runs along the real axis, but avoids the poles of the integrand at

$$\omega = \pm \frac{kc}{\sqrt{\varepsilon\mu}}$$

with infinitesimal half circles in the upper complex half plane. The contour is complemented by a half circle with infinite radius enclosing the upper half plane. This part of the contour does not contribute to the value of the integral and can be added without changing the result. The contour C_1 does not include any singularities of the integrand. According the theorem of Cauchy, the value of this contour integral is zero.

The contour C_2, which should be chosen for $(t - t') > 0$, differs from the contour C_1 by closure with a very large half circle in the lower complex half plane (Fig. 1.17b). Two poles of the integrand are then included within the contour. The integral can be evaluated after a partial fraction decomposition

$$\frac{1}{\left(k^2 - (\omega^2/c_{med}^2)\right)} = \frac{c_{med}}{2k} \left[\frac{1}{(\omega + c_{med}k)} - \frac{1}{(\omega - c_{med}k)} \right]$$

with the Cauchy integral formula. One finds

$$I_{complex} = \frac{2\pi\, c_{med}}{k} \sin(c_{med}k(t - t')) \,.$$

This is to be followed by integration over the space of wave numbers (d^3k), which yields the retarded Green's function

$$G^{(+)}(\mathbf{r} - \mathbf{r}', t - t') = \begin{cases} 0 & t - t' < 0 \\ \dfrac{\delta\left(t - t' - |\mathbf{r} - \mathbf{r}'|/c_{med}\right)}{|\mathbf{r} - \mathbf{r}'|} & t - t' > 0 \end{cases}. \quad (1.83)$$

It satisfies the differential equation and the condition of causality. A charge, which is moved at the time t' at the position r' causes a disturbance at the time

$$t = t' + \frac{|r - r'|}{c_{med}} \geq t'$$

at the position r. The time $|r - r'|/c_{med}$ is exactly the time, which a wave signal needs to cover the distance $|r - r'|$ with the velocity c_{med}. The simple denominator is an indication, that this Green's function describes a case with simple boundary conditions.

The retarded potential $V^{(+)}$ of the inhomogeneous wave equation

$$V^{(+)}(r, t) = \frac{k_e}{\varepsilon} \int dt' \iiint d^3 r' \, G^{(+)}(r, t, r', t') \rho_{tr}(r', t') \tag{1.84}$$

can be obtained explicitly by evaluation of the time integral. This potential is

$$V^{(+)}(r, t) = \frac{k_e}{\varepsilon} \iiint d^3 r' \frac{\rho_{tr}(r', t - |r - r'|/c_{med})}{|r - r'|} . \tag{1.85}$$

With a similar manner one calculates the retarded vector potential

$$A^{(+)}(r, t) = \mu k_m \iiint d^3 r' \frac{j_{tr}(r', t - |r - r'|/c_{med})}{|r - r'|} . \tag{1.86}$$

These formulae are a special form of the **Kirchhoff representation** of the solution of the inhomogeneous wave equation with simple boundary conditions for a given time dependent current and charge distribution. They allow the calculation of the four potentials involved and hence the corresponding fields.

1.7 Details

1.7.1 The Transformation of Faraday's Law

The aim is to prove that

$$\frac{d}{dt} B(r, t) = \frac{\partial}{\partial t} B(r, t) + \nabla \times (B \times v)$$

is valid. The magnetic field B is dependent on r and t, so we have to use the chain rule

$$\frac{d}{dt} = \frac{\partial}{\partial t} + v \cdot \nabla$$

1.7 Details

to rewrite the term

$$\frac{d}{dt} B(r,t) = \frac{\partial}{\partial t} B(r,t) + (v \cdot \nabla) B \ .$$

With the product rule for the application of the ∇-operator (see Appendix C.3)

$$\nabla \times (A \times B) = A(\nabla \cdot B) - B(\nabla \cdot A) + (B \cdot \nabla)A - (A \cdot \nabla)B$$

$$\Longleftrightarrow$$

$$(A \cdot \nabla)B = A(\nabla \cdot B) - B(\nabla \cdot A) + (B \cdot \nabla)A - \nabla \times (A \times B)$$

one obtains for $(v \cdot \nabla)B$

$$(v \cdot \nabla)B = -B(\nabla \cdot v) + (B \cdot \nabla)v - \nabla \times (v \times B) \ . \tag{1.87}$$

The term $v(\nabla \cdot B)$ vanishes because of $\nabla \cdot B = \mathrm{div}\, B = 0$.
The detailed form of the first two terms of the right hand side of (1.87) is

$$B(\nabla \cdot v) = B \left(\frac{\partial v_x}{\partial x} + \frac{\partial v_y}{\partial y} + \frac{\partial v_z}{\partial z} \right)$$

and

$$(B \cdot \nabla)v = \left(B_x \frac{\partial}{\partial x} + B_y \frac{\partial}{\partial y} + B_z \frac{\partial}{\partial z} \right) v$$

$$= \left(B_x \frac{\partial}{\partial x} + B_y \frac{\partial}{\partial y} + B_z \frac{\partial}{\partial z} \right) v_x \, e_x$$

$$+ \left(B_x \frac{\partial}{\partial x} + B_y \frac{\partial}{\partial y} + B_z \frac{\partial}{\partial z} \right) v_y \, e_y$$

$$+ \left(B_x \frac{\partial}{\partial x} + B_y \frac{\partial}{\partial y} + B_z \frac{\partial}{\partial z} \right) v_z \, e_z \ .$$

Each term of these equations disappears because $(x = x_1, \ y = x_2, \ z = x_3)$

$$\frac{\partial}{\partial x_i} v_j = \frac{\partial}{\partial x_i} \frac{d}{dt} x_j = \frac{d}{dt} \frac{\partial}{\partial x_i} x_j = 0 \qquad (i = 1, 2, 3) \ .$$

The result is

$$\frac{d}{dt} B(r,t) = \frac{\partial}{\partial t} B(r,t) - \nabla \times (v \times B)$$

$$= \frac{\partial}{\partial t} B(r,t) + \nabla \times (B \times v) \, .$$

1.7.2 Some Properties of Plane Electromagnetic Waves

The free Maxwell equations for the electromagnetic fields lead to wave equations. One can gain additional information on the wave solutions, if these are reintroduced into the original field equations. One example is the transverse character of the waves. Insertion into the Maxwell equations, which contain the operator $\nabla \times$

$$\nabla \times E(r,t) + k_f \frac{\partial}{\partial t} B(r,t) = 0$$

$$\nabla \times B(r,t) - \varepsilon\mu \frac{k_m}{k_e} \frac{\partial}{\partial t} E(r,t) = 0$$

yields, in the notation of Chap. 1.3.3,

$$i \left[(k \times e_1) E_0 - \omega k_f e_2 B_0 \right] e^{i(k \cdot r - \omega t)} = 0 \qquad (1.88)$$

$$i \left[(k \times e_2) B_0 + \varepsilon\mu \frac{\omega k_m}{k_e} e_1 E_0 \right] e^{i(k \cdot r - \omega t)} = 0 \, . \qquad (1.89)$$

From Eq. (1.88) one extracts directly

$$E_0 (k \times e_1) = \omega k_f B_0 e_2 \, .$$

This vector equation can only be satisfied, if $(k \times e_1)$ and e_2 point in the same direction. This fact can be expressed in the form

$$\frac{1}{k} (k \times e_1) = e_2$$

for the two unit vectors. As a consequence, the relation

$$E_0 (k \times e_1) = \omega k_f B_0 e_2 \quad \Longleftrightarrow \quad E_0 = \omega \frac{k_f}{k} B_0 \qquad (1.90)$$

1.7 Details

can be established between the field amplitudes E_0 and B_0. For the vector product $(k \times e_2) = \frac{1}{k}(k \times (k \times e_1))$ one finds because of

$$(k \times (k \times e_1)) = (k \cdot e_1)k - (k \cdot k)e_1 = -k^2 e_1$$

the corresponding relation

$$\frac{1}{k}(k \times e_2) = -e_1 ,$$

and therefore with (1.89)

$$E_0 = \frac{k k_e}{\varepsilon \mu \omega k_m} B_0 . \tag{1.91}$$

If Eqs. (1.90) and (1.91) are compared, one finds with (1.41)

$$\left(\frac{\omega}{k}\right)^2 = \frac{1}{\varepsilon \mu} \frac{k_e}{k_f k_m} = \frac{c^2}{\varepsilon \mu} = v^2 .$$

Equation (1.90) (and (1.91)) is then equivalent to the relation

$$E_0 = \frac{c k_f}{\sqrt{\varepsilon \mu}} B_0 .$$

With the choice of the direction of the propagation $k = (0, 0, k)$ the E-field takes the form (complex notation)

$$E(r,t) = \left(E_x\, e^{i(kz-\omega t)},\ E_y\, e^{i(kz-\omega t)},\ 0 \right) .$$

The complex amplitudes can be written as real magnitude and phase

$$E_x = \tilde{E}_x\, e^{i\alpha_x} \qquad E_y = \tilde{E}_y\, e^{i\alpha_y} \qquad (\tilde{E}_a, \alpha_a\ \text{real}) .$$

The real part of E takes, because of

$$E_a\, e^{i(kz-\omega t)} = \tilde{E}_a\, e^{i(kz-\omega t+\alpha_a)} \qquad a = x,\ y$$

the form

$$E_{\text{real}}(r,t) = \left\{ \tilde{E}_x \cos(kz-\omega t+\alpha_x),\ \tilde{E}_y \cos(kz-\omega t+\alpha_y),\ 0 \right\} .$$

The real part of the **B**-field can, corresponding to (1.47) in Chap. 1.3.3

$$\boldsymbol{B}(\boldsymbol{r},t) = \frac{1}{\omega k_f}(\boldsymbol{k} \times \boldsymbol{E}(\boldsymbol{r},t)) \, ,$$

be calculated with the result

$$\boldsymbol{B}_{\text{real}}(\boldsymbol{r},t) = \frac{1}{\omega k_f}\begin{pmatrix} 0 \\ 0 \\ k \end{pmatrix} \times \begin{pmatrix} \tilde{E}_x \cos(kz - \omega t + \alpha_x) \\ \tilde{E}_y \cos(kz - \omega t + \alpha_y) \\ 0 \end{pmatrix}$$

$$= \frac{k}{\omega k_f}\begin{pmatrix} -\tilde{E}_y \cos(kz - \omega t + \alpha_y) \\ \tilde{E}_x \cos(kz - \omega t + \alpha_x) \\ 0 \end{pmatrix}.$$

1.7.3 Illustration of the Dispersion of Wave Packets

A wave packet, which is often used for the illustration of the dispersion, is a Gauss packet. At the initial time ($t = 0$) it has the form

$$\Psi(x) = \frac{a}{\sqrt{\pi}} e^{-a^2 x^2} .$$

This function represents a curve, which is called a bell-shape curve (Fig. 1.18). It is symmetric with respect to the maximum at the point $x = 0$. The bell is broader for smaller values of a. The prefactor is chosen, that the area under the bell has the value 1.

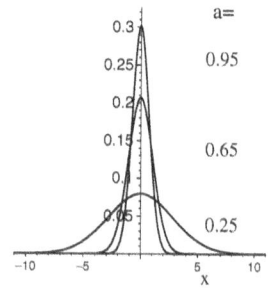

Fig. 1.18 Bell-shaped curves $\Psi(x)$ for different parameters a

1.7 Details

In order to verify this statement, one needs the integral

$$\int_{-\infty}^{\infty} dx\, e^{-x^2} .$$

The evaluation can be achieved, by taking the square, writing the product as a double integral

$$\left(\int_{-\infty}^{\infty} dx\, e^{-x^2}\right)^2 = \int_{-\infty}^{\infty} dx \int_{-\infty}^{\infty} dy\, e^{-(x^2+y^2)}$$

and by introducing plane polar coordinates

$$= \int_0^{2\pi} d\varphi \int_0^{\infty} e^{-r^2} r\, dr = 2\pi \int_0^{\infty} e^{-r^2} r\, dr .$$

The radial integral can be obtained via the substitution $z = r^2$

$$\int_0^{\infty} dr\, r\, e^{-r^2} = \frac{1}{2} \int_0^{\infty} dz\, e^{-z} = \frac{1}{2}$$

so that

$$\int_{-\infty}^{\infty} dx\, e^{-x^2} = \sqrt{\pi} .$$

An often used extension is

$$\int_{-\infty}^{\infty} dx\, e^{-(\alpha x^2 + 2\beta x + \gamma)} = \sqrt{\frac{\pi}{\alpha}}\, e^{(\beta^2 - \alpha\gamma)/\alpha} ,$$

where one has to assume, that the real part of α is larger than zero ($\text{Re}(\alpha) > 0$) and β, γ can be chosen as an arbitrary complex number. It is necessary to use a quadratic complement of the exponent

$$-(\alpha x^2 + 2\beta x + \gamma) = -\alpha\left(x^2 + 2\frac{\beta}{\alpha}x + \frac{\beta^2}{\alpha^2}\right) + \frac{\beta^2}{\alpha} - \gamma$$

and the substitution

$$z = \sqrt{\alpha}\left(x + \frac{\beta}{\alpha}\right) \qquad dx = \frac{dz}{\sqrt{\alpha}} ,$$

in order to obtain

$$\int_{-\infty}^{\infty} dx\, e^{-(\alpha x^2 + 2\beta x + \gamma)} = \frac{1}{\sqrt{\alpha}} e^{(\beta^2 - \alpha\gamma)/\alpha} \int_{-\infty}^{\infty} dz\, e^{-z^2}$$

and hence the result quoted.

The initial Gauss packet can be represented by a Fourier integral

$$\Psi(x) = \int_{-\infty}^{\infty} dk\, A(k)\, e^{ikx}\ . \tag{1.92}$$

The calculation of the Fourier coefficients $A(k)$ can be started with

$$\int_{-\infty}^{\infty} dx\, \Psi(x)\, e^{-ik'x} = \int_{-\infty}^{\infty} dx \int_{-\infty}^{\infty} dk\, A(k)\, e^{i(k-k')x}\ .$$

By interchanging the sequence of the integration on the right hand side, using the properties of the δ-function, one finds

$$\int_{-\infty}^{\infty} dk\, A(k) \int_{-\infty}^{\infty} dx\, e^{i(k-k')x} = 2\pi \int_{-\infty}^{\infty} dk\, A(k)\, \delta(k-k') = 2\pi A(k')\ .$$

The formula quoted can be used on the right hand side

$$\frac{a}{\sqrt{\pi}} \int_{-\infty}^{\infty} dx\, e^{-a^2 x^2} e^{-ik'x} = e^{-k'^2/4a^2}\ .$$

The Fourier amplitudes of the Gauss packet are obtained by assembling all factors

$$A(k) = \frac{1}{2\pi} e^{-k^2/4a^2}\ .$$

The Fourier amplitude is also a bell-shape curve with a width, which is characterised by $1/a^2$. If the bell-shaped curve is wide in space, the Fourier amplitude is localised in momentum (wave number) space (see Figs. 1.19 and 1.20).

The discussion of the time development of the Gauss packet can be calculated with an expansion in terms of plane waves

$$\Phi_k(x, t) = e^{i(kx - \omega(k)t)}\ .$$

1.7 Details

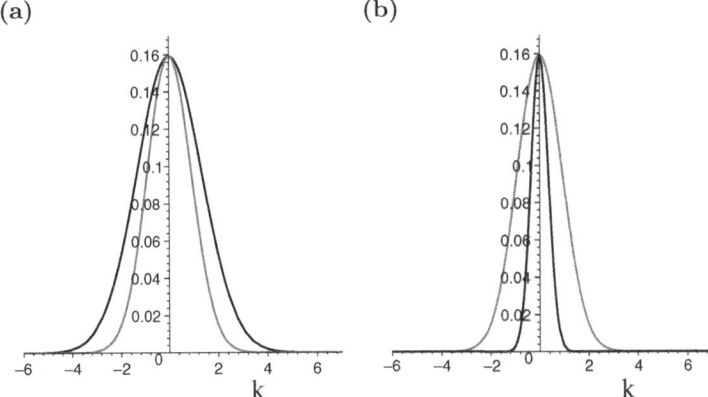

Fig. 1.19 Fourier amplitude $A(k)$ for different parameters a. (**a**) $a = 0.95$ (black), $a = 0.65$ (grey). (**b**) $a = 0.65$ (grey), $a = 0.25$ (black)

The frequency $\omega(k)$ can be given in the form $\omega(k) = ck$ (free packet) respectively $\omega(k) = ck + hk^2$ (dispersive packet). The calculation of the time dependent packets requires

$$\Psi(x,t) = \int_{-\infty}^{\infty} dk\, A(k)\, e^{i(kx - \omega(k)t)},$$

where $A(k)$ is equal to the initial coefficients (see (1.92)).

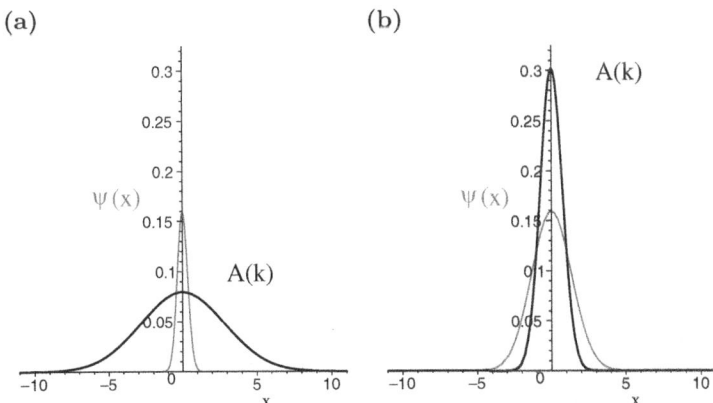

Fig. 1.20 The Fourier amplitude $A(k)$ (black) in comparison with $\Psi(x)$ (grey) for different values of a. (**a**) $a = 0.25$. (**b**) $a = 0.95$

For the case $\omega(k) = ck$ one has to evaluate the integral

$$I_1 = \int_{-\infty}^{\infty} dk\, e^{-k^2/4a^2} e^{ik(x-ct)} \ .$$

This corresponds to the integral considered above

$$\alpha = \frac{1}{4a^2} \qquad \beta = -\frac{i}{2}(x-ct) \ ,$$

so that one can write

$$I_1 = 2a\sqrt{\pi}\, e^{-a^2(x-ct)^2}$$

respectively

$$\Psi_1(x,t) = \frac{a}{\sqrt{\pi}} e^{-a^2(x-ct)^2} \ .$$

The result is a Gauss packet, which moves without dispersion uniformly to the right (Fig. 1.21).
For $\omega(k) = ck + hk^2$, the integral

$$I_2 = \int_{-\infty}^{\infty} dk\, e^{-k^2(1/4a^2 + iht)} e^{ik(x-ct)}$$

with

$$\alpha = \left(\frac{1}{4a^2} + iht\right) \qquad \beta = -\frac{i}{2}(x-ct)$$

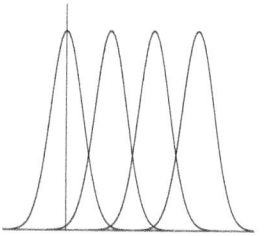

Fig. 1.21 Gauss packet moving without dispersion ($t=0, \Delta t, 2\Delta t, 3\Delta t$)

1.7 Details

has to be calculated

$$I_2 = \frac{\sqrt{\pi}}{\left[\frac{1}{4a^2} + iht\right]^{1/2}} \exp\left[-\frac{(x-ct)^2}{4\left[\frac{1}{4a^2} + iht\right]}\right].$$

Instead of the determination of the real part of this function, one can calculate the simpler form

$$|\Psi(x,t)|^2 = \Psi^*(x,t)\Psi(x,t),$$

in accordance with the usage of quantum mechanics. This function conveys also an idea of the time development of the packet. One finds

$$|I_2|^2 = \frac{\pi}{\left[\frac{1}{16a^4} + h^2t^2\right]^{1/2}} \cdot$$

$$\exp\left[-\frac{(x-ct)^2}{4\left[\frac{1}{16a^4} + h^2t^2\right]}\left(\left[\frac{1}{4a^2} - iht\right] + \left[\frac{1}{4a^2} + iht\right]\right)\right]$$

$$= \frac{4a^2\pi}{\left[1 + 16a^4h^2t^2\right]^{1/2}} \exp\left[-2a^2\frac{(x-ct)^2}{\left[1 + 16a^4h^2t^2\right]}\right]$$

respectively after sorting all factors

$$|\Psi_2(x,t)|^2 = \frac{a^2}{\pi\left[1 + 16a^4h^2t^2\right]^{1/2}} \exp\left[-2a^2\frac{(x-ct)^2}{\left[1 + 16a^4h^2t^2\right]}\right].$$

This function can be compared with

$$|\Psi_1(x,t)|^2 = \frac{a^2}{\pi} \exp\left[-2a^2(x-ct)^2\right].$$

The centre of the wave packet moves also with the velocity c to the right. The prefactor and the additional terms in the exponent cause, however, a broadening of the packet (Fig. 1.22).

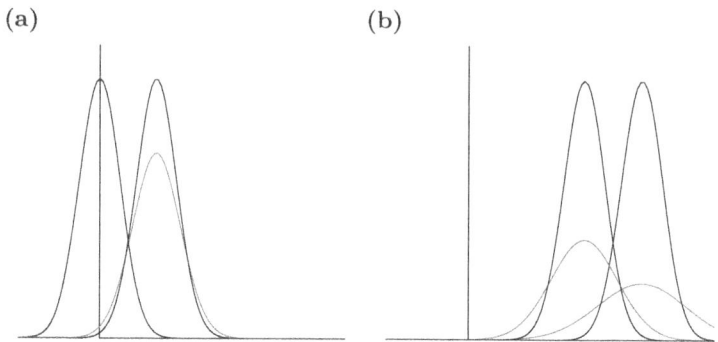

Fig. 1.22 Comparison of $|\Psi_1(x,t)|^2$ (black) and $|\Psi_2(x,t)|^2$ (grey). (**a**) $t = 0, \Delta t$. (**b**) $t = 2\Delta t, 3\Delta t$

1.7.4 The Temporal Mean Value of the Poynting Vector

Often only the mean value over an oscillatory period $T = 2\pi/\omega$ in the case of a sufficiently large frequency of a plane electromagnetic wave is of interest, as the action of the individual phases in a point can not be resolved. The calculation of this mean value is not difficult. In addition, the validity of the formula (1.61) will be shown to be valid.

The claim, that the temporal mean value of the Poynting vector for an electromagnetic plane wave in every point of space has the same value (1.60), can be demonstrated in the following fashion: With the starting point (1.59)

$$S_{\text{pw}}(r) = \frac{c}{4\pi k_e} \sqrt{\frac{\varepsilon}{\mu}} (E(r,t) \cdot E(r,t)) \frac{k}{k}$$

and the real part

$$\text{Re}(E(r,t)) = \tilde{E} \cos(kr - \omega t + \varphi)$$

one finds for

$$E(r,t) \cdot E(r,t) = |\tilde{E}|^2 \cos^2(kr - \omega t + \varphi) .$$

For the calculation of the mean value one uses the addition theorem

$$\int_0^T \cos^2(kr - \omega t + \varphi) \, dx =$$

$$\int_0^T dt \left[\cos^2(kr + \varphi) \cos^2(\omega t) \right.$$

1.7 Details

$$+2\cos(\boldsymbol{kr}+\varphi)\sin(\boldsymbol{kr}+\varphi)\sin\omega t \cos\omega t$$

$$+\sin^2(\boldsymbol{kr}+\varphi)\sin^2(\omega t)\Big].$$

The time integrals needed are (use $T = 2\pi/\omega$)

$$\int_0^T \sin^2(\omega t)\,dt = \left(\frac{1}{2}t + \frac{1}{4\omega}\sin 2\omega t\right)\Big|_0^T = \frac{T}{2}$$

$$\int_0^T \cos^2(\omega t)\,dt = \left(\frac{1}{2}t - \frac{1}{4\omega}\sin 2\omega t\right)\Big|_0^T = \frac{T}{2}$$

$$\int_0^T \sin(\omega t)\cos(\omega t)\,dt = \frac{1}{2\omega}\sin^2\omega t\Big|_0^T = 0,$$

so that the result for

$$\bar{S}_{\text{pw}}(\boldsymbol{r}) = \frac{1}{T}\int_0^T S_{\text{pw}}(\boldsymbol{r},t)\,dt$$

is

$$\bar{S}_{\text{pw}}(\boldsymbol{r}) = \bar{S}_{\text{pw}} = \frac{c}{8\pi k_e}\sqrt{\frac{\varepsilon}{\mu}}|\tilde{\boldsymbol{E}}|^2\frac{\boldsymbol{k}}{k}.$$

The alternative form

$$\bar{S}_{\text{pw}}(\boldsymbol{r},t) = \frac{1}{8\pi k_h}\left[\boldsymbol{E}(\boldsymbol{r},t)\times\boldsymbol{H}^*(\boldsymbol{r},t)\right]$$

results in

$$\boldsymbol{H}^*(\boldsymbol{r},t) = \sqrt{\frac{\varepsilon}{\mu}}\frac{k_h}{ckk_f k_m}\left(\boldsymbol{k}\times\boldsymbol{E}^*(\boldsymbol{r},t)\right)$$

and

$$\boldsymbol{E}(\boldsymbol{r},t) = \boldsymbol{E}_0\,e^{i(\boldsymbol{k}'\cdot\boldsymbol{r}-\omega't)} \qquad \boldsymbol{E}_0 \text{ real}$$

$$\bar{S}_{\text{pw}}(\boldsymbol{r}) = \frac{1}{8\pi k_f k_m}\sqrt{\frac{\varepsilon}{\mu}}\frac{1}{ck}\left[\boldsymbol{E}_0(\boldsymbol{r},t)\times(\boldsymbol{k}\times\boldsymbol{E}_0)^*\right]$$

$$= \frac{c}{8\pi k_e}\sqrt{\frac{\varepsilon}{\mu}}(\boldsymbol{E}_0 \cdot \boldsymbol{E}_0^*)\frac{\boldsymbol{k}}{k} \quad \left(\text{employ } \sqrt{\frac{k_e}{k_f k_m}} = c\right).$$

The difference of the complex amplitude $\boldsymbol{E}_0$ with respect to the real amplitude is only a phase

$$E_{0x} = \tilde{E}_{0x}\, e^{i\alpha_x},$$

so that

$$(\boldsymbol{E}_0 \cdot \boldsymbol{E}_0^*) = \tilde{\boldsymbol{E}} \cdot \tilde{\boldsymbol{E}}$$

holds.

1.7.5 The Elements of the Maxwell Stress Tensor

The discussion of the conservation of the momentum in electrodynamics leads to some non-transparent combinations of field vectors. These have to be analysed in detail in order to obtain the standard definition of the Maxwell stress tensor.

One is asked to show, that the validity of the simple material equations

$$\boldsymbol{D} = \varepsilon \frac{k_d}{k_e}\boldsymbol{E} \qquad \boldsymbol{B} = \mu \frac{k_m}{k_h}\boldsymbol{H}$$

leads to the validity of the relations

$$\left[(\nabla \cdot \boldsymbol{D})\boldsymbol{E} - \boldsymbol{D} \times (\nabla \times \boldsymbol{E})\right]_i = \frac{\varepsilon k_d}{4\pi k_e}\left[(\nabla \cdot \boldsymbol{E})\boldsymbol{E} - \boldsymbol{E} \times (\nabla \times \boldsymbol{E})\right]_i \tag{1.93}$$

$$= \frac{\varepsilon k_d}{4\pi k_e}\sum_{k=1}^{3}\frac{\partial}{\partial x_k}\left(E_i E_k - \frac{1}{2}E^2 \delta_{i,k}\right)$$

and

$$\boldsymbol{B} \times (\nabla \times \boldsymbol{H}) = \frac{k_f}{4\pi \mu k_m}\sum_{k=1}^{3}\frac{\partial}{\partial x_k}\left(B_i B_k - \frac{1}{2}B^2 \delta_{i,k}\right). \tag{1.94}$$

In order to discuss (1.94) one writes the vector form on the left hand side in detail as

$$(\nabla \cdot \boldsymbol{E})\boldsymbol{E} - \boldsymbol{E} \times (\nabla \times \boldsymbol{E}) =$$

1.7 Details

$$(\partial_1 E_1 + \partial_2 E_2 + \partial_3 E_3) \begin{pmatrix} E_1 \\ E_2 \\ E_3 \end{pmatrix} - \begin{pmatrix} E_1 \\ E_2 \\ E_3 \end{pmatrix} \times \begin{pmatrix} \partial_2 E_3 - \partial_3 E_2 \\ \partial_3 E_1 - \partial_1 E_3 \\ \partial_1 E_2 - \partial_2 E_1 \end{pmatrix} =$$

$$(\partial_1 E_1 + \partial_2 E_2 + \partial_3 E_3) \begin{pmatrix} E_1 \\ E_2 \\ E_3 \end{pmatrix}$$

$$- \begin{pmatrix} E_2 (\partial_1 E_2 - \partial_2 E_1) - E_3 (\partial_3 E_1 - \partial_1 E_3) \\ E_3 (\partial_2 E_3 - \partial_3 E_2) - E_1 (\partial_1 E_2 - \partial_2 E_1) \\ E_1 (\partial_3 E_1 - \partial_1 E_3) - E_2 (\partial_2 E_3 - \partial_3 E_2) \end{pmatrix}.$$

For the verification of (1.94) one needs only to examine one component e.g. the first component, which can be arranged in the form

$$E_1 \partial_1 E_1 + E_1 \partial_2 E_2 + E_1 \partial_3 E_3 + E_2 \partial_2 E_1 + E_3 \partial_3 E_1$$
$$- E_2 \partial_1 E_2 - E_3 \partial_1 E_3 \,.$$

Add now and subtract a term with

$$E_1 \partial_1 E_1 \,,$$

and combine the first three terms as

$$E_1 \sum_k \partial_k E_k \,,$$

the next two plus the added term as

$$\sum_k E_k \partial_k E_1$$

and the remaining terms as

$$-\frac{1}{2} \partial_1 \sum_k E_k^2 \,.$$

The result is exactly the expression

$$\sum_{k=1}^{3} \frac{\partial}{\partial x_k}\left(E_1 E_k - \frac{1}{2}E^2 \delta_{1,k}\right).$$

The same procedure can be used for the components with $i = 2, 3$.

The treatment of the magnetic contributions follows exactly this pattern, if one uses $\nabla \cdot \boldsymbol{B} = 0$. With a corresponding addendum and the simple material equations one finds indeed

$$\boldsymbol{B} \times (\nabla \times \boldsymbol{H}) + (\nabla \cdot \boldsymbol{B})\boldsymbol{H} = \frac{k_f}{4\pi \mu k_m} \sum_{k=1}^{3} \frac{\partial}{\partial x_k}\left(B_i B_k - \frac{1}{2}B^2 \delta_{i,k}\right).$$

Electrodynamics: Applications 2

Electrodynamics provided the basis for the technological developments of our age. The first generators and electric motors were already developed in the nineteenth century. Telegraphy (with the construction of senders and receivers at an ever higher standard) followed. Optical questions, as the complete understanding of diffraction phenomena could be approached on a quantitative level. The examination of the radiation of moving point charges lead to the discovery of X-rays and the construction of particle accelerators (starting with bevatrons up to high energy machines), which opened a general field of research. A few examples of the many facets of practical electrodynamics will be introduced in this chapter. The CGS-system of units will be used throughout.

2.1 Technical Implementation of Induction

A direct application of induction is the alternative current motor and in reversal the alternate current (AC) generator. M.H. von Jacobi developed the first electric motor in 1834, shortly after the discovery of induction. First dynamo machines were in use by 1853. The necessary technology has been extended further by W. von Siemens (1866). With the generation of currents and the use of currents for many purposes a need to adjust the voltage of the currents to particular conditions arose. The transformer, which has been developed for this purpose, is yet another application of the laws of induction. The principle behind this variety of appliances will be addressed briefly in the next sections.

Fig. 2.1 AC generator. (a) Schematic presentation. (b) Orientation of the surface vector

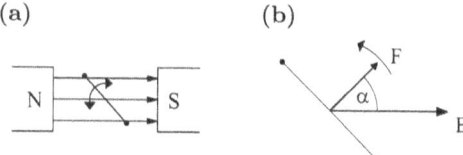

2.1.1 The AC Generator

The magnetic flux of a plane current loop with the area F is rotated in a uniform magnetic B-field (Fig. 2.1a). The magnetic flux changes with time according to

$$\phi_B = BF \cos\alpha(t).$$

The angle $\alpha(t)$ describes the variation of the orientation of the surface vector of the loop with respect to the magnetic field (Fig. 2.1b). A uniform rotation $\alpha(t) = \omega t$ produces according to the law of induction (practical form, (1.2)) an induced voltage

$$U_{ind} = -\frac{1}{c}\frac{d\phi_B}{dt} = \frac{BF}{c}\omega \sin\omega t = U_0 \sin\omega t. \quad (2.1)$$

An AC voltage with a time dependence in sine form with the frequency $f = \omega/2\pi$ is induced, which can be picked off with a suitable mechanism. The maximum amplitude U_0 can e.g. be increased by using a coil with N windings instead of one loop. The flux and the voltage U_0 are (in good approximation) enlarged by a factor N.

2.1.2 The Transformer

The principle behind the transformer is the mutual induction of two current circuits. In the primary circuit (with a resistance R_1) an AC voltage

$$U(t) = U_0 \cos\omega t$$

is applied (Fig. 2.2).

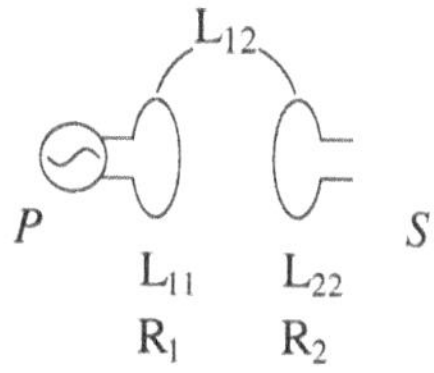

Fig. 2.2 Transformer: Principle

2.1 Technical Implementation of Induction

The corresponding AC current $i_1(t)$ in this circuit acts on itself by self-induction (with the coefficient of self-induction L_{11}) and on a secondary circuit per mutual induction (coefficient of mutual induction L_{12}, with a resistance R_2). The current $i_2(t)$, which is induced in the secondary circuit, acts back on the first circuit and on itself by self-induction (coefficient of self-induction L_{22}). The voltage in the primary circuit $U_1(t)$ is therefore composed of three contributions: the applied voltage $U(t)$ and the voltages $U_1^s(t)$ and $U_1^m(t)$ induced by self- (s) and mutual (m) induction. With the relations (1.5) in Chap. 1.1.2 between induced voltage and induced current one can write

$$U_1(t) = U(t) + U_1^s(t) + U_1^m(t) = R_1 i_1(t)$$

$$= U(t) - L_{11}\frac{d}{dt}i_1(t) - L_{12}\frac{d}{dt}i_2(t) = R_1 i_1(t). \qquad (2.2)$$

For the secondary circuit one finds in a similar fashion

$$U_2(t) = U_2^s(t) + U_2^m(t) = R_2 i_2(t)$$

$$= -L_{22}\frac{d}{dt}i_2(t) - L_{21}\frac{d}{dt}i_1(t) = R_2 i_2(t). \qquad (2.3)$$

The coefficients for the mutual induction can be calculated with the relation (1.8)

$$L_{21} = L_{12} = \frac{1}{c^2}\oint_{K_1}\oint_{K_2}\frac{d\mathbf{r}_1 \cdot d\mathbf{r}_2}{r_{12}}.$$

The differential equations (2.2) and (2.3) are a set of differential equations for the currents $i_1(t)$ and $i_2(t)$. The first differential equation is inhomogeneous, the voltage $U(t)$ is applied to this circuit. One can solve the differential equation by inserting

$$i_1(t) = i_{10}\cos(\omega t - \varphi_1) \qquad i_2(t) = i_{20}\cos(\omega t - \varphi_2)$$

or better the complex form[1]

$$i_1(t) = i_{10}\,e^{i(\omega t - \varphi_1)} \qquad i_2(t) = i_{20}\,e^{i(\omega t - \varphi_2)}$$

into the differential equations and calculate the four constants, the amplitudes of the currents (i_{10} and i_{20}) and the phase shifts (φ_1 and φ_2) of the currents with respect to the applied AC voltage by comparison of the coefficients (see Detail 2.5.1.1).

One result, which is much used in practical applications, can be obtained with simpler means. It relies on the neglect of the resistance of the primary circuit

---

[1] One has to use the complex form also for $U(t)$.

Fig. 2.3 Transformer: Realisation

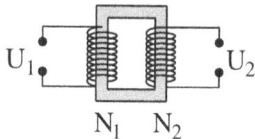

($R_1 \ll \omega L_{11}$) and on the assumption of an ideal coupling of the circuits. The last assumption can be realised approximately by connecting the circuits (actually coils) with an iron core and by minimising the loss of flux with a layer structure of the core (Fig. 2.3). If the ansatz for the complex currents, as given above, is used, one finds for the equations of the transformer

$$i\omega \{L_{11} i_1(t) + L_{12} i_2(t)\} = U(t)$$
$$i\omega \{L_{12} i_1(t) + L_{22} i_2(t)\} = -U_2(t) ,$$
(2.4)

which can be used to find the ideal transformer equation (see Detail 2.5.1.2)

$$U_2(t) = -\frac{N_2}{N_1} U(t) .$$
(2.5)

The ratio of secondary voltage to the applied voltage can be chosen by the ratio of the number of windings.

2.2 Wave Propagation

Optical phenomena can often be understood with simple means, e.g. with a ray construction or with Huygens principle. If one aims, however at the calculation of intensity distributions, one has to revert to wave optics. This chapter deals with three optical problems: crystal optics, metal optics and the diffraction of electromagnetic waves. It is probably useful to start with a definition of the mechanisms, which will be needed in covering these topics:

- **Diffraction** addresses the deflection of waves (not only electromagnetic ones) by obstructions of any kind.
- **Refraction** involves the change of the direction and the speed of propagation due e.g. to a change of the media, which the wave transverses, and therefore the index of refraction.
- **Reflection**—alternatively in older English: **Reflexion**—indicates a return of waves (or particles).

2.2.1 Reflection and Refraction in Crystal Optics

It is easy to observe the refraction of a light beam for the transition from air into water. This phenomenon can be understood by analysing the plane wave solution of the Maxwell equations. It can be observed, if one concentrates on the plane interface of two different materials, e.g. the x–y plane ($z = 0$). The material constants are ε_1, μ_1 respectively ε_2, μ_2. A monochromatic plane wave (wave number vector $\boldsymbol{k}_1$) impinges on the interface in the first material under the angle α. The wave is in part reflected (angle β, wave number vector $\boldsymbol{k}'_1$), in part it continues as a refracted wave in the second material (angle γ, wave number vector $\boldsymbol{k}_2$). All angles are measured with respect to the normal of the interface (Fig. 2.4). The task is the calculation of the relation between the angle of incidence and the angles of the reflected and the refracted beams, as well as the amplitudes of the corresponding electromagnetic waves.

The homogeneous Maxwell equations and the plane wave solutions are supposed to hold in the two materials

$$\nabla \cdot \boldsymbol{B} = \nabla \cdot \boldsymbol{E} = 0$$

$$\nabla \times \boldsymbol{E} = -\frac{1}{c}\frac{\partial \boldsymbol{B}}{\partial t} \qquad \nabla \times \boldsymbol{B} = \frac{n_i^2}{c}\frac{\partial \boldsymbol{E}}{\partial t} \qquad (i = 1, 2)\,.$$

This special variant of the Maxwell equations defines the regime of **crystal optics**. There should be no true charges or currents in the two materials or at the interface. The material constants define the index of refraction in the materials $n_i = \sqrt{\varepsilon_i \mu_i}$.

Fig. 2.4 Kinematical aspects of reflection and diffraction at the plane $z = 0$

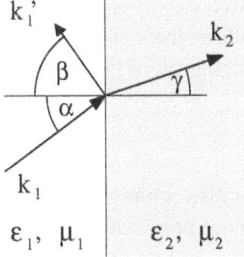

The three waves can be characterised by the following fields

incident wave:
$$E_1 = E_{10} e^{i(k_1 \cdot r - \omega_1 t)} \qquad \omega_1 = \frac{c}{n_1} k_1$$
$$B_1 = \frac{n_1}{k_1} (k_1 \times E_1)$$

refracted wave:
$$E_2 = E_{20} e^{i(k_2 \cdot r - \omega_2 t)} \qquad \omega_2 = \frac{c}{n_2} k_2$$
$$B_2 = \frac{n_2}{k_2} (k_2 \times E_2)$$

reflected wave:
$$E'_1 = E'_{10} e^{i(k'_1 \cdot r - \omega'_1 t)} \qquad \omega'_1 = \frac{c}{n_1} k'_1$$
$$B'_1 = \frac{n_1}{k'_1} (k'_1 \times E'_1) \, .$$

The well known geometric laws for

$$\text{reflection:} \quad \alpha = \beta \quad \text{and} \quad \text{refraction:} \quad \frac{\sin \alpha}{\sin \gamma} = \frac{n_2}{n_1}$$

follow from a direct kinematic argumentation: It is necessary, that the state of oscillation of the three waves be the same for all points of the interface. This implies

$$k_1 \cdot r - \omega_1 t = k_2 \cdot r - \omega_2 t = k'_1 \cdot r - \omega'_1 t$$

for all vectors $r = (x, y, 0)$ pointing to or coming from the interface at all times. A trivial phase difference of $2m\pi$ is excluded, as the conditions should also be valid for $r = 0$ and $t = 0$.

If one regards in particular the origin and arbitrary times $r = 0$, $t \neq 0$, one obtains

$$\omega_1 = \omega'_1 = \omega_2 \equiv \omega \, .$$

All three waves oscillate with the same frequency. For the wave lengths respectively the magnitude of the wave lengths one finds

2.2 Wave Propagation

Fig. 2.5 Projection of the wave number vectors

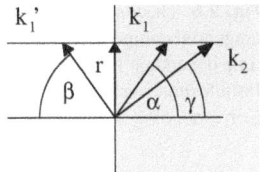

$$k_1 = \frac{\omega_1}{v_1} = \frac{n_1}{c}\omega$$

$$k_1' = \frac{\omega_1'}{v_1'} = \frac{n_1}{c}\omega$$

$$\Rightarrow k_1 = k_1'.$$

The two waves in the medium (1) have (as expected) the same wave length. In the second medium one has

$$k_2 = \frac{\omega_2}{v_2} = \frac{n_2}{c}\omega.$$

The ratio of the wave length in the two media is therefore

$$\frac{\lambda_2}{\lambda_1} = \frac{k_1}{k_2} = \frac{n_1}{n_2} \longrightarrow \lambda_2 = \frac{n_1}{n_2}\lambda_1.$$

The wave length in medium (2) is smaller, if the index of refraction is larger in this medium than in medium 1.

If one considers points with $r \neq 0$ (but $z = 0$) and arbitrary values of t, one finds

$$\boldsymbol{k}_1 \cdot \boldsymbol{r} = \boldsymbol{k}_2 \cdot \boldsymbol{r} = \boldsymbol{k}_1' \cdot \boldsymbol{r}.$$

The explicit form of the scalar products is (see Fig. 2.5)

$$k_1 r \cos(90° - \alpha) = k_1' r \cos(90° - \beta) = k_2 r \cos(90° - \gamma).$$

Use of the relation $\cos(90° - \alpha) = \sin\alpha$ and the results for the wave numbers, lead to the law of reflection

$$\sin\alpha = \sin\beta \longrightarrow \alpha = \beta. \qquad (2.6)$$

The associated law of refraction was discovered already around 1620 by the Dutch scientist Snellius and by Descartes

$$\frac{\sin\alpha}{\sin\gamma} = \frac{k_2}{k_1} = \frac{n_2}{n_1} = \frac{\lambda_1}{\lambda_2} = \frac{v_1}{v_2}. \qquad (2.7)$$

The answer concerning the question of the distribution of intensities takes a bit more space. It is a question, which addresses the dynamics of the way, in which the amplitudes of the three waves are related. The answer requires a consideration of the continuity of the time dependent fields at the interface. These conditions follow,

Fig. 2.6 Discontinuity of the tangential component of the time dependent fields. (**a**) Infinitesimal surface. (**b**) Corresponding Stokes loop

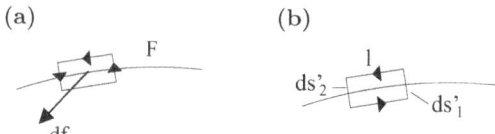

as in the stationary situation from the field equation, that is in this case the Maxwell equations.

The equation $\nabla \cdot \boldsymbol{B} = 0$ states, that the normal components should satisfy the relation $B_{1n} + B'_{1n} = B_{2n}$. As a charge density is not present in the case of crystal optics ($\rho_{tr} = 0$), the source equation for the $\boldsymbol{D}$-field is $\nabla \cdot \boldsymbol{D} = 0$, so that the statements $D_{1n} + D'_{1n} = D_{2n}$ or $\varepsilon_1 (E_{1n} + E'_{1n}) = \varepsilon_2 E_{2n}$ are also valid at the interface.

The relations for the tangential components differ from those in the stationary case. The relevant equation for the $\boldsymbol{E}$-field is

$$\nabla \times \boldsymbol{E} = -\frac{1}{c}\frac{\partial}{\partial t}\boldsymbol{B} \ .$$

This relation can be exploited by integration both sides of the equation over a rectangular area F perpendicular to the interface (Fig. 2.6a)

$$\iint_F (\nabla \times \boldsymbol{E}) \cdot \mathbf{d}\boldsymbol{f} = -\frac{1}{c} \iint_F \frac{\partial \boldsymbol{B}}{\partial t} \cdot \mathbf{d}\boldsymbol{f} \ .$$

The left hand side is transformed with the theorem of Stokes (Fig. 2.6b)

$$\oint_{R(F)} \boldsymbol{E} \cdot \mathbf{d}\boldsymbol{s} = -\frac{1}{c} \iint_F \frac{\partial \boldsymbol{B}}{\partial t} \cdot \mathbf{d}\boldsymbol{f} \ .$$

The flux integral vanishes, if an infinitely narrow rectangle is chosen. The line integral on the left hand side over two lines of length L for the contributions of the remaining sides yields

$$(E_{1t} + E'_{1t} - E_{2t})L = 0 \ .$$

The line integral along the interface (I) is

$$(E_{1t} + E'_{1t})\big|_\mathrm{I} = E_{2t}\big|_\mathrm{I} \ ,$$

even if the fields are not vortex free. The fourth Maxwell equation states

$$\nabla \times \boldsymbol{H} = \frac{1}{c}\frac{\partial \boldsymbol{D}}{\partial t} \ ,$$

2.2 Wave Propagation

Fig. 2.7 Continuity of linearly polarised plane waves. (**a**) E perpendicular to the k-plane. (**b**) E in the k-plane

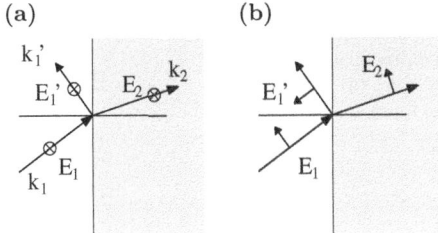

so that one finds

$$(H_{1t} + H'_{1t})\big|_I = H_{2t}\big|_I \quad \text{resp.} \quad \frac{1}{\mu_1}(B_{1t} + B'_{1t})\big|_I = \frac{1}{\mu_2} B_{2t}\big|_I ,$$

if one uses a corresponding argument. These relations can be used for the discussion of the optical problem at hand. It is sufficient to consider two special cases of linearly polarised plane waves. A situation of arbitrarily circularly or elliptically polarised waves could be treated by superposition. The E-field oscillates in the following manner in the two special cases:

- in a plane perpendicular to the wave vectors k_1, k'_1, k_2 (Fig. 2.7a)
- or in the plane of the three wave vectors (Fig. 2.7b).

The discussion of some of the details will be restricted to the first case. The discussion of the second case is similar enough, so that only the results will be cited.

The states of oscillation for all waves are, as mentioned above, equal in the interface. A consequence is the fact that the discontinuities concern only the amplitudes. Electric fields, which oscillate perpendicularly to k-plane do not have a normal component with respect to the interface

$$E_{1n} = 0 \quad E_{1'n} = 0 \quad E_{2n} = 0 ,$$

the tangential components are

$$E_{1t} = E_{10} \quad E'_{1t} = E'_{10} \quad E_{2t} = E_{20} .$$

The B-vectors are perpendicular to the E- and the k-vectors. The magnetic induction has a normal and a tangential component. For the incident beam they are (see Fig. 2.8a)

$$B_{1n} = -B_{10} \sin\alpha \quad B_{1t} = B_{10} \cos\alpha .$$

Fig. 2.8 Connection of the magnetic fields of a linearly polarised plane wave ($E \perp k$). (a) Reflected beam. (b) Refracted beam

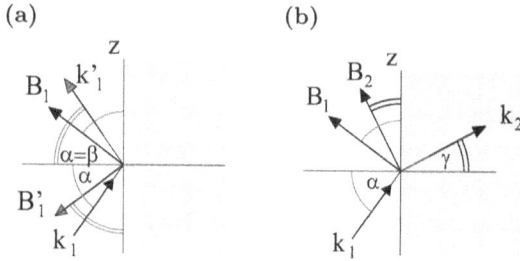

If, in addition, the relation (1.48) is used in the form $B = nE$, one can write

$$B_{1n} = -n_1 E_{10} \sin \alpha \qquad B_{1t} = n_1 E_{10} \cos \alpha .$$

For the **B**-vectors of the reflected and the refracted wave corresponding statements are valid (Fig. 2.8a and b)

$$B'_{1n} = -n_1 E'_{10} \sin \alpha \qquad B'_{1t} = -n_1 E'_{10} \cos \alpha$$

$$B_{2n} = -n_2 E_{20} \sin \gamma \qquad B_{2t} = n_2 E_{20} \cos \gamma .$$

At this stage one has to apply the discontinuity conditions. The normal components of the **B**-field are continuous. This statement is valid for the complete **B**-field in (1), which in turn is equal to the **B**-field in (2)

$$B_{1n} + B'_{1n} = B_{2n} .$$

Insert here the relations for the components

$$-n_1 E_{10} \sin \alpha - n_1 E'_{10} \sin \alpha = -n_2 E_{20} \sin \gamma$$

and use the law of diffraction to write

$$E_{10} + E'_{10} = E_{20} .$$

The condition for the tangential components of the **E**-fields is the same as for the normal components of the **B**-fields. This confirms the consistency of the discontinuity condition extracted from the law of induction. The discontinuity condition for the normal components of the **E**-fields need not be considered as these components vanish. The conditions for the tangential components of the **B**-fields are

$$\frac{1}{\mu_1}(B_{1t} + B'_{1t}) = \frac{1}{\mu_2} B_{2t} .$$

2.2 Wave Propagation

Insertion of the three components leads to the relation

$$\frac{n_1}{\mu_1}(E_{10} - E'_{10})\cos\alpha = \frac{n_2}{\mu_2} E_{20}\cos\gamma, \tag{2.8}$$

which can be treated with the law of refraction in order to eliminate the ratio of the numbers of refraction

$$E_{10} - E'_{10} = \frac{\mu_1 \tan\alpha}{\mu_2 \tan\gamma} E_{20}.$$

The discontinuity conditions have been used to find two linear relations between the amplitudes of the electric fields. They can be used for the determination of the ratios E'_{10}/E_{10} and E_{20}/E_{10} of the amplitudes. There is a good reason, why this is possible. The intensity (resp. the amplitude) of the incident wave can be chosen freely. With the abbreviation

$$x = \frac{\mu_1 \tan\alpha}{\mu_2 \tan\gamma}$$

one can write for the linear equations

$$\frac{E'_{10}}{E_{10}} - \frac{E_{20}}{E_{10}} = -1 \qquad \frac{E'_{10}}{E_{10}} + x\frac{E_{20}}{E_{10}} = 1.$$

Their solution is

$$\frac{E'_{10}}{E_{10}} = \frac{1-x}{1+x} \qquad \frac{E_{20}}{E_{10}} = \frac{2}{1+x}. \tag{2.9}$$

These formulae are known under the name **Fresnel's relations**. They allow the determination of the quantity x, if the angle of incidence and the material constants are given, provided the angle γ is obtained from the law of refraction. The quantity x determines the ration of the amplitudes.

The materials in crystal optics are, in good approximation, characterised by $\mu_1 = \mu_2 = 1$. One then obtains[2] the Fresnel formulae

$$\frac{E'_{10}}{E_{10}} = \frac{\tan\gamma - \tan\alpha}{\tan\gamma + \tan\alpha} = \frac{\sin(\gamma - \alpha)}{\sin(\gamma + \alpha)}$$

$$\frac{E_{20}}{E_{10}} = \frac{2\tan\gamma}{\tan\gamma + \tan\alpha} = \frac{2\sin\gamma\cos\alpha}{\sin(\gamma + \alpha)}.$$
(2.10)

---

[2] The tools for the conversion of the trigonometric functions can be found in any collection of their properties.

The case of perpendicular incidence $\alpha = 0$ has to be treated separately. Equation (2.8) yields for $\alpha = \gamma = 0$ with $\mu_1 = \mu_2 = 1$ the relation

$$E_{10} - E'_{10} = \frac{n_2}{n_1} E_{20} \qquad \left(x = \frac{n_2}{n_1} \right),$$

which leads to

$$\frac{E'_{10}}{E_{10}} = \frac{n_2 - n_1}{n_1 + n_2} \qquad \frac{E_{20}}{E_{10}} = \frac{2n_1}{n_1 + n_2}. \qquad (2.11)$$

The same result can be obtained with the Fresnel formulae using the argument

$$\lim_{\alpha,\gamma \to 0} \frac{\tan \alpha}{\tan \gamma} = \lim_{\alpha,\gamma \to 0} \frac{\sin \alpha}{\sin \gamma} = \frac{n_2}{n_1}.$$

For perpendicular incidence and $n_2 > n_1$ (a transition from a thinner optical medium into a denser medium, e.g. from air into water), one finds $E'_{10}/E_{10} < 0$. The electric vector of the reflected wave is turned by 180° with respect to the electric vector of the incident wave. If $n_1 = n_2$ (no interface exists), one finds

$$E'_{10} = 0 \qquad E_{20} = E_{10}.$$

The plane wave just continues.

The calculation for the case, that the E-vectors oscillate in the plane of the wave vectors, differs only in the decomposition of the components. The final formulae in the approximation with $\mu_1 = \mu_2 = 1$ are in general respectively for perpendicular incidence

$$\frac{E'_{10}}{E_{10}} = \frac{\tan(\alpha - \gamma)}{\tan(\alpha + \gamma)} \longrightarrow \frac{n_2 - n_1}{n_1 + n_2}$$

$$\frac{E_{20}}{E_{10}} = \frac{2 \sin \gamma \cos \alpha}{\sin(\gamma + \alpha) \cos(\gamma - \alpha)} \longrightarrow \frac{2n_1}{n_1 + n_2}. \qquad (2.12)$$

(The necessary calculations for this and additional situations can be found in Detail 2.5.2.1). A special situation has to be pointed out though. For the angles with $\alpha + \gamma = \pi/2$, one obtains because of $\tan(\pi/2) \to \infty$ the result $E'_{10}/E_{10} = 0$. A reflected wave does not exist. The angle, for which one finds only a refracted wave, is determined by

$$n_1 \sin \alpha_B = n_2 \sin \left(\frac{\pi}{2} - \alpha_B \right) = n_2 \cos \alpha_B \longrightarrow \tan \alpha_B = \frac{n_2}{n_1}.$$

2.2 Wave Propagation

Fig. 2.9 The Brewster angle for the transition from air into glass

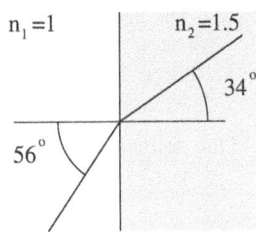

It is called the **Brewster angle**. Its value is e.g. $\alpha_B \approx 56°$ for the transition from air ($n_1 = 1$) into glass ($n_2 = 1.5$) (Fig. 2.9).

An arbitrarily polarised plane wave can be decomposed into components parallel and perpendicular to this plane

$$\boldsymbol{E}(\boldsymbol{r}, t) = \left(E_{1,\|}\boldsymbol{e}_\| + E_{2,\perp}\boldsymbol{e}_\perp\right) e^{i(\boldsymbol{k}\cdot\boldsymbol{r} - \omega t)} ,$$

with

$$E_{1,\|} = \tilde{E}_{1,\|}\, e^{i\varphi_\|} \qquad E_{2,\perp} = \tilde{E}_{2,\perp}\, e^{i\varphi_\perp} .$$

The vector $\boldsymbol{e}_\|$ lies in the $\boldsymbol{k}$-plane, the vector $\boldsymbol{e}_\perp$ is perpendicular to this plane. The reflected beam has only a component perpendicular to the wave vector plane, if a wave with these specifications is incident on an interface under the Brewster angle. This property can be used for the production of electromagnetic waves with a definite direction of the polarisation.

The ratios of the amplitudes, which have been calculated, can be used to determine the flux of energy through the interface of two materials. No charge is moved nor does the field energy change on the average, if a 'flat' volume is arranged about the interface (Fig. 2.10). This shows that only the Poynting vector can contribute to the energy balance. This balance is

$$F\boldsymbol{e}_F \cdot \bar{\boldsymbol{S}}_{in} + F\boldsymbol{e}_F \cdot \bar{\boldsymbol{S}}_{refl} + F\boldsymbol{e}'_F \cdot \bar{\boldsymbol{S}}_{refr} = 0 ,$$

as only the temporal mean value can be measured in an experiment. The quantity F is the surface of the base of this volume, the unit vectors characterise the normal of this surface and the quantities $\bar{S}$ are the time averaged Poynting vectors of the three

Fig. 2.10 Transport of field energy for reflection and refraction

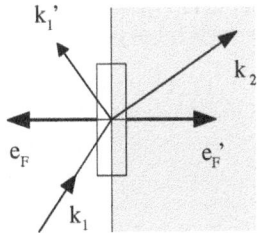

waves. The geometry is indicated in Fig. 2.10. It leads to the energy balance

$$-\cos\alpha\, \bar{S}_{in} + \cos\alpha\, \bar{S}_{refl} + \cos\gamma\, \bar{S}_{refr} = 0 \, .$$

One defines the quantities

$$\text{reflection coefficient:} \qquad R = \frac{\bar{S}_{refl}}{\bar{S}_{in}}$$

$$\text{transmission coefficient:} \qquad T = \frac{\cos\gamma}{\cos\alpha}\frac{\bar{S}_{refr}}{\bar{S}_{in}}$$

(2.13)

and writes the flux in the compact form

$$R + T = 1 \, . \tag{2.14}$$

The result for the case of an oscillation of E perpendicular to the wave vector plane is (up to irrelevant factors)

$$\bar{S}_{in} \propto \frac{n_1}{\mu_1} E_{10}^2 \, , \qquad \bar{S}_{refl} \propto \frac{n_1}{\mu_1} E_{10}^{\prime 2} \, , \qquad \bar{S}_{refr} \propto \frac{n_2}{\mu_2} E_{20}^2 \, .$$

This yields the statement for the flux of the energy

$$\left(\frac{E'_{10}}{E_{10}}\right)^2 + \frac{n_2 \mu_1 \cos\gamma}{n_1 \mu_2 \cos\alpha}\left(\frac{E_{20}}{E_{10}}\right)^2 = 1 \, .$$

The ratios of the amplitudes, which are given by the Fresnel formulae, are in accord with this statement on energy conservation (Detail 2.5.2.2). As an example for the application of these relations, one can look at the transition of light through a glass pane for perpendicular incidence (Fig. 2.11). At the two interfaces (air → glass, glass → air) one has as a consequence of the symmetry of the Fresnel formulae

$$R = \left(\frac{n_1 - n_2}{n_1 + n_2}\right)^2 \qquad\qquad T = \frac{4 n_1 n_2}{(n_1 + n_2)^2} \, .$$

Fig. 2.11 Transition of light through a glass pane

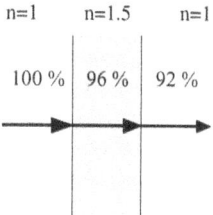

With the values

$$n_{\text{air}} = 1 \qquad n_{\text{glass}} \approx 1.5$$

one finds for each interface

$$R \approx 0.04 \quad \text{and} \quad T \approx 0.96 \; .$$

96% of the energy enters into the pane, 4% is reflected directly. At the second interface another 4% is reflected, so that about 92% of the incident field energy passes through the pane. Glass is a transparent medium.

2.2.2 Wave Propagation in Metals

The propagation of plane electromagnetic waves in isotropic, uniform, dielectric materials has been discussed in the last section. The question, of the propagation of electromagnetic waves in metals is the topic of this section. Formulated in a more direct fashion, one could ask: What is the difference if light falls on a metal plate rather than on a glass pane? Normally, the density of the true charges in a metal block is zero, at least on the average, $\bar{\rho}_{tr} \to \rho_e = 0$. This implies that the divergence of the D-field (resp. of the E-field) vanishes

$$\varepsilon \nabla \cdot E(r,t) = 0 \; .$$

The true current density is, however, not equal to zero ($j_{tr} \neq 0$). Through the action of the time dependent electromagnetic fields on the free charges, they can start to move and produce a current. For the discussion of the situation one has to work with the complete law of Ampère

$$\nabla \times H(r,t) = \frac{1}{c} \frac{\partial}{\partial t} D(r,t) + \frac{4\pi}{c} j_{tr}(r,t) \; .$$

The current is produced by the electric field. The relation, which connects the current density with this field, is Ohm's law

$$j_{tr}(r,t) = \sigma E(r,t) \; .$$

The additional Maxwell equations, that need to be considered, are

$$\nabla \cdot B(r,t) = 0 \qquad \nabla \times E(r,t) = -\frac{1}{c} \frac{\partial}{\partial t} B(r,t) \; .$$

The use of the simple material equations for the magnetic fields is adequate, if there are no strong effects of magnetisation. It is possible in this case to obtain a free wave equation with the relation

$$\nabla \times (\nabla \times E(r,t)) = -\frac{1}{c}\frac{\partial}{\partial t}\nabla \times B(r,t)$$

(see Detail 2.5.3) and, with the same arguments as for the derivation of the free wave equation in Chap. 1.3.2, the differential equation[3]

$$\Delta E(r,t) - \frac{\varepsilon\mu}{c^2}\frac{\partial^2 E(r,t)}{\partial t^2} - \left(\frac{4\pi}{c^2}\mu\sigma\right)\frac{\partial E(r,t)}{\partial t} = 0. \qquad (2.15)$$

The corresponding magnetic equation can be found by using $\nabla \times (\nabla \times B$. It has the same structure

$$\Delta B(r,t) - \frac{\varepsilon\mu}{c^2}\frac{\partial^2 B(r,t)}{\partial t^2} - \left(\frac{4\pi}{c^2}\mu\sigma\right)\frac{\partial B(r,t)}{\partial t} = 0. \qquad (2.16)$$

These equations have the name **telegraph equations**. They differ from the free wave equations (1.39) and (1.41) by the presence of a first derivative of the fields. This term is similar to the term, which is present in the damped oscillator problem of mechanics. One can expect a solution of the telegraph equations in the form of a plane wave, damped in space. As the present differential equations are linear, this means, that it is a superimposition of damped plane waves.

The simplest solution can be found by inserting the ansatz

$$\left.\begin{array}{l} E(r,t) = E_0 \\ B(r,t) = B_0 \end{array}\right\} \cdot e^{i(\kappa \cdot r - \omega t)} \qquad \text{with } \kappa = \text{complex}$$

into the differential equations. The resulting quadratic equation

$$\kappa^2 - \left(\frac{\varepsilon\mu\omega^2}{c^2} + 4\pi i\frac{\mu\omega\sigma}{c^2}\right) = 0$$

can be solved by using

$$\kappa = k + i\beta \qquad k, \beta = \text{real}.$$

In the resulting coupled equations with

$$k^2 - \beta^2 = \frac{\varepsilon\mu\omega^2}{c^2} \quad \text{and} \quad 2k\beta = 4\pi\frac{\mu\omega\sigma}{c^2},$$

---

[3] Please note: Eq. (2.15) can also be obtained in the case that $\rho_{tr} \neq 0$ and $\nabla \rho_{tr} = \mathbf{0}$, that is a uniform charge density in the metal.

2.2 Wave Propagation

one can eliminate β in order to have a quadratic equation for k^2

$$(k^2)^2 - \frac{\varepsilon\mu\omega^2}{c^2}(k^2) = 4\pi^2 \frac{\mu^2\omega^2\sigma^2}{c^4}.$$

The real solution for k, which goes over into a dispersion relation for a plane wave in the limit $\sigma \longrightarrow 0$, is

$$k = \sqrt{\varepsilon\mu}\,\frac{\omega}{c}\left[\frac{1}{2}\left(\left[1 + \left(\frac{4\pi\sigma}{\varepsilon\omega}\right)^2\right]^{1/2} + 1\right)\right]^{1/2}. \quad (2.17)$$

From the equation

$$\beta^2 = k^2 - \frac{\varepsilon\mu\omega^2}{c^2}$$

one extracts then (with the choice of the sign, that corresponds to damping)

$$\beta = \sqrt{\varepsilon\mu}\,\frac{\omega}{c}\left[\frac{1}{2}\left(\left[1 + \left(\frac{4\pi\sigma}{\varepsilon\omega}\right)^2\right]^{1/2} - 1\right)\right]^{1/2}. \quad (2.18)$$

In order to have a compact form of the particular solutions of the telegraph equations, one defines the vectors $\boldsymbol{k} = k\boldsymbol{e}$ and $\boldsymbol{\beta} = \beta\boldsymbol{e}$, where the unit vector $\boldsymbol{e}$ is a vector in the direction of the propagation. The particular solution is then

$$\left.\begin{array}{l} \boldsymbol{E}(\boldsymbol{r},t) = \boldsymbol{E}_0 \\ \boldsymbol{B}(\boldsymbol{r},t) = \boldsymbol{B}_0 \end{array}\right\} \cdot e^{-\boldsymbol{\beta}\cdot\boldsymbol{r}}\, e^{i(\boldsymbol{k}\cdot\boldsymbol{r} - \omega t)}. \quad (2.19)$$

The magnitude of the vector $\boldsymbol{\beta}$ determines the degree of damping. The parameter σ has for a good conductor, e.g. copper, a value of the order of magnitude 10^{17} s^{-1}. This means, that (ignoring details of the material and the dependence on the frequency) the damping parameter $4\pi\sigma/(\varepsilon\omega)$ is much larger than 1, so that the depth of the penetration of a wave hitting a metal (defined as $x = 1/\beta$), is only a few millimeters.

The inhomogeneous versions of the telegraph equations will not be discussed in detail. They lead to a coupling of the $\boldsymbol{B}$- and the $\boldsymbol{E}$-fields. The simplest discussion of the coupling effects analyses the action of all Maxwell equations on the damped plane waves with the complex wave number vector $\boldsymbol{\kappa} = \boldsymbol{k} + i\boldsymbol{\beta}$ in a metal. The divergence equations

$$\nabla \cdot \boldsymbol{E}(\boldsymbol{r},t) = \nabla \cdot \boldsymbol{B}(\boldsymbol{r},t) = 0$$

yield

$$\boldsymbol{\kappa} \cdot \boldsymbol{E} = \boldsymbol{\kappa} \cdot \boldsymbol{B} = 0$$

and

$$\boldsymbol{k} \cdot \boldsymbol{E}(\boldsymbol{r}, t) = \boldsymbol{k} \cdot \boldsymbol{B}(\boldsymbol{r}, t) = 0 \,.$$

The electromagnetic waves in a metal are transverse, too. The law of induction

$$\nabla \times \boldsymbol{E}(\boldsymbol{r}, t) = -\frac{1}{c}\frac{\partial}{\partial t}\boldsymbol{B}(\boldsymbol{r}, t)$$

leads to the relation

$$\boldsymbol{\kappa} \times \boldsymbol{E}_0 = \frac{\omega}{c}\boldsymbol{B}_0$$

which states, that the vectors $\boldsymbol{\kappa}$, $\boldsymbol{E}$, $\boldsymbol{B}$ form a right handed tripod.

The fact, that the wave number is a complex number, is the reason for a phase shift between the $\boldsymbol{E}$- and the $\boldsymbol{B}$-fields. In order to demonstrate this statement, it is convenient to use the representation of complex numbers in the form of magnitude times phase

$$\kappa = |\kappa| \, e^{i\alpha} \,,$$

with

$$|\kappa| = [k^2 + \beta^2]^{1/2} = \sqrt{\varepsilon\mu}\,\frac{\omega}{c}\left[1 + \left(\frac{4\pi\sigma}{\varepsilon\omega}\right)^2\right]^{1/4}$$

and

$$\tan\alpha = \frac{\beta}{k} \,,$$

where the angle α is best obtained from

$$\tan 2\alpha = \frac{2\tan\alpha}{1 - \tan^2\alpha} = \frac{2\beta k}{k^2 - \beta^2} = \frac{4\pi\sigma}{\varepsilon\omega} \,.$$

The relation between the amplitudes of the fields are then

$$\boldsymbol{B}_0 = \frac{c}{\omega}[(k + i\beta) \times \boldsymbol{E}_0] = \sqrt{\varepsilon\mu}\left[1 + \left(\frac{4\pi\sigma}{\varepsilon\omega}\right)^2\right]^{1/4} e^{i\alpha} \left(\frac{\boldsymbol{k}}{k} \times \boldsymbol{E}_0\right).$$

(2.20)

2.2 Wave Propagation

The quantity $(4\pi\sigma)/(\omega\varepsilon)$ is very large in a metal with a high conductivity. This is the reason why the magnetic field in a metal is much stronger than the electric field and, in addition, phase shifted by up to $\alpha = 45°$.

The discussion of surface effects is of particular interest for metals, as there is not much action *in* metals from the point of view of electromagnetism. The main question concerns the propagation of electromagnetic waves in conductors: How is it possible, that wave propagation takes place in metals? The answer is: Waves do not propagate in a conductor but on its surface. The most effective conductors of electromagnetic waves are **wave guides**. The reflection-diffraction properties for the transition of an electromagnetic wave into a metal are such, that waves can propagate along a wave guide practically without a loss. A detailed explanation of this statement will be offered in the next section.

2.2.3 Wave Guides and Wire Waves

Electromagnetic waves can propagate with minimal loss in wave guides, which are hollow metal tubes with a uniform cross section and conducting inner surfaces (Fig. 2.12). The wave guides can be filled with a dielectric material. This is usually thought to be characterised by simple material equations with the constants ε and μ. In the discussion of wave guides it is (nearly) always possible to assume, that the inner walls constitute an *ideal* conductor. This implies that the normal components of the **B**-field and the tangential **E**-field vanish on the inner surface. The assumption is therefore

$$e_n \cdot B(r,t)|_{\text{edge}} = 0$$
$$e_n \times E(r,t)|_{\text{edge}} = 0 \tag{2.21}$$

where e_n represents the normal of the inner surface of the tube. In the interior of the wave guide there should be no true charges or true currents, so that the source-free Maxwell equations are responsible for the description of the situation within the guide

$$\nabla \cdot E(r,t) = 0 \tag{2.22}$$

$$\nabla \cdot B(r,t) = 0 \tag{2.23}$$

Fig. 2.12 Model of a wave guide with a uniform cross section

$$\nabla \times \boldsymbol{E}(\boldsymbol{r},t) = -\frac{1}{c}\frac{\partial \boldsymbol{B}(\boldsymbol{r},t)}{\partial t} \tag{2.24}$$

$$\nabla \times \boldsymbol{B}(\boldsymbol{r},t) = \frac{\varepsilon\mu}{c}\frac{\partial \boldsymbol{E}(\boldsymbol{r},t)}{\partial t} \,. \tag{2.25}$$

The assumed, simple material equations have been used in writing the last two equations, (2.24) and (2.25). The divergence of these equations

$$\nabla \cdot (\nabla \times \boldsymbol{E}(\boldsymbol{r},t)) = 0 = -\frac{1}{c}\frac{\partial}{\partial t}(\nabla \cdot \boldsymbol{B}(\boldsymbol{r},t))$$

shows, that Eqs. (2.22) and (2.23) are satisfied automatically. The Maxwell equations lead to the wave equations (compare Chap. 1.3.2)

$$\left(\Delta - \frac{\varepsilon\mu}{c^2}\frac{\partial^2}{\partial t^2}\right)\left\{\begin{array}{c}\boldsymbol{E}(\boldsymbol{r},t) \\ \boldsymbol{B}(\boldsymbol{r},t)\end{array}\right\} = 0\,. \tag{2.26}$$

For the discussion, which follows, one can—depending on the cross section of the wave guide—use either Cartesian or cylindrical coordinates. Here cylindrical coordinates z and the transverse coordinates r and ϕ will be used. A basic solution, which is well suited for such a wave guide, are wave solutions with the form

$$\left\{\begin{array}{c}\boldsymbol{E}(\boldsymbol{r},t) \\ \boldsymbol{B}(\boldsymbol{r},t)\end{array}\right\} = \left\{\begin{array}{c}\boldsymbol{E}(r,\varphi) \\ \boldsymbol{B}(r,\varphi)\end{array}\right\} \exp\{i(\pm p(\omega)z - \omega t)\}\,. \tag{2.27}$$

These solutions describe the propagation of a monochromatic (ω fixed) wave in the direction of the wave guide. The frequency ω is provided by the monochromatic source. Arbitrary wave forms can be generated by superposition, as in the case of free electromagnetic waves. The 'wave number' $p(\omega)$ is determined by the boundary conditions and does not correspond to (see below) plane waves.

The Maxwell equations (2.22) to (2.25) yield, with the ansatz (2.27), a wave propagating along the wave guide in the positive direction ($+pz$), which is characterised by a system of equations for the six components of the electromagnetic field[4]

$$\frac{1}{r}\frac{\partial E_z}{\partial \varphi} - ip\, E_\varphi = ik\, B_r$$

$$ip\, E_r - \frac{\partial E_z}{\partial r} = ik\, B_\varphi$$

---

[4] The wave number k is defined by the relation $k = \omega/c$.

2.2 Wave Propagation

$$\frac{1}{r}\frac{\partial(rE_\varphi)}{\partial r} - \frac{1}{r}\frac{\partial E_r}{\partial \varphi} = ik\, B_z \qquad (2.28)$$

$$\frac{1}{r}\frac{\partial B_z}{\partial \varphi} - ip\, B_\varphi = -i\varepsilon\mu k\, E_r$$

$$ip\, B_r - \frac{\partial B_z}{\partial r} = -i\varepsilon\mu k\, E_\varphi$$

$$\frac{1}{r}\frac{\partial(rB_\varphi)}{\partial r} - \frac{1}{r}\frac{\partial B_r}{\partial \varphi} = -i\varepsilon\mu k\, E_z \,.$$

These components of the electromagnetic fields satisfy a wave equation of the form

$$\left[(\Delta - \frac{\partial^2}{\partial z^2}) + (\varepsilon\mu k^2 - p^2)\right]K_i(r, \varphi) = 0 \qquad (2.29)$$

where K_i stands for B_r, B_φ etc. An often used, alternative notation for this differential equation in two dimensions is

$$\left[\Delta_t + (\varepsilon\mu k^2 - p^2)\right]K_i(r, \varphi) = 0 \,.$$

Three different modes of solutions of the Maxwell or the wave equations of the wave guide problem can be distinguished:

Name	Abbreviation	Characterisation
Transverse magnetic	TM	$B_z = 0$
Transverse electric	TE	$E_z = 0$
Transverse electromagnetic	TEM	$E_z = B_z = 0$

The z-components of the **B**-field or the **E**-field vanish for all points of the wave guide in case of the two normal modes TM and TE. Besides the two basic types a third version of solutions can occur. A TEM wave is present, if the z-components of the two fields vanish for all points. The three basic types of wave guides shall be discussed in some detail, beginning with the TEM waves.

2.2.3.1 TEM-Waves
If one starts with the TEM waves, one has to consider the Maxwell equations

$$-ip\, E_\varphi = ik\, B_r$$

$$ip\, E_r = ik\, B_\varphi$$

$$-ip\, B_\varphi = -i\varepsilon\mu k\, E_r \qquad (2.30)$$

$$ip B_r = -i\varepsilon\mu k E_\varphi$$

$$0 = \frac{1}{r}\frac{\partial(rE_\varphi)}{\partial r} - \frac{1}{r}\frac{\partial E_r}{\partial \varphi}$$

$$0 = \frac{1}{r}\frac{\partial(rB_\varphi)}{\partial r} - \frac{1}{r}\frac{\partial B_r}{\partial \varphi} .$$

The first four equations indicate that a non-trivial solution can only exist, if the conditions

$$\varepsilon\mu k^2 = p^2 \quad \text{or} \quad p = \sqrt{\varepsilon\mu}\,\frac{\omega}{c} \tag{2.31}$$

with e.g.

$$E_\varphi = -\frac{k}{p}B_r = \frac{\varepsilon\mu k^2}{p^2}E_\varphi$$

are satisfied. TEM waves propagate in a wave guide in the same fashion as plane waves, if the quantities ε and μ are constant. They are not influenced by the restrictive metallic mantle. One obtains with the simple dispersion relation (2.31)

$$B_r = -\sqrt{\varepsilon\mu}\,E_\varphi \quad \text{and} \quad B_\varphi = \sqrt{\varepsilon\mu}\,E_r . \tag{2.32}$$

These equations show, that for TEM waves the relation

$$\mathbf{B}\cdot\mathbf{E} = 0$$

holds, as for plane waves. The magnetic field and the electric field are perpendicular.

The remaining two Maxwell equations can be rewritten with the aid of (2.32) in the form

$$\frac{\partial(rE_\varphi)}{\partial r} - \frac{\partial E_r}{\partial \varphi} = 0 \tag{2.33}$$

$$\frac{\partial(rE_r)}{\partial r} + \frac{\partial(rE_\varphi)}{\partial \varphi} = 0 \tag{2.34}$$

(or a corresponding form for the magnetic field). Equation (2.33) is satisfied, if the components of the electric field can be represented by the gradient of a function $V(r,\varphi)$ in planar polar coordinates. For

$$\left(E_r,\,E_\varphi\right) = -\nabla_2 V(r,\varphi) = \left(-\frac{\partial V}{\partial r},\,-\frac{1}{r}\frac{\partial V}{\partial \varphi}\right) \tag{2.35}$$

2.2 Wave Propagation

one finds (provided that V can be differentiated twice with a continuous result)

$$\frac{\partial(rE_\varphi)}{\partial r} - \frac{\partial E_r}{\partial \varphi} = -\left(\frac{\partial^2 V}{\partial r \partial \varphi} - \frac{\partial^2 V}{\partial \varphi \partial r}\right) = 0.$$

The last of the equations (2.34) leads to an equation for the determination of $V(r, \varphi)$

$$-\left(\frac{\partial}{\partial r}\left(r\frac{\partial V}{\partial r}\right) + \frac{1}{r}\frac{\partial^2 V}{\partial \varphi^2}\right) = 0,$$

or after multiplication with $1/r$

$$\Delta_2 V(r, \varphi) = -\left(\frac{1}{r}\frac{\partial}{\partial r}\left(r\frac{\partial V}{\partial r}\right) + \frac{1}{r^2}\frac{\partial^2 V}{\partial \varphi^2}\right) = 0.$$

The potential function $V(r, \varphi)$ is determined by a Laplace equation (in two dimensions).

The boundary condition, that the tangential component of the electric field vanishes on the interior surface of the wave guide, is satisfied, if the function $V(r, \varphi)$ is constant on the edge of the cross section (Fig. 2.13)

$$E_t = \left.\frac{\partial V}{\partial t}\right|_{\text{edge}} = 0 \quad \longrightarrow \quad V_{\text{edge}} = \text{const}.$$

If one prefers to discuss the problem from the point of view of the B-field, one can represent the field also as the gradient of a scalar function of r and φ

$$(B_r, B_\varphi) = -\nabla_2 U(r, \varphi).$$

This function $U(r, \varphi)$ satisfies a corresponding Laplace equation

$$\Delta_2 U(r, \varphi) = 0.$$

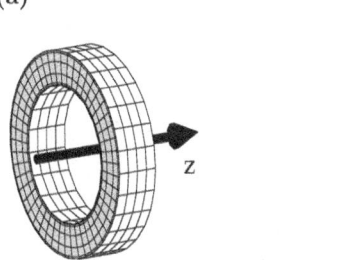
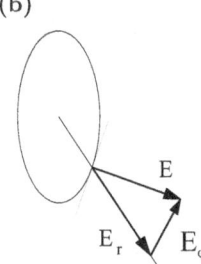

Fig. 2.13 Illustration of the boundary condition for the E-field of a TEM wave (a) (b)

The boundary condition, which has to be satisfied in this case, is: The normal component of the B-field has to vanish on the inner edge of the wave guide

$$B_n = \frac{\partial U}{\partial n}\bigg|_{edge} = 0 .$$

It can be shown, that in case of a simply connected edge of the wave guide the boundary condition for the E- and the B-fields can only be satisfied, if both the potentials V and U are constant in the complete interior area. No electromagnetic field exists in the wave guide. TEM waves can only exist in multiply connected wave guides (as e.g. a wave guide consisting of two coaxial cylinders).

2.2.3.2 TM-Waves
The individual Maxwell equations for TM waves are

$$\frac{1}{r}\frac{\partial E_z}{\partial \varphi} - ip\, E_\varphi = ik\, B_r$$

$$ip\, E_r - \frac{\partial E_z}{\partial r} = ik\, B_\varphi$$

$$\frac{1}{r}\frac{\partial (r E_\varphi)}{\partial r} - \frac{1}{r}\frac{\partial E_r}{\partial \varphi} = 0 \qquad (2.36)$$

$$ip\, B_\varphi = i\varepsilon\mu k\, E_r$$

$$ip\, B_r = -i\varepsilon\mu k\, E_\varphi$$

$$\frac{1}{r}\frac{\partial (r B_\varphi)}{\partial r} - \frac{1}{r}\frac{\partial B_r}{\partial \varphi} = -i\varepsilon\mu k\, E_z .$$

The number of independent components is reduced in this case as well. If one uses

$$B_r = -\varepsilon\mu \frac{k}{p} E_\varphi \qquad B_\varphi = \varepsilon\mu \frac{k}{p} E_r$$

for the elimination of two of the components of the B-field, one finds four equations, in which the components of the E-field are connected

$$\frac{1}{r}\frac{\partial E_z}{\partial \varphi} + \frac{i}{p}\kappa^2 E_\varphi = 0 \qquad (2.37)$$

$$\frac{\partial E_z}{\partial r} + \frac{i}{p}\kappa^2 E_r - = 0 \qquad (2.38)$$

$$\frac{1}{r}\frac{\partial (r E_\varphi)}{\partial r} - \frac{1}{r}\frac{\partial E_r}{\partial \varphi} = 0 \qquad (2.39)$$

2.2 Wave Propagation

$$\frac{1}{r}\frac{\partial(rE_r)}{\partial r} + \frac{1}{r}\frac{\partial E_\varphi}{\partial r} = -ipE_z . \qquad (2.40)$$

The quantity κ^2 stands for

$$\kappa^2 = \varepsilon\mu k^2 - p^2 .$$

Equations (2.37) and (2.38) state, that E_r and E_φ can be represented, up to a factor, by a gradient of E_z

$$(E_r, E_\varphi) = i\frac{p}{\kappa^2}\left(\frac{\partial E_z}{\partial r}, \frac{1}{r}\frac{\partial E_z}{\partial \varphi}\right) . \qquad (2.41)$$

One finds, that this equation is satisfied identically, if (2.41) is inserted into (2.39). If Eq. (2.41) is inserted into (2.40), one obtains (compare (2.29)) an equation for the determination of E_z

$$\Delta_2 E_z + \kappa^2 E_z = \left\{\frac{1}{r}\frac{\partial}{\partial r}\left(r\frac{\partial E_z}{\partial r}\right) + \frac{1}{r^2}\frac{\partial^2 E_z}{\partial \varphi^2}\right\} + \kappa^2 E_z = 0 . \qquad (2.42)$$

The solution has to satisfy the boundary condition

$$E_z|_{\text{edge}} = 0 ,$$

so that, as demanded in (2.21), the tangential component of the electric field vanishes on the edge of the wave guide. This boundary condition can not be satisfied for all values of κ^2 (and hence of p^2). The determination of acceptable values of κ^2, enforced by the boundary conditions, shows, that Eq. (2.42) presents an eigenvalue problem.

2.2.3.3 TE-Waves
The equations for the discussion of the TE waves can be obtained by the transformation[5]

$$E_{\text{TM}} = H_{\text{TE}} \quad \text{and} \quad H_{\text{TM}} = -E_{\text{TE}} .$$

The eigenvalue problem for these waves is therefore (suppress the indices)

$$\Delta_2 B_z + \kappa^2 B_z = 0 .$$

---

[5] This transformation is known as the Fitzgerald-transformation.

The boundary condition (2.21)

$$\mathbf{e}_n \cdot \mathbf{B}\Big|_{\text{edge}} = \mathbf{e}_n \cdot (B_r \mathbf{e}_r + B_\varphi \mathbf{e}_\varphi)\Big|_{\text{edge}} = 0$$

requires for a wave guide with arbitrary cross section

$$B_r\Big|_{\text{edge}} = B_\varphi\Big|_{\text{edge}} = 0.$$

The condition is satisfied because of

$$(B_r, B_\varphi) = i\frac{p}{\kappa^2}\left(\frac{\partial B_z}{\partial r}, \frac{1}{r}\frac{\partial B_z}{\partial \varphi}\right),$$

if the differential equation for B_z is solved with the boundary condition

$$\frac{\partial B_z}{\partial n}\Big|_{\text{edge}} = 0.$$

The normal derivative of B_z has to vanish at the edge of the cross section.

As both modes, TM and TE, satisfy different boundary conditions, one obtains in general different eigenvalues for the two modes. Both TE and TM modes can propagate in wave guides in the case of a simply connected edge.

2.2.3.4 Wire Waves

While the assumption of an ideal conductor yields acceptable results for wave guides, it is necessary to take into account the finite penetration depth for the discussion of **wire waves**. They describe the propagation of electromagnetic waves along an extended, cylindrical metallic body. In this case one has to use the complete Ampère law

$$\nabla \times \mathbf{H} = \frac{4\pi}{c}\mathbf{j}_{tr} + \frac{1}{c}\frac{\partial}{\partial t}\mathbf{D}$$

with the specification that the relations

$$\mathbf{j}_{tr} = \sigma \mathbf{E} \qquad \mathbf{D} = \varepsilon \mathbf{E}$$

have to be used in the interior of the conductor. The differential equation of Ampère involves in this case a time dependent factor as for monochromatic waves $\exp^{i\omega t}$ in the form

$$\nabla \times \mathbf{H} = \left(\frac{4\pi}{c}\sigma + ik\varepsilon\right)\mathbf{E}.$$

2.2 Wave Propagation

The complex wave number, which replaces the purely imaginary number for wave guides, does not change the formal structure of the relevant equations but their physical content. The boundary conditions are in this case: The electromagnetic fields H and E have to go to zero for points, which are infinitely far away from the wire. Different conditions for the connection of the fields on the surface of the wire has to be considered as well.

2.2.4 Diffraction

It is possible to observe diffraction patterns of an electromagnetic wave, e.g. a monochromatic plane wave, which has impinged on and passed a surface with one or more apertures. One requirement for the appearance of such patterns is the fact that the diameter of the opening must be small in comparison with the wave length of the radiation. An example is the diffraction of light by a slit, for which one is able to observe a pattern of bright and dark strips. A corresponding, but more developed pattern can be found for the diffraction by a grid of parallel slits, a less prominent pattern for the diffraction by an obstacle, which is e.g. produced by a wire.

A heuristic explanation of diffraction patterns is provided by the **principle of Huygens**, which states that each point of the aperture is the starting point of an elementary spherical wave. Interference of these spherical waves produces the observed pattern. It is possible to derive a formula for maximum interference (Fig. 2.14a and b) in the case of the diffraction by a grid. This formula is based on the optical path difference of spherical waves (or more simply rays), which emanate from neighbouring slits of the grid. Maximal interference, that is bright strips, is observed under an angle α, which satisfies the condition

$$d \sin \alpha = n\lambda \qquad (n = 0, 1, 2, \ldots).$$

The grid has a grid constant d (the regular distance between neighbouring slits), the number n is the order of the diffraction and λ is the wave length of the light, that is used.

The task of theory is the justification of the principle of Huygens and the provision of tools for the quantitative calculation of the intensity distributions of the

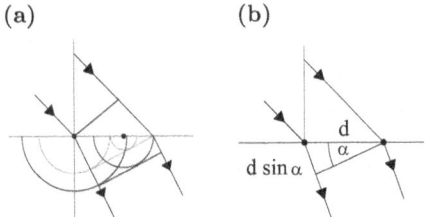

Fig. 2.14 Principle of Huygens. (**a**) Elementary waves. (**b**) Condition for interference

electromagnetic radiation after the passage of openings or obstacles. It is necessary to distinguish two domains of the space for this purpose:

- The Fresnel diffraction zone, which is characterised by a moderate distance from the system, where diffraction is initiated.
- The Fraunhofer diffraction zone, which covers the area of larger distances.

The theory of diffraction is based on the adaption of the solutions of the d'Alembert equations (1.72) and (1.73), for electromagnetic waves with the aid of physically motivated conditions, in order to find a form, which can be applied more easily. The first step is the extraction of an integral representation of the solutions. The derivation is usually based on a consideration of scalar waves, which is sufficient to cover the main aspects of the diffraction. The scalar wave function $\psi(r, t)$ is supposed to be a solution of the differential equation (1.73)

$$\Box \psi(r,t) = \Delta \psi(r,t) - \frac{1}{c^2}\frac{\partial^2}{\partial t^2}\psi(r,t) = -4\pi g(r,t) \qquad (2.43)$$

in the vacuum. The source term g is not specified at this stage. The discussion of the solution of this differential equation begins with a time dependent extension of Green's theorem (Dreizler and Lüdde, 2024, Electrostatics and Magnetostatics (Springer Berlin Heidelberg), Eq. (4.24))

$$\iiint_V [\psi(r)\Delta\phi(r) + \nabla\psi(r)\cdot\nabla\phi(r)]\,dV = \oiint_S \psi(r)\frac{\partial\phi(r)}{\partial n}\,df.$$

$\Longrightarrow$

$$\int_{t_i}^{t_f} dt' \iiint_V [\phi(r',t')\Delta'\psi(r',t') - \psi(r',t')\Delta'\phi(r',t')]\,dV'$$
(2.44)
$$= \int_{t_i}^{t_f} dt' \oiint_S \left[\phi(r',t')\frac{\partial\psi(r',t')}{\partial n'} - \psi(r',t')\frac{\partial\phi(r',t')}{\partial n'}\right]\,df',$$

where S is the surface of a volume V. The time t' is integrated over an interval starting at the initial time t_i. This involves the steps:

- Replace the function ϕ by the retarded Green's function $G^{(+)}(r,t;r',t')$ in Eq. (1.83), so that the second time variable t lies in the interval $t_i \leq t \leq t_f$. Afterwards one uses (in analogy to the treatment of the corresponding stationary problem in Dreizler and Lüdde, 2024, Electrostatics and Magnetostatics (Springer Berlin Heidelberg), Chap. 4.3) in the volume integral the d'Alembert

2.2 Wave Propagation

equation (1.77) for $G^{(+)}$ as well as Eq. (2.43) for ψ. The volume integral (VI) in (2.44) is then

$$\text{VI} = \int_{t_i}^{t_f} dt' \iiint_V \left[4\pi \delta(t-t')\delta(\mathbf{r}-\mathbf{r}')\psi(\mathbf{r}',t') \right.$$

$$- 4\pi G^{(+)}(\mathbf{r},t;\mathbf{r}',t')g(\mathbf{r}',t') + \frac{1}{c^2}\left(G^{(+)}(\mathbf{r},t;\mathbf{r}',t')\frac{\partial^2 \psi(\mathbf{r}',t')}{\partial t'^2}\right.$$

$$\left.\left. - \psi(\mathbf{r}',t')\frac{\partial^2 G^{(+)}(\mathbf{r},t;\mathbf{r}',t')}{\partial t'^2}\right)\right] dV'.$$

- The contribution with the time derivatives can be integrated partially using

$$G^{(+)}\frac{\partial^2 \psi}{\partial t'^2} = \frac{\partial}{\partial t'}\left(G^{(+)}\frac{\partial \psi}{\partial t'}\right) - \frac{\partial G^{(+)}}{\partial t'}\frac{\partial \psi}{\partial t'}$$

and the corresponding expression for the second term (note the abbreviated notation). As the Green's function and its derivative vanish at the upper limit $t' = t_f$, one finds

$$\int_{t_i}^{t_f} dt' \left(G^{(+)}\frac{\partial^2 \psi}{\partial t'^2} - \psi\frac{\partial^2 G^{(+)}}{\partial t'^2}\right) = \left(G^{(+)}\frac{\partial \psi}{\partial t'} - \psi\frac{\partial G^{(+)}}{\partial t'}\right)_{t'=t_i}.$$

The result of this first step is a representation of the wave function $\psi(\mathbf{r},t)$ for points within the volume V, which is confined by the surface S. In addition, the specification of the sources, that is of the initial values of the wave function and of the values of the wave function on the surface are required.

$$\psi(\mathbf{r},t) = \int_{t_i}^{t_f} dt' \left[\iiint_V dV' \, G^{(+)}g \right.$$

$$+ \frac{1}{4\pi} \oiint_S \left(G^{(+)}\frac{\partial \psi}{\partial n'} - \psi\frac{\partial G^{(+)}}{\partial n'}\right) df' \right]$$

$$+ \frac{1}{4\pi c^2} \iiint_V dV' \left(G^{(+)}\frac{\partial \psi}{\partial t'} - \psi\frac{\partial G^{(+)}}{\partial t'}\right)_{t'=t_i}.$$

The integral representation of the solution can be found by the introduction of a first set of additional conditions, which are based on the geometry indicated in Fig. 2.15, as introduced by Kirchhoff. The volume V, which contains the point of observation, is confined by two closed surfaces S_1 and S_2. The source of the radiation lies in the

Fig. 2.15 Geometry used in Kirchhoff's representation of diffraction

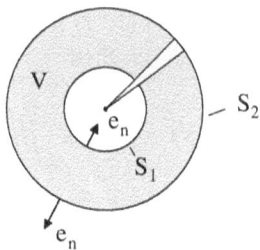

interior the surface S_1. The objects, which produce the diffraction, are situated on this surface. The assumptions are

- There are no sources within the volume V. This is expressed by

$$g(r', t') = 0 .$$

- The initial values of the experiment do not play any role. One assumes, that the experiment is arranged, so that the initial wave function is confined to the interior of S_1 and that one can use

$$\psi(r, t_i) = 0 \quad \text{and} \quad \left.\frac{\partial \psi(r, t)}{\partial t}\right|_{t=t_i} = 0 .$$

The remaining equation, which only contains the contribution of the total surface $S = S_1 + S_2$

$$\psi(r, t) = \frac{1}{4\pi} \int_{t_i}^{t_f} dt' \oint_S \left[G^{(+)}(r, t; r', t') \frac{\partial \psi(r', t')}{\partial n'} \right.$$

$$\left. - \psi(r', t') \frac{\partial G^{(+)}(r, t; r', t')}{\partial n'} \right] df' \quad (2.45)$$

can be processed further with the formula (1.83) for the Green's function. The normal derivative is represented for this purpose by the gradient

$$\frac{\partial}{\partial n'} = e_n \cdot \nabla' .$$

The vector e_n, which represents the direction of the normal, points out of the closed volume V. This means into the volume, which is enclosed by the surface S_1 and into the volume excluded by the surface S_2. For the gradient of the Green's function

$$G^{(+)}(r, t; r', t') = G^{(+)}(r - r', t - t') = \frac{\delta(R/c + t' - t)}{R}$$

2.2 Wave Propagation

($R = r - r'$) one obtains the result

$$\nabla' G^{(+)}(R, t - t') = \frac{\partial G^{(+)}(R, t - t')}{\partial R} \nabla' R$$

$$= \frac{R}{R^3} \delta(R/c + t' - t) - \frac{R}{cR^2} \delta'(R/c + t' - t).$$

The time integration in (2.45) can be executed. (The individual steps for the transition from (2.45) to (2.46) can be found in Detail 2.5.4.1). The final result, a general variant of **Kirchhoff's integral representation**, has the form

$$\psi(r, t) = \frac{1}{4\pi} \oiint_S df' \, e_n \cdot \left[\frac{\nabla' \psi(r', t')}{R} \right. \tag{2.46}$$

$$\left. - \frac{R\psi(r', t')}{R^3} - \frac{R}{cR^2} \frac{\partial \psi(r', t')}{\partial t'} \right]_{t' = t - R/c}.$$

This result corresponds to the principle of Huygens: The wave function in the volume V is determined by the wave function on the surface S (the starting point of the spherical waves of Huygens). The information, necessary for the evaluation of this formula (knowledge of the wave function and its time derivative on the surface S) is, however, known only after the solution of the complete problem (an initial value problem with the specification of the sources of the radiation). In order to avoid this step, it is necessary to introduce in addition to the postulates above, further approximations for the solution of the diffraction problem.

- The first one is the assumption, that one is dealing with a monochromatic wave function[6]

$$\psi(r, t) = e^{-i\omega t} \psi(r).$$

With the time derivative

$$\left. \frac{\partial \psi(r', t')}{\partial t'} \right|_{t' = t - R/c} = -i\omega \, e^{-i\omega(t - R/c)} \psi(r)$$

one obtains with (2.46) for the space part ($\omega = ck$) of the wave function

$$\psi(r) = \frac{1}{4\pi} \oiint_S df' \, \frac{e^{ikR}}{R} e_n \cdot \left[\nabla' \psi(r') + ik \left(1 + \frac{i}{kR} \right) \frac{R\psi(r')}{R} \right].$$

---

[6] The discussion of wave packets is possible.

- One uses a free wave (plane or spherical) as a wave function and its derivative for points of the opening. This assumption is based on the (not quite correct) idea, that this wave is propagated from a source to the opening without being influenced by the opening. A corresponding remark applies in the case of an obstacle.
- The surface S_2 is assumed to be the surface of a sphere with a sufficiently large radius, so that it does not contribute to the surface integral. This requirement can not be met directly, as the potential of a spherical wave (see Chap. 2.3) decreases only as $1/r'$. One argues for this reason, that a wave front can not reach the surface, which is far away, and can therefore not contribute to the surface integral on S_2. The remaining Kirchhoff integral is usually cited in the form

$$\psi(r) = -\frac{1}{4\pi} \iint_{O(S_1)} \frac{e^{ikR}}{R} \bar{e}_n \cdot \left[\nabla' \psi(r') \right.$$

$$\left. + ik \left(1 + \frac{i}{kR} \right) \frac{R\psi(r')}{R} \right] df'. \qquad (2.47)$$

The term $O(S_1)$ indicates the opening or the obstacle on the surface S_1. The normal $\bar{e}_n$ points now, according to an often used convention, into the volume V between the two surfaces. This requires a sign change.
- The last assumption restricts the discussion to **Fraunhofer diffraction**, which is generally of greater interest. The point of observation r is usually sufficiently well removed from the opening (compare the discussion of the 'radiation zone' in Chap. 2.3.1), so that one can use the approximation

$$\frac{1}{R} \approx \frac{1}{r}, \quad \frac{1}{R^2} \approx 0, \quad \frac{R}{R} \approx e_r \quad \text{and} \quad e^{ikR} = e^{ikr} e^{-ik \cdot r'}.$$

The wave number vector is defined by $k = k e_r$. It points in the direction of the point of observation. For the discussion of **Fresnel diffraction** the inclusion of higher order terms in the expansions indicated have to be included.

The Kirchhoff integral for Fraunhofer diffraction, which is generally applied, finally takes the form

$$\psi(r) = -\frac{e^{ikr}}{4\pi r} \iint_{O(S_1)} df' \, e^{-ik \cdot r'} \bar{e}_n \cdot \left[\nabla' \psi(r') + ik \psi(r') \right]. \qquad (2.48)$$

The prefactor represents a spherical wave emanating from the origin. This is modified by the integral over the surface with the objects, which are responsible for the diffraction.

An example, which was treated by G. Airy already in 1835—though with different methods—is the pattern of diffraction, which arises if a plane wave impacts

2.2 Wave Propagation

Fig. 2.16 Geometry of the diffraction on a circular opening

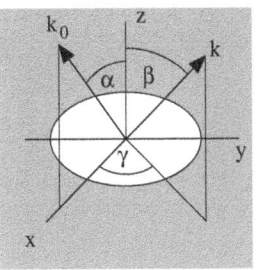

at an angle α on a plane, circular opening with radius a. The corresponding geometry and the choice of the coordinate system are presented in Fig. 2.16. The z-axis points in the direction of the normal ($\bar{e}_n = e_z$), the circle is located in the x–y plane. The x-axis is arranged in such a way, that the projection of the wave number vector k_0 of the incoming plane wave coincides in the x–y plane with this axis. The direction to the point of observation is characterised by the polar angle β and the azimuthal angle γ. For the calculation of the integral over the spherical surface polar coordinates have been used. The following quantities (vectors, wave function and scalar products) have been employed:

- The wave number vector of the incoming radiation (with the choice of the coordinate system)

$$k_0 = k\,(\cos\alpha\; e_z + \sin\alpha\; e_x)\ .$$

- The wave number vector $k = k e_r$, respectively the direction of the point of observation

$$e_r = \cos\beta\; e_z + \sin\beta\,(\cos\gamma\; e_x + \sin\gamma\; e_y)\ .$$

- The components of the vector r', which represents the surface of the circle

$$r' = \rho'\,(\cos\varphi'\,e_x + \sin\varphi'\,e_y)\ .$$

- The gradient of the incoming plane wave

$$\psi(r') = \psi_0\,e^{ik_0\cdot r'} \quad \text{is} \quad \nabla'\psi(r') = ik_0\,\psi_0\,e^{ik_0\cdot r'}\ .$$

- The list of the necessary scalar products

$$k_0 \cdot r' = k\rho'\sin\alpha\;\cos\varphi'$$
$$k_0 \cdot e_n = k\cos\alpha$$

$$\mathbf{k} \cdot \mathbf{e}_n = k \cos \beta$$
$$\mathbf{k} \cdot \mathbf{r}' = k\rho' \sin \beta \left(\cos \gamma \cos \varphi' + \sin \gamma \sin \varphi' \right) = k\rho' \sin \beta \cos(\varphi' - \gamma) \,.$$

The integral, which has to be evaluated, is therefore

$$\psi(\mathbf{r}) = -\frac{ik\psi_0 \, e^{ikr}}{4\pi r} (\cos \alpha + \cos \beta) \int_0^a \rho' d\rho'$$

$$\cdot \int_0^{2\pi} d\varphi' \, e^{ik\rho'(\sin \alpha \cos \varphi' - \sin \beta \cos(\varphi' - \gamma))} \,.$$

For this calculation one uses the conversion of the argument in the exponent (the integral is discussed in Detail 2.5.4.2)

$$A \cos \varphi' + B \sin \varphi' = \sqrt{(A^2 + B^2)}$$

$$\cdot \left(\frac{A}{\sqrt{(A^2 + B^2)}} \cos \varphi' + \frac{B}{\sqrt{(A^2 + B^2)}} \sin \varphi' \right),$$

and defines the functions

$$\cos \varphi_0 = -\frac{A}{\sqrt{(A^2 + B^2)}} \qquad \sin \varphi_0 = -\frac{B}{\sqrt{(A^2 + B^2)}} \,,$$

which can be combined to

$$A \cos \varphi' + B \sin \varphi' = -\sqrt{(A^2 + B^2)} \cos(\varphi' - \varphi_0) \,.$$

The resulting angular integral

$$\text{WI} = \int_0^{2\pi} d\varphi' \, e^{ik\rho'(\sin \alpha \cos \varphi' - \sin \beta \cos(\varphi' - \gamma))} = \int_0^{2\pi} d\varphi' \, e^{-ik\sigma\rho' \cos(\varphi' - \varphi_0)}$$

can be abbreviated in the form

$$\sigma = \sqrt{(A^2 + B^2)} = \left[\sin^2 \alpha + \sin^2 \beta - 2 \sin \alpha \sin \beta \cos \gamma \right]^{1/2}$$

and, with the periodicity of the cosine, written as

$$\text{WI} = \int_0^{2\pi} d\varphi' \, e^{-ik\sigma\rho' \cos \varphi'} \,.$$

2.2 Wave Propagation

This integral is not elementary. It is, up to a factor, the integral representation of the function J_0, a **Bessel function of the first kind** with integer order

$$WI = 2\pi J_0(k\sigma\rho') .$$

In the evaluation of the integral one encounters a second formula with Bessel functions. The derivative of the Bessel function J_1 is

$$zJ_0(z) = \frac{d}{dz}(zJ_1(z)) ,$$

so that the ρ'-integral yields after a substitution, the final result

$$\psi(r) = -\frac{ik\psi_0 e^{ikr}}{2r} a^2 (\cos\alpha + \cos\beta) \frac{J_1(ak\sigma)}{ak\sigma} . \quad (2.49)$$

The distribution of the intensity of the scalar wave can be interpreted in the spirit of the Poynting theorem (Chap. 1.4.1) as

$$\frac{dP}{d\Omega} = \frac{c}{8\pi}|\psi|^2 = \frac{c(ka^2)^2|\psi_0|^2}{32\pi r^2}(\cos\alpha + \cos\beta)^2 \left|\frac{J_1}{ak\sigma}\right|^2 . \quad (2.50)$$

Figure 2.17a shows the variation of the essential factor with the variable σ, which depends on the incident angle and the angle of observation

$$f(\eta) = |J_1(\eta)/\eta|^2 \qquad \eta = ak\sigma ,$$

as well as the oscillatory structure starting at the first zero of this function

$$\eta^2 f(\eta) = |J_1(\eta)|^2 .$$

The true diffraction pattern (for $\alpha = 0$, Fig. 2.17b) consists of a bright, central disk, which is surrounded by concentric dark and bright rings (these statements

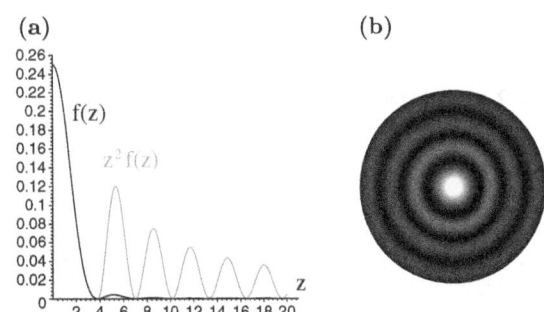

Fig. 2.17 Diffraction pattern of a circular opening. (**a**) Analytic representation ($z = \eta$). (**b**) Reality

concerning the diffraction pattern are justified in Detail 2.5.4.3). The intensity of the bright rings decreases quickly with the radius. The distance between neighbouring rings is approximately (const.)/(ka) (according to the difference of the minima of the function $|J_1(\eta)/\eta|^2$), so that one observes for $ka \gg 1$ practically only the image of the opening and minimal diffraction effects. One can observe a distinct diffraction pattern for $ka \approx 1$, as the Bessel function changes slowly with the angle β. For a wave length, that is large in comparison with the radius of the opening $ka \ll 1$, one can not meet the assumption of an undisturbed wave function in the opening. The real situation does not correspond to the expression (2.50).

Two remarks have to be added:

- The scalar approach to diffraction does not include one property of electromagnetic waves, the polarisation. For the inclusion of polarisation a formulation of the Kirchhoff integral in terms of vector functions is necessary. The high frequency of light ($f = c/\lambda \approx 10^{14}$ s^{-1}) allows, however, only the measurement of time-averaged quantities, so that polarisation effects can only be observed in a limited way. The theoretical description of experimental results is actually quite adequate in terms of the scalar version.
- One can ask, what is the difference between the diffraction patterns of a circular opening and of an impenetrable circular disk (Fig. 2.18).

  The answer to this question can be obtained with the following argument: The wave functions of the disk ψ_S and the opening ψ_O in the point of observation are determined by integration over two complementary domains. The sum of the wave functions corresponds to a value, which can be found by integration over the complete plane on which the diffracting objects are placed

$$\psi_{\text{total}}(r) = \psi_S(r) + \psi_O(r).$$

This relation is known by the name **Babinet's principle**. There is no diffraction by the complete plane, only *one* wave (e.g. a plane ψ_{total}), which passes through the plane. It is characterised by the same wave number vector as the incident wave k_0 with $\psi_{\text{total},k \neq k_0} = 0$.

For the diffraction by the two objects follows

$$\psi_S(r) = -\psi_O(r).$$

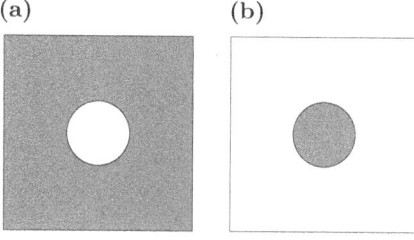

Fig. 2.18 Babinet's principle. (**a**) Circular opening. (**b**) Circular disk

The pattern of diffraction of complementary objects, as the circular disk and the circular opening, are equal, as the pattern is determined by $|\psi|^2$.

2.3 Generation of Waves: Transmitters

The problem of sending and receiving electromagnetic waves played an important role from modest beginnings up to the creation of a worldwide net for the exchange of information. This section starts with a standard example, the dipole approximation. The discussion of transmission problems is continued with two topics: The contributions of the next multipoles and the complete multipole expansion. The section is concluded with the consideration of a special transmitting problem, which can be solved exactly.

2.3.1 Specification of Transmitters

A simple charge distribution, which could be used for the evaluation of the formulae of Kirchhoff ((1.85) and (1.86)) is a harmonically oscillating charge distribution

$$\rho_{tr}(\boldsymbol{r},t) = \rho(\boldsymbol{r})\, e^{-i\omega t} \ . \tag{2.51}$$

The space part should be confined to a distribution about the origin of a coordinate system. The transmitter with an oscillating frequency ω is monochromatic. The real part of the factor depending on time represents the actual time dependence (e.g. $\cos \omega t$).

If the current density vector $\boldsymbol{j}_{tr}(\boldsymbol{r},t)$, with the harmonic time part

$$\boldsymbol{j}_{tr}(\boldsymbol{r},t) = \boldsymbol{j}(\boldsymbol{r})\, e^{-i\omega t}$$

is inserted into the continuity equation

$$\nabla \cdot \boldsymbol{j}_{tr}(\boldsymbol{r},t) = -\frac{\partial \rho_{tr}(\boldsymbol{r},t)}{\partial t} = i\omega \rho(\boldsymbol{r})\, e^{-i\omega t} \ ,$$

one can observe: This equation is satisfied, if the space part of the current density satisfies the relation

$$\nabla \cdot \boldsymbol{j}(\boldsymbol{r}) = i\omega \rho(\boldsymbol{r}) \tag{2.52}$$

with the same density. If the expression (1.85) for a scalar potential is evaluated with the charge distribution (2.51), one finds

$$V(\boldsymbol{r},t) = \iiint dV' \frac{\rho(\boldsymbol{r}')}{|\boldsymbol{r}-\boldsymbol{r}'|} e^{-i\omega(t-(1/c)|\boldsymbol{r}-\boldsymbol{r}'|)} \ .$$

Using the standard definition of the wave number $k = \omega/c = 2\pi/\lambda$ one can express the scalar part of the emitted wave as a product of a function depending on the space coordinates times a harmonic time part

$$V(\boldsymbol{r},t) = V(\boldsymbol{r})\,e^{-i\omega t}$$

$$V(\boldsymbol{r}) = \iiint dV' \frac{\rho(\boldsymbol{r}')}{|\boldsymbol{r}-\boldsymbol{r}'|}\,e^{ik|\boldsymbol{r}-\boldsymbol{r}'|}\,. \tag{2.53}$$

The representation of the space part in the form

$$V(\boldsymbol{r}) = \iiint dV'\, G^{(+)}(\boldsymbol{r},\boldsymbol{r}')\rho(\boldsymbol{r}')$$

yields the space part of the retarded Green's function of the transmitter. (A more detailed discussion of this function is found in Chap. 1.6.) The corresponding expression for the vector potential, which characterises this scalar potential, is

$$\boldsymbol{A}(\boldsymbol{r},t) = \boldsymbol{A}(\boldsymbol{r})\,e^{-i\omega t}$$

$$\boldsymbol{A}(\boldsymbol{r}) = \frac{1}{c}\iiint dV' \frac{\boldsymbol{j}(\boldsymbol{r}')}{|\boldsymbol{r}-\boldsymbol{r}'|}\,e^{ik|\boldsymbol{r}-\boldsymbol{r}'|}\,. \tag{2.54}$$

The retardation manifests itself, in comparison with the stationary case, by an exponential function under the integration symbol. The potentials and hence the fields have the same time function as the source. A chromatic transmitter emits a monochromatic electromagnetic wave.

The task, that one faces, is the evaluation of the space integrals for a given charge distribution, which corresponds to the solution of the inhomogeneous **Helmholtz equation**, as e.g.

$$\Delta V(\boldsymbol{r}) + k^2 V(\boldsymbol{r}) = -4\pi\rho(\boldsymbol{r})\,. \tag{2.55}$$

This differential equation with the specification (2.51) and the ansatz (2.53) is obtained from the d'Alembert equation.

For the evaluation of the integrals one discerns, with the assumption of a small transmitter (L_S), three domains of space:

- The near (or local) zone. Here the distance of a point in space from the centre of the transmitter is large in comparison to the extent of the transmitter but smaller than the wave length (λ) of the emitted wave ($L_S \ll r_N \ll \lambda$).
- The intermediate zone. Distances up to the order of the magnitude of the wave length play a role in this zone ($L_S \ll r_N \leq r_Z \approx \lambda$).
- The far (or radiation) zone, which is characterised by $L_S \ll \lambda \approx r_Z \ll r_S$.

2.3 Generation of Waves: Transmitters

The evaluation of the integrals relies on different approximations in the three sections of space. In the near zone one can replace the exponential function $\exp(ik|r - r'|)$ by 1, as the value of the product $\exp(ik|r - r'|)$ is quite small. The space part is therefore essentially equal to the stationary result. In the intermediate zone one has to expand the total integrand in a consistent fashion (compare Chap. 2.3.4). Most interest, for practical reasons, is given to the **radiation zone**. In the region with $L_S \ll \lambda \approx r_Z \ll r_S$ an expansion of both factors of the integrand is required. The function of the separation is expanded as

$$|r - r'| = r\left[1 - 2\frac{r' \cdot e_r}{r} + \left(\frac{r'}{r}\right)^2\right]^{1/2}$$

$$= r - r' \cdot e_r + \frac{1}{r}\frac{(r'^2 - (r' \cdot e_r))}{2} + \dots$$

and the inverse as

$$|r - r'|^{-1} = \frac{1}{r}\left[1 - 2\frac{r' \cdot e_r}{r} + \left(\frac{r'}{r}\right)^2\right]^{-1/2}$$

$$= \frac{1}{r} + \frac{1}{r^2}.(r' \cdot e_r)) + \dots .$$

In this zone, a consistent treatment of the contributions up to the order $1/r$ is essential. This means, that it suffices to expand the argument of the exponential function only up to the first two terms and the expansion of the inverse separation only to first order. A consistent approximation of the space part, e.g. for the vector potential, is therefore in the radiation zone

$$A_{RZ}(r) \approx \frac{1}{c}\iiint dV' \, j(r')\frac{e^{ik(r-r'\cdot e_r)}}{r} = \frac{e^{ikr}}{cr}\iiint dV' \, j(r')e^{-ik\cdot r'} .$$

The definition of the wave number vector in the radial direction is $k = k\, e_r$. The time dependent vector potential (and the corresponding scalar potential) in the radiation zone have therefore the form

$$A_{RZ}(r,t) = A_{RZ}(k)\left[\frac{1}{r} e^{i(kr-\omega t)}\right] \tag{2.56}$$

$$A_{RZ}(k) = \frac{1}{c}\iiint dV' \, j(r')e^{-ik\cdot r'} . \tag{2.57}$$

The factor in the square bracket, in which the wave number is multiplied with the distance from the centre of the transmitter, represents a spherical wave, which

Fig. 2.19 Spherical wave with a modification of the angle (dimension of the transmitter $L_S \ll r$)

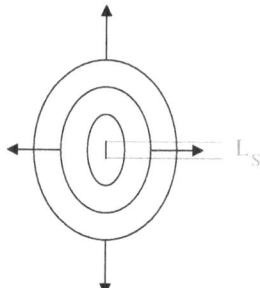

is emitted from this source. The wave fronts are spherical surfaces around the oscillating charge distribution (Fig. 2.19). The factor of the amplitude, $A_{RZ}(k)$, which depends on the wave number k, modifies the spherical wave in different directions. If the product of a wave number with the extension of the transmitter is small compared to 1

$$k \cdot r' \leq kL_S < 1 \, ,$$

a more complete expansion is required. The amplitude function e.g. of the vector potential $A_{RZ}(k)$ is expanded as

$$A_{RZ}(k) = \frac{1}{c} \iiint dV' \, j(r') \sum_{n=0}^{\infty} \frac{1}{n!}(-i k \cdot r')^n = \sum_{n=0}^{\infty} A^{(n)}(k) \, . \qquad (2.58)$$

This expansion is useful, if the dimension of the source is small in comparison with the wave length of the emitted radiation

$$2\pi \, L_S < \lambda \, .$$

In the simplest approximation, the extreme **long wave length approximation**, only the first term of this expansion is used. The next section contains the proof, that the radiation of this approximation has dipole character.

2.3.2 The Hertz Dipole

The integral

$$A^{(0)}(k) \approx \frac{1}{c} \iiint dV' \, j(r')$$

is to be evaluated in the simplest approximation. It turns out, that it is sufficient to consider the vector potential, as both the magnetic as well as the electric field

2.3 Generation of Waves: Transmitters

(in this approximation) can be calculated with this potential. The integral can be transformed, using the relation (see Detail 2.5.5.1)

$$\iiint \mathrm{d}V' \, \boldsymbol{j}(\boldsymbol{r}') = - \iiint \mathrm{d}V' \, \boldsymbol{r}' \left(\boldsymbol{\nabla}' \cdot \boldsymbol{j}(\boldsymbol{r}') \right) .$$

The term $\boldsymbol{\nabla} \cdot \boldsymbol{j}$ in the resulting expression

$$\boldsymbol{A}^{(0)}(\boldsymbol{k}) = -\frac{1}{c} \iiint \boldsymbol{r}' \left(\boldsymbol{\nabla}' \cdot \boldsymbol{j}(\boldsymbol{r}') \right) \mathrm{d}V'$$

can be replaced with (2.52) by ρ

$$\boldsymbol{A}^{(0)}(\boldsymbol{k}) = -\frac{\mathrm{i}\omega}{c} \iiint \boldsymbol{r}' \rho(\boldsymbol{r}') \, \mathrm{d}V' ,$$

so that one obtains the electric dipole moment of the charge distribution

$$\boldsymbol{p} = \iiint \boldsymbol{r}' \rho(\boldsymbol{r}') \mathrm{d}V' .$$

The final result for the approximation of the vector potential is thus (use $\omega/c = k$)

$$\boldsymbol{A}^{(0)}(\boldsymbol{r}, t) = -\boldsymbol{p} \left(\frac{\mathrm{i}k}{r} \right) \mathrm{e}^{\mathrm{i}(kr - \omega t)} . \qquad (2.59)$$

A corresponding formula is valid for the scalar potential. The dominant contribution in the radiation zone is a potential in the direction of the dipole moment of the source. The time-space structure is a spherical wave. This radiation pattern is known as a **Hertz dipole radiation**. The standard nomenclature is $E1$ (electric, $l = 1$).

The corresponding magnetic field is

$$\boldsymbol{B}_{(E1)}(\boldsymbol{r}, t) = \boldsymbol{\nabla} \times \boldsymbol{A}^{(0)}(\boldsymbol{r}, t) .$$

It is sufficient to include only the action of the nabla-operator on the wave part, as the inverse separation has been expanded to the order $1/r$

$$\boldsymbol{\nabla} \mathrm{e}^{\mathrm{i}(kr - \omega t)} = \mathrm{i} \boldsymbol{e}_r \cdot k \, \mathrm{e}^{\mathrm{i}(kr - \omega t)} = \mathrm{i} \boldsymbol{k} \, \mathrm{e}^{\mathrm{i}(kr - \omega t)} .$$

The magnetic field in lowest order is therefore

$$\boldsymbol{B}_{(E1)}(\boldsymbol{r}, t) = (\boldsymbol{k} \times \boldsymbol{p}) \frac{k}{r} \mathrm{e}^{\mathrm{i}(kr - \omega t)} . \qquad (2.60)$$

It is perpendicular to the direction of propagation and the dipole moment of the transmitter (Fig. 2.20).

Fig. 2.20 Orientation of the fields of the dipole radiation

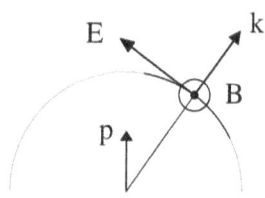

For the calculation of the E-field in the far zone, for which $\boldsymbol{j}_{tr}(\boldsymbol{r}) = \boldsymbol{0}$ is valid, one can use Ampère's law (in the vacuum)

$$\frac{1}{c}\frac{\partial \boldsymbol{E}}{\partial t} = \nabla \times \boldsymbol{B} \,.$$

If the time dependence of the E-field is the same as that of the B-field, the relation

$$-\mathrm{i}\frac{\omega}{c}\boldsymbol{E} = \mathrm{i}\,(\boldsymbol{k} \times \boldsymbol{B})$$

follows. If this equation is inserted into the result for $\boldsymbol{B}$, one finds after sorting[7]

$$\boldsymbol{E}_{(E1)}(\boldsymbol{r},t) = ((\boldsymbol{k} \times \boldsymbol{p}) \times \boldsymbol{k})\,\frac{1}{r}\,\mathrm{e}^{\mathrm{i}(kr-\omega t)} \,. \tag{2.61}$$

The E-field is, as the B-field, a modified spherical wave, which is perpendicular to the B-field and the direction of propagation $\boldsymbol{k} = k\boldsymbol{e}_r$. The relative orientation of the electromagnetic fields corresponds to the situation for plane waves.

The electromagnetic wave emitted by a Hertz dipole is illustrated by a snap shot of electric field lines in Fig. 2.21. Figure 2.21a shows the pattern of field lines over a larger spatial area, which corresponds to approximately 8 wave lengths. The near zone is shown in Fig. 2.21b.

The energy situation of the Hertz dipole transmitter can be discussed by considering the flux of energy from the source. This is given by the averaged Poynting vector

$$\bar{\boldsymbol{S}} = \frac{c}{8\pi}\left[\boldsymbol{E}_{(E1)}(\boldsymbol{r},t) \times \boldsymbol{B}^{*}_{(E1)}(\boldsymbol{r},t)\right] \,.$$

With the fields (2.60) and (2.61) one finds in the first step

$$\bar{\boldsymbol{S}} = -\frac{ck}{8\pi}\,[(\boldsymbol{k} \times (\boldsymbol{k} \times \boldsymbol{p})) \times (\boldsymbol{k} \times \boldsymbol{p})]\,\frac{1}{r^2} \,.$$

---

[7] This result can be confirmed by an explicit calculation of the electric field with the scalar potential (Detail 2.5.5.2).

2.3 Generation of Waves: Transmitters

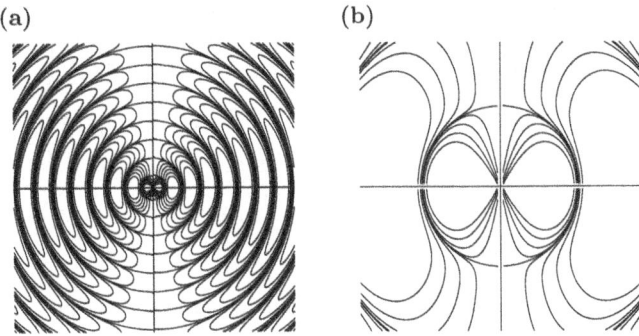

Fig. 2.21 Snap shots of the electric field lines of a Hertz dipole. (**a**) Intermediate and far zone. (**b**) Near and intermediate zone

The vector product

$$(k \times a) \times a$$

can be changed with the standard formula

$$(k \times a) \times a = (a \cdot k) a - (a \cdot a) k = - (a \cdot a) k ,$$

as the vector $a = k \times p$ is perpendicular to the vector k. The final result is then

$$\bar{S} = \frac{c}{8\pi} |(k \times p)|^2 \frac{k}{r^2} . \qquad (2.62)$$

The flux of energy is, as expected, in the radial direction and decreases due to the spherical geometry with the $1/r^2$ law.

The scalar product of the Poynting vector and the surface element yields for the spherical geometry

$$\bar{S} \cdot df = \bar{S} \cdot e_r \, r^2 d\Omega .$$

This quantity has the dimension energy divided by time, that is power P. The emitted power per solid angle is therefore

$$\frac{dP}{d\Omega} = \frac{ck^2}{8\pi} |(k \times p)|^2 . \qquad (2.63)$$

This result can also be expressed by the angle θ between the direction of the dipole p and the radial direction $k = k\, e_r$

$$\frac{dP}{d\Omega} = \frac{ck^4}{8\pi} |p|^2 \sin^2 \theta . \qquad (2.64)$$

Fig. 2.22 Radiation pattern of the Hertz dipole

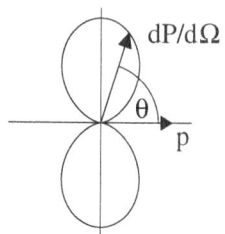

The angular dependence is usually presented in a polar diagram, in which the quantity $dP/d\Omega$ is plotted as the length of a ray. Such a plot leads to the characteristic radiation pattern of a Hertz dipole, which is presented in Fig. 2.22. The quantity power per solid angle does not depend on the polar angle φ, the radiation pattern shows cylindrical symmetry. The energy is mainly radiated in the direction of the dipole moment. The total emitted power is in the average

$$P = \int dP = \frac{ck^4}{8\pi} |p|^2 \iint \sin^2\theta \, d\Omega = \frac{1}{3} ck^4 |p|^2 \,. \tag{2.65}$$

The power (in the long wave length approximation) is inversely proportional to the fourth power of the wave length.

2.3.3 Contributions of Higher Multipoles

The next term in the expansion of the amplitude function (2.58) in the radiation zone is

$$\boldsymbol{A}^{(1)}(\boldsymbol{r}) = \frac{-i\,e^{ikr}}{cr} \iiint dV'\, \boldsymbol{j}(\boldsymbol{r}')(\boldsymbol{k}\cdot\boldsymbol{r}')\,. \tag{2.66}$$

The analysis of the physical content of this expression requires, as the sorting of the multipole contributions in the stationary problem, a transcription of the integrand. For this purpose one uses (see Detail 2.5.6.1) a split of the integrand into contributions, which are symmetric respectively antisymmetric with respect to the interchange of $\boldsymbol{r}'$ and $\boldsymbol{j}$

$$(\boldsymbol{k}\cdot\boldsymbol{r}')\boldsymbol{j} = \frac{1}{2}[(\boldsymbol{k}\cdot\boldsymbol{r}')\boldsymbol{j} + (\boldsymbol{k}\cdot\boldsymbol{j})\boldsymbol{r}'] + \frac{1}{2}(\boldsymbol{r}'\times\boldsymbol{j})\times\boldsymbol{k}\,. \tag{2.67}$$

The symmetric term leads, as shown below, to the electric and the magnetic field of the electric quadrupole radiation, which is identified by the code $E2$. The antisymmetric term leads to the electromagnetic fields of the magnetic dipole radiation ($M1$).

2.3 Generation of Waves: Transmitters

The volume integral over the term in square brackets can be summed up with the formula

$$\nabla \cdot (f(r) \, b(r)) = b(r) \cdot (\nabla f(r)) + f(r)(\nabla \cdot b(r))$$

and yields after partial integration (see Detail 2.5.6.2)

$$\iiint dV' \, [(k \cdot r') j(r') + (k \cdot j(r')) r'] = - \iiint dV' \, r'(k \cdot r')(\nabla' \cdot j(r')) \, .$$

The divergence of the current density can be replaced by the density via the equation of continuity (2.52) $\nabla \cdot j(r) = i\omega \rho(r)$. The complete contribution of the symmetric part of the vector potential is

$$A^{(1)}_{(\text{sym})}(r) = \frac{-k \, e^{ikr}}{2r} \iiint dV' \, r'(k \cdot r') \rho(r') \, .$$

The second moment of the charge distribution, which features here, shows that it is a electric quadrupole contribution. It is possible, to define the vector

$$Q(k) = 3 \iiint dV' \, r'(k \cdot r') \rho(r') \tag{2.68}$$

with components, which are composed of the elements of a quadrupole tensor

$$Q_i = \sum_{l=1}^{3} Q_{il} k_l = \sum_{l=1}^{3} \left[3 \iiint dV' \, x'_i x'_l \rho(r') \right] k_l \qquad i = 1, 2, 3$$

and write the result in the form

$$A^{(1)}_{(\text{sym})}(r) = -\frac{k}{6} Q(k) \frac{e^{ikr}}{r} \, . \tag{2.69}$$

One can recognise in the decomposition (2.67) the antisymmetric part of the magnetic moment of the charge distribution (Dreizler and Lüdde, 2024, Electrostatics and Magnetostatics (Springer Berlin Heidelberg), Eq. (5.33))

$$m = \frac{1}{2c} \iiint dV' (r' \times j(r')) \, .$$

The corresponding contribution to the vector potential is for this reason

$$A^{(1)}_{(\text{asym})}(r) = i(k \times m) \frac{e^{ikr}}{r} \, . \tag{2.70}$$

The electromagnetic fields in the radiation zone can be calculated on the basis of the two parts of the vector potential, as in the case of the dipole approximation, with the formulae

$$B = i(k \times A) \qquad E = \frac{i}{k}((k \times A) \times k) .$$

The results for the space parts of the magnetic dipole fields ($A = A_{(\mathrm{asym})}$) are

$$B_{(M1)}(r) = ((k \times m) \times k) \frac{1}{r} e^{i\,kr}$$

$$E_{(M1)}(r) = -k(k \times m) \frac{1}{r} e^{i\,kr} .$$

(2.71)

For the extraction of the result for the E-field it is necessary to apply a formula for the reduction of multiple vector products. The results for the magnetic dipole radiation show, that (up to a sign) the role of the fields are exchanged in comparison with the electric dipole radiation.

- The magnetic field of the magnetic dipole radiation $B_{(M1)}$ lies in the plane spanned by the vectors k and m. The electric field of the electric dipole radiation $E_{(E1)}$ (2.61) lies in the plane of k and p.
- The magnetic field of the electric dipole radiation $M_{(E1)}$ (2.60) is perpendicular to the plane spanned by the vectors k and p. The electric field $E_{(M1)}$ of the magnetic dipole radiation is perpendicular to k and m.

The situation is illustrated in Fig. 2.23. A consequence of this symmetry is: The radiation pattern for purely magnetic dipole radiation is not different from the pattern of purely electric dipole radiation.

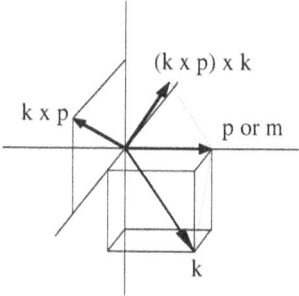

Fig. 2.23 Orientation of multipole fields (($k \times$ moment) $\longrightarrow B_{(E1)}$, $E_{(M1)}$, (($k \times$ moment)$\times k$) $\longrightarrow B_{(M1)}$, $E_{(E1)}$)

2.3 Generation of Waves: Transmitters

The electric quadrupole fields are

$$B_{(E2)}(r) = -\frac{ik}{6}(k \times Q(k)) \frac{1}{r} e^{ikr}$$

$$E_{(E2)}(r) = -\frac{i}{6}((k \times Q(k)) \times k) \frac{1}{r} e^{ikr}.$$
(2.72)

This fields have a similar form as the fields in the electric dipole approximation ((2.60) and (2.61)), if the electric dipole moment is replaced by the more complicated quadrupole vector $-i\,Q/6$. The radiation pattern of the pure electric quadrupole radiation is, except for different factors, described by the formula of the electric dipole

$$\frac{dP}{d\Omega} = \frac{ck^2}{288\pi} |(k \times Q(k))|^2.$$
(2.73)

A typical quadrupole radiation pattern is already obtained for a quadrupole radiating in a simple way. If the quadrupole can be described by

$$Q_{il} = \delta_{il} Q_{ii} \quad \text{with} \quad Q_{11} = Q_{22} = -\frac{1}{2}Q_{33} = \frac{1}{2}Q_0,$$

one finds for this spheroidal quadrupole (see Detail 2.5.6.3)

$$\frac{dP}{d\Omega} = \frac{ck^6}{128\pi} Q_0^2 \sin^2\theta \cos^2\theta.$$

The polar diagram (without the prefactors) with the typical four-leaf clover pattern is presented in Fig. 2.24a. The spatial distribution is produced by a rotation of the picture about the axis indicated. Figure 2.24b shows the functional dependence in the standard representation.

Fig. 2.24 Angular distribution of the quadrupole radiation ($E2$) as a function of $x = \theta$. (a) Polar diagram. (b) Standard representation

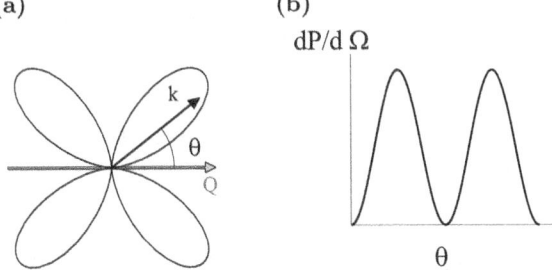

2.3.4 The Complete Multipole Expansion

A systematic access to the radiation can be obtained with a discussion of the spatial part of the retarded Green's function, which describes the radiation process (compare (2.53), (2.54))

$$G^{(+)}(\mathbf{r}, \mathbf{r}') = \frac{e^{ik|\mathbf{r}-\mathbf{r}'|}}{|\mathbf{r}-\mathbf{r}'|}.$$

This function satisfies the Helmholtz equation

$$(\Delta + k^2) G^{(+)}(\mathbf{r}, \mathbf{r}') = -4\pi\, \delta(\mathbf{r} - \mathbf{r}'),$$

the symmetry condition

$$G^{(+)}(\mathbf{r}, \mathbf{r}') = G^{(+)}(\mathbf{r}', \mathbf{r})$$

and the boundary conditions:

- The function has to be finite at the origin of the coordinate system

$$G^{(+)}(\mathbf{r}, \mathbf{r}') \xrightarrow{r' \to 0} \text{finite} \quad (r > r'),$$

  so that integration over a given charge density yields reasonable results.
- The function has to turn into an outgoing wave in the asymptotic region

$$G^{(+)}(\mathbf{r}, \mathbf{r}') \xrightarrow{r' \to \infty} \text{const.} \frac{e^{ikr}}{r} \quad (r < r').$$

The discussion of this function is based on an expansion with spherical harmonics (compare in Dreizler and Lüdde, 2024, Electrostatics and Magnetostatics (Springer Berlin Heidelberg), Chap. 4.3)

$$G^{(+)}(\mathbf{r}, \mathbf{r}') = \sum_{l,m} g_l(r, r') Y_{lm}^*(\Omega') Y_{lm}(\Omega).$$

If this expansion is inserted into the differential equation and the dependence on the angles is eliminated via integration, one finds an equation for the determination of the radial functions g_l

$$\left(\frac{d^2}{dr^2} + \frac{2}{r}\frac{d}{dr} + k^2 - \frac{l(l+1)}{r^2} \right) g_l(r, r') = -\frac{1}{r^2} \delta(r - r'). \tag{2.74}$$

2.3 Generation of Waves: Transmitters

The solutions of this inhomogeneous differential equation can be found via the solutions of the homogeneous differential equation. The corresponding homogeneous differential equation for functions of one variable is the differential equation for the **spherical Bessel functions**.

The solution of the inhomogeneous differential equation, that one is looking for, should satisfy the conditions for the Green's function stated above. In addition, it should reproduce the singularity at the position $r = r'$. The solutions, which are assembled in Detail 2.5.7, are products of a spherical Bessel function j_l with a spherical Hankel function $h_l^{(+)}$

$$g_l(r, r') = 4\pi i k j_l(kr_<) h_l^{(+)}(kr_>) .$$

The symbol $r_<$ applies to the smaller value of r- respectively r'-values, the symbol $r_>$ to the larger one. If this expansion

$$\frac{e^{ik|r-r'|}}{|r-r'|} = 4\pi i k \sum_{l,m} j_l(kr_<) h_l^{(+)}(kr_>) Y_{lm}^*(\Omega') Y_{lm}(\Omega) \qquad (2.75)$$

is inserted into the relation for the vector product

$$A(r) = \frac{1}{c} \iiint dV' \frac{j(r')}{|r-r'|} e^{ik|r-r'|} ,$$

the transmitter problem is solved in principle, even if the actual calculation of the multipole expansion of the electric and magnetic fields still requires some effort.

Thus one has for points with $r > r'$ for the vector potential

$$A(r) = \frac{4\pi i k}{c} \sum_{l,m} h_l^{(+)}(kr) Y_{lm}(\Omega) A_{lm}(k) , \qquad (2.76)$$

where the coefficient functions, which are sorted with respect to the multipolarity, are to be calculated by

$$A_{lm}(k) = \iiint dV' j_l(kr') Y_{lm}^*(\Omega') \, j(r') . \qquad (2.77)$$

The results are valid for *all* points, which are not lying in the current and the density distributions, even in the intermediate zone. It is sufficient to use the asymptotic form of the Hankel function

$$h_l^{(+)}(kr) \xrightarrow{r \to \infty} (-i)^{l+1} \frac{e^{i kr}}{kr}$$

in the radiation zone $r \gg r'$. In this case a spherical wave with angular modifications results

$$A_{RZ}(r) = \frac{4\pi}{c} \frac{e^{ikr}}{r} \sum_{l,m} (-i)^l Y_{lm}(\Omega) A_{lm}(k) \,. \qquad (2.78)$$

If one restricts, in addition, the attention to the long wave approximation

$$j_l(kr') \xrightarrow{kr' \to 0} \frac{(kr')^l}{(2l+1)!!} \,,$$

one can make a connection with the results obtained directly in Chaps. 2.3.2 and 2.3.3 or one can extend these results.

There exist two options for the calculation of the multipole expansion of the electromagnetic fields. One of them is to use, as before, the standard relations (with the assumptions of harmonic time dependence, considering only the area outside the specified charge and current distributions)

$$\boldsymbol{B}(r) = \nabla \times \boldsymbol{A}(r) \quad \text{and} \quad \boldsymbol{E}(r) = \frac{i}{k} (\nabla \times \boldsymbol{B}(r)) \,.$$

The second one is a direct multipole expansion of the electromagnetic fields in terms of the fundamental solutions of the Bessel differential equation e.g. for the $\boldsymbol{B}$-field

$$\boldsymbol{B}(r) = \sum_{l,m} \left(C^{(1)}_{l,m} h^{(+)}_l(kr) + C^{(2)}_{l,m} h^{(-)}_l(kr) \right) Y_{lm}(\Omega) \,.$$

The coefficients have to be determined by insertion into the Maxwell equations.

In both options one is faced with the application of the operator product $\nabla \times$ on the multipole expansion. This is a tedious affair. With the decomposition of the rotation operator in spherical coordinates (compare p. 311), one has to work out the results of the action of the derivatives with respect to the angular coordinates of the spherical harmonics. These results leads to a tensor classification of the electric and the magnetic parts based on the **Gradient Formula**.[8]

In this respect one can also mention, that, as indicated in Chap. 2.3.3, the classification of the vector potential (and the associated electric and magnetic multipole fields) can be generalised. The parts of the vector potential, which are antisymmetric under spatial reflections (the replacement of r by $-r*$) are magnetic parts

$$\boldsymbol{A}_{(M)}(-r) = -\boldsymbol{A}_{(M)}(r) \,.$$

---

[8] Interested readers may consult e.g. the standard book of M.E. Rose, Elementary Theory of Angular Momentum.

2.3 Generation of Waves: Transmitters

Electric parts of the vector potential (E1, E2, ...) are symmetric

$$A_{(E)}(-r) = +A_{(E)}(r).$$

2.3.5 An Exactly Solvable Transmitter Problem

A problem, for which an analytic solution can be given, is the symmetric, linear antenna with central feed. The current density vector with an orientation with respect to the z-axis can be stated as

$$j(r,t) = \begin{cases} i_0 \sin\left(\frac{kd}{2} - k|z|\right) e^{-i\omega t} \delta(x)\delta(y) e_z & \text{for } |z| < \frac{d}{2} \\ 0 & \text{for } |z| \geq \frac{d}{2}. \end{cases} \quad (2.79)$$

The arms of the antenna have each the length $d/2$. The local amplitude is a sine function with a maximum of $\sin(kd)/2$ at the central position of the rod at $z = 0$ and the value zero at the edge of the rod at $z = \pm d/2$.

This distribution is modulated with the time dependent factor $\cos \omega t$. Figure 2.25 shows the initial phase of the current distribution for two values of the parameter $a = kd/2$. The specified spatial pattern is modified in a simple manner by the time factor. Such an arrangement can be viewed as an idealisation of a real antenna with a finite extension in the x- and y-directions.

The calculation of the vector potential in the *far zone* requires the evaluation of the integral (2.66) for the specification above

$$A(r) = \frac{i_0 e^{ikr}}{cr} \int_{-d/2}^{d/2} dz' \sin\left(\frac{kd}{2} - k|z'|\right) e^{-ikz'\cos\theta} e_z.$$

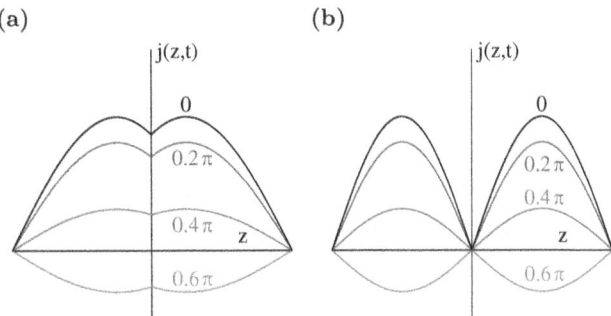

Fig. 2.25 Snap shots of the current distribution $\text{Re}(j(z,t))$: The numbers indicate ωt. (a) $a = kd/2 = 2\pi/3$. (b) $a = kd/2 = \pi$

For this purpose one splits the integral into two contributions ($[-d/2, 0]$ and $[0, d/2]$), sorts the sine function with the addition theorem and calculates the resulting integrals of trigonometric functions multiplied by an exponential function (see Detail 2.5.8.1). In this way one finds

$$A(r) = \frac{2i_0 \, e^{ikr}}{ckr} \left[\frac{\cos\left(\frac{kd}{2}\cos\theta\right) - \cos\left(\frac{kd}{2}\right)}{\sin^2\theta} \right] e_z . \tag{2.80}$$

For the calculation of the magnetic induction and the electric field one uses spherical coordinates

$$A(r) = A(r,\theta)(\cos\theta \, e_r - \sin\theta \, e_\theta) = A_r(r,\theta)e_r + A_\theta(r,\theta)e_\theta ,$$

as the vector potential is represented by the variables r and θ. The application of the operator $\nabla \times$ in spherical coordinates

$$\nabla \times A(r) = \frac{1}{r\sin\theta}\left\{\frac{\partial(\sin\theta \, A_\varphi)}{\partial \theta} - \frac{\partial A_\theta}{\partial \varphi}\right\} e_r$$

$$+ \frac{1}{r}\left\{\frac{\partial(rA_\theta)}{\partial r} - \frac{\partial A_r}{\partial \theta}\right\} e_\varphi + \frac{1}{r}\left\{\frac{1}{\sin\theta}\frac{\partial A_r}{\partial \varphi} - \frac{\partial(rA_\varphi)}{\partial r}\right\} e_\theta$$

is definitely simpler, if one restricts the attention to the radiation zone, that is the sole consideration of terms of the order of $1/r$. For the magnetic field one has to consider in this order only the term with

$$B(r) = \nabla \times A(r) \approx \frac{1}{r}\frac{\partial(rA_\theta)}{\partial r} e_\varphi$$

$$= -2i\left(\frac{i_0}{cr}\right) e^{ikr} \frac{\left(\cos\left(\frac{kd}{2}\cos\theta\right) - \cos\left(\frac{kd}{2}\right)\right)}{\sin\theta} e_\varphi .$$

For the electric field one finds with the Ampère law in the vacuum and for harmonic excitation

$$E(r) = \frac{i}{k}(\nabla \times B(r)) \approx -\frac{i}{kr}\frac{\partial(rB_\varphi)}{\partial r} e_\theta$$

$$= -2i\frac{i_0}{cr} e^{ikr} \frac{\left(\cos\left(\frac{kd}{2}\cos\theta\right) - \cos\left(\frac{kd}{2}\right)\right)}{\sin\theta} e_\theta .$$

Both fields satisfy the relation

$$B(r) = e_r \times E(r) ,$$

2.3 Generation of Waves: Transmitters

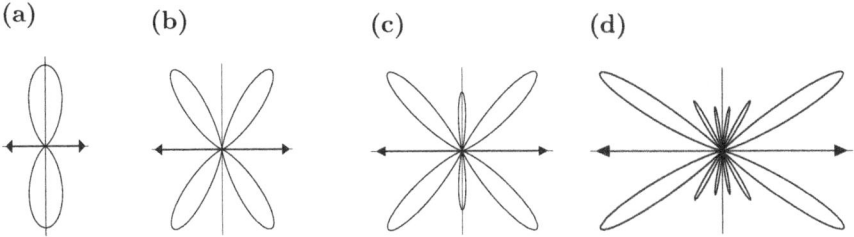

Fig. 2.26 Radiation pattern of the centrally fed antenna. (**a**) $a = \pi$. (**b**) $a = 2\pi$. (**c**) $a = 3\pi$. (**d**) $a = 6\pi$

which shows that they form together with the vector $\boldsymbol{k} = k\boldsymbol{e}_r$ a tripod. They have the same magnitude in the vacuum. The time averaged Poynting vector (1.61) respectively the corresponding emitted power per solid angle is therefore

$$\bar{\boldsymbol{S}}(\boldsymbol{r}) = \frac{c}{8\pi} \left(\boldsymbol{E}(\boldsymbol{r}) \times \boldsymbol{B}(\boldsymbol{r})^* \right) = \frac{i_0^2}{2\pi c r^2} \frac{\left(\cos\left(\frac{kd}{2}\cos\theta\right) - \cos\left(\frac{kd}{2}\right) \right)^2}{\sin^2\theta} \boldsymbol{e}_r$$

$$\frac{dP}{d\Omega} = \frac{i_0^2}{2\pi c} \frac{\left(\cos\left(\frac{kd}{2}\cos\theta\right) - \cos\left(\frac{kd}{2}\right) \right)^2}{\sin^2\theta} = \frac{i_0^2}{2\pi c} f(\theta) \qquad (2.81)$$

The wave number k is determined by the specification of the oscillation frequency of the transmitter, the distribution of the radiation by the product $a = kd/2$. Polar diagrams of the exact radiation pattern (shown is the combination of the trigonometric functions in (2.81) without the prefactor) for the parameter values $a = \pi, 2\pi, 3\pi, 16\pi$ are illustrated in Fig. 2.26a–d. With the increase of the parameter a (which corresponds to an increase of the length of the antenna and/or the reduction of the wave length of the radiation) one observes a transition from the dipole pattern to radiation patterns with increasingly higher multipoles.

An approximation of the exact results for the radiation pattern can be obtained by a direct expansion (see Detail 2.5.8.2) of the cosine functions in (2.81) in powers of $a = kd/2$

$$\frac{dP}{d\Omega} = \frac{i_0^2}{8\pi c} a^4 \sin^2\theta \left\{ 1 - \frac{1}{6} a^2 \left(1 + \cos^2\theta \right) \right. $$
$$\left. + \frac{1}{720} a^4 \left(9 + 14 \cos^2\theta + 9 \cos^4\theta \right) \dots \right\}.$$

This expansion should be useful for small values of the parameter a. A comparison of the results of this expansion to the order $(a)^6$ respectively $(a)^8$ is found in Detail 2.5.8.2. A better approximation can be found in Detail 2.5.8.3 with the multipole expansion, which has been developed in Chap. 2.3.4. The starting point is the representation of the vector potential in the radiation zone[9]

$$A_{RZ}(r) = \frac{4\pi}{c} \frac{e^{ikr}}{r} \sum_{l,m} (-i)^l Y_{lm}(\Omega) A_{lm}(k)$$

with the multipole coefficients

$$A_{lm}(k) = \iiint dV' j_l(kr') Y^*_{lm}(\Omega') \, \boldsymbol{j}(r') \, .$$

A representation of the current density vector (2.79) in spherical coordinates gives

$$\boldsymbol{j}(r,t) = \begin{cases} \dfrac{i_0}{2\pi r^2} \sin\left(\dfrac{kd}{2} - k|z|\right) (\delta(\cos\theta - 1) - \delta(\cos\theta - 1)) \, \boldsymbol{e}_r \\ 0 \end{cases} ,$$

where the upper line is valid for $|z| < d/2$, the lower one for $|z| \geq d/2$. One obtains for the expansion coefficients A_{lm} with even l in this case

$$A_{lm}(k) = \delta_{m0} A_l(k, a)$$

$$= \delta_{m0} \left\{ \frac{i_0}{k} \sqrt{\frac{(2l+1)}{\pi}} \int_0^a dt \, \sin(a-t) j_l(t) \right\} \boldsymbol{e}_z \, ,$$

and for odd values

$$A_{lm}(k) = 0 \, .$$

The remaining radial integrals

$$I_l(a) = \int_0^a dt \, \sin(a-t) j_l(t)$$

can be represented by special functions or be obtained with numerical methods. The results of a numerical integration for $l = 0, 2, 4$ over the range of values $0 \leq a \leq 2\pi$ are displayed in Fig. 2.27. In Fig. 2.27a one can observe, that the integrals alone would contribute with increasing l only for higher values of the parameter a, but Fig. 2.27b shows that different multipole contributions dominate in different regions of this parameter, if the different weights of the expansion are included.

---

[9] A treatment of the intermediate zone would be possible.

2.4 Generation of Waves: The Radiation of Moving Point Charges

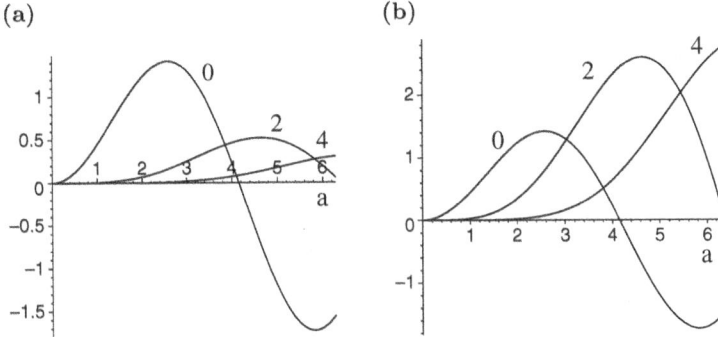

Fig. 2.27 Radial integrals $I_l(a)$ for $l = 0, 2, 4$. (**a**) The integral $I_l(a)$. (**b**) The terms $(2l+1)I_l(a)$

The multipole expansion of the vector potential is thus

$$A_{RZ}(r) = \frac{2i_0}{c} \frac{e^{ikr}}{kr} \sum_{l=\text{even}} (-i)^l (2l+1) I_l(a) P_l(\cos\theta) e_z .$$

The calculation of the electromagnetic fields and of the emitted power per solid angle and unit time follows the pattern of the exact solution, so that one obtains in the end

$$\frac{dP}{d\Omega} = \frac{i_0^2}{2\pi c} \sin^2\theta \left| \sum_{l=\text{even}} (-i)^l (2l+1) I_l(a) P_l(\cos\theta) \right|^2$$

$$= \frac{i_0^2}{2\pi c} g(a,\theta) . \tag{2.82}$$

Results of the multipole expansion with $l_{\max} = 2$ and $l_{\max} = 4$ are illustrated in Figs. 2.28 and 2.29. One notes that the results are definitely better than those obtained with the simple expansion. The reason is obvious: The integrals $I_l(a)$ contain more detailed information as the pure powers a^l. In addition one sees, that the multipole expansion with only two terms reproduced the dipole pattern correctly, but that it is not able to describe the transition to a quadrupole pattern correctly. An expansion with three terms corrects this error, but has difficulties to reproduce the fully developed quadrupole pattern.

2.4 Generation of Waves: The Radiation of Moving Point Charges

The last example for the evaluation of the formulae (1.85) and (1.85) with the electromagnetic potentials is the production of radiation by moving point charges. The explicit examples, the bremsstrahlung and the Čerenkov radiation, will be

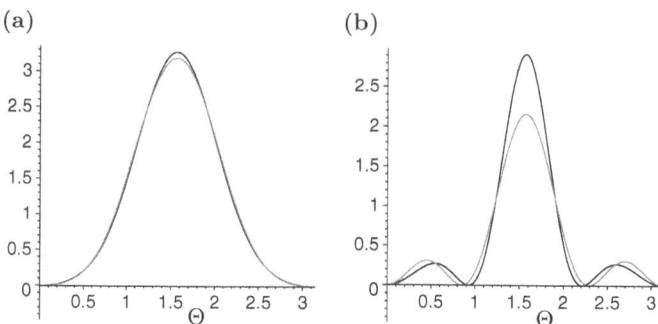

Fig. 2.28 Radiation pattern of the centrally fed antenna: Comparison of the exact solution (black) with an approximate multipole expansion ($l_{\max} = 2$, grey). (**a**) $a = 0.8\pi$. (**b**) $a = 1.25\pi$

presented briefly after a general discussion of the generation of electric and magnetic fields by moving charges.

2.4.1 The Liénard-Wiechert Potentials

The starting point for the remarks, that follow, are the solutions of the inhomogeneous wave equations (1.85) and (1.86) in the form (1.76)

$$V^{(+)}(\mathbf{r}, t) = \frac{1}{\varepsilon} \int dt' \iiint d^3r' \, G^{(+)}(\mathbf{r}, t; \mathbf{r}', t') \rho_{tr}(\mathbf{r}', t')$$

$$A^{(+)}(\mathbf{r}, t) = \frac{\mu}{c} \int dt' \iiint d^3r' \, G^{(+)}(\mathbf{r}, t; \mathbf{r}', t') \mathbf{j}_{tr}(\mathbf{r}', t') \,, \tag{2.83}$$

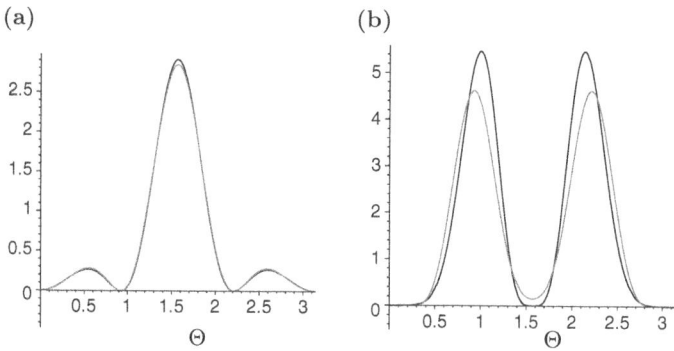

Fig. 2.29 Radiation pattern for the centrally fed antenna: Comparison of the exact solution (black) with a multipole expansion ($l_{\max} = 2, 4$, grey). (**a**) $a = 1.25\pi$. (**b**) $a = 2\pi$

2.4 Generation of Waves: The Radiation of Moving Point Charges

which can be represented in terms of the retarded Green's function (1.83)

$$G^{(+)}(\mathbf{r}-\mathbf{r}', t-t') = \frac{\delta\left(t - t' - |\mathbf{r}-\mathbf{r}'|/c_{\text{med}}\right)}{|\mathbf{r}-\mathbf{r}'|} \qquad t-t' > 0.$$

The velocity of light in a medium is written as $c_{\text{med}} = c/\sqrt{\varepsilon\mu}$.

A point charge q, moving along a trajectory $\mathbf{s}(t)$ with the velocity $\mathbf{v}(t)$ is characterised by the charge density

$$\rho_{tr}(\mathbf{r}, t) = q\,\delta(\mathbf{r} - \mathbf{s}(t))$$

and the current density

$$\mathbf{j}_{tr}(\mathbf{r}, t) = q\,\mathbf{v}(t)\,\delta(\mathbf{r} - \mathbf{s}(t)) \qquad \text{with} \quad \mathbf{v}(t) = \dot{\mathbf{s}}(t).$$

As one is dealing with a point charge, an additional δ-function has appeared, so that the integration over the primed space coordinates can be done directly. The result is

$$V^{(+)}(\mathbf{r}, t) = \frac{q}{\varepsilon} \int dt'\, \frac{\delta\left(t' - t + |\mathbf{r} - \mathbf{s}(t')|/c_{\text{med}}\right)}{|\mathbf{r} - \mathbf{s}(t')|} \qquad (2.84)$$

$$\mathbf{A}^{(+)}(\mathbf{r}, t) = \frac{q\mu}{c} \int dt'\, \mathbf{v}(t')\frac{\delta\left(t' - t + |\mathbf{r} - \mathbf{s}(t')|/c_{\text{med}}\right)}{|\mathbf{r} - \mathbf{s}(t')|}. \qquad (2.85)$$

In order to execute the remaining time integration, the substitution

$$f(t') = t' + |\mathbf{r} - \mathbf{s}(t')|/c_{\text{med}} \qquad (2.86)$$

with

$$\frac{df(t')}{dt'} = 1 - \frac{(\mathbf{r} - \mathbf{s}(t'))\cdot \mathbf{v}(t')}{|\mathbf{r} - \mathbf{s}(t')|\,c_{\text{med}}} \qquad (2.87)$$

can be applied. The integration over f leads to the **Liénard-Wiechert potentials**

$$V^{(+)}(\mathbf{r}, t) = \frac{q}{\varepsilon} \left[\frac{1}{|\mathbf{r} - \mathbf{s}(t')| - (\mathbf{r} - \mathbf{s}(t'))\cdot \frac{\mathbf{v}(t')}{c_{\text{med}}}}\right]_{f(t')-t=0} \qquad (2.88)$$

$$\mathbf{A}^{(+)}(\mathbf{r}, t) = \frac{q\mu}{c} \left[\frac{\mathbf{v}(t')}{|\mathbf{r} - \mathbf{s}(t')| - (\mathbf{r} - \mathbf{s}(t'))\cdot \frac{\mathbf{v}(t')}{c_{\text{med}}}}\right]_{f(t')-t=0}. \qquad (2.89)$$

These potentials are reduced to the Coulomb potential ($V^{(+)} = V_{\text{Coul}}$, $A^{(+)} = 0$), if the point charge is at rest ($v(t') = 0$, $s(t') = r'$). The index states, that these results have to be processed further. It is necessary to find the solution(s) at $t' = t_{\text{ret}}$ of the transcendent equations $f(t') - t = 0$ and insert them into Eqs. (2.88) and (2.89).

In the vacuum one has $c_{\text{med}} = c$. Equation (2.87) allows the estimate

$$1 - \frac{v(t')}{c} \leq \frac{df(t')}{dt'} \leq 1 + \frac{v(t')}{c},$$

from which follows, as one has $v/c < 1$, that the derivative of the function $f(t')$ is positive definite. The strictly monotonically increasing function f can have at most one zero t_{ret}, which can be calculated, if the path of the charge has been specified.

A different situation is found, if the radiation takes place in a medium (as compared to the vacuum), as multiple zeros of the equation $f(t') - t = 0$ can occur. E.g. for a uniform motion with

$$v(t') = v_0 \qquad s(t') = r'_0 + v_0 t'$$

one obtains from $f(t') - t = 0$ the relation

$$c_{\text{med}}^2 (t - t')^2 = (r - r_0 - v_0 t')^2 . \tag{2.90}$$

In order to continue the argumentation, one introduces the vector

$$X(t) = r - r_0 - v_0 t , \tag{2.91}$$

which describes the distance of the point of observation r from the *momentary* position of the charge $r_0 + v_0 t$ (see Fig. 2.30, replace $R(t)$ in the Figure by X(t)).

Fig. 2.30 Motion of a point charge: Geometry for the calculation of the potentials

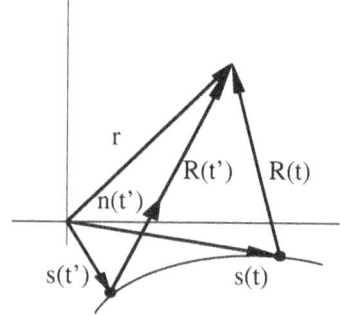

2.4 Generation of Waves: The Radiation of Moving Point Charges

Add and subtract in Eq. (2.90) in the bracket $v_0 t$ on the right hand side, one obtains a quadratic equation in $(t - t')$

$$c_{med}^2 (t - t')^2 = X^2(t) + 2v_0 \cdot X(t)(t - t') + v_0^2 (t - t')^2$$

with the solutions

$$(t - t') = \frac{\{v_0 \cdot X(t) \pm [(v_0 \cdot X(t))^2 + (c_{med}^2 - v_0^2) X^2(t)]^{1/2}\}}{(c_{med}^2 - v_0^2)} . \qquad (2.92)$$

Only real, positive roots are admitted due to the causality $t > t'$. The root is real and larger than the product $v_0 \cdot X(t)$ for $v_0 < c_{med}$. In this case, there exists only one causal solution for the retarded time. For $c_{med} < v_0$ (which is possible in a medium), it is possible to have two causal solutions. This radiation will be addressed in Chap. 2.4.3.

The evaluation of the fields associated with the Liénard-Wiechert potentials can be executed in two ways:[10] One differentiates the representations (2.83) according to

$$B(r, t) = \nabla \times A^{(+)}(r, t)$$

and

$$E(r, t) = -\nabla V^{(+)}(r, t) - \frac{1}{c} \frac{\partial A^{(+)}(r, t)}{\partial t}$$

and carries out the time integration with the intermediate results obtained in this way. An alternative method is the direct differentiation of the potentials (2.88) and (2.89). The following abbreviations are introduced in order to present the result in a compact form:

- The difference vector from the position of the point charge at the retarded time to the point of observation (see Fig. 2.30) is

$$R(r, t') \equiv R(t') = r - s(t') . \qquad (2.93)$$

The dependence of the quantities on the coordinates of the point charge are suppressed in order to simplify the notation.
- A unit vector in the direction of $R(t')$ defined as

$$n(r, t') \equiv n(t') = \frac{R(t')}{R(t')} . \qquad (2.94)$$

---

[10] The somewhat longer calculations for both options are shown in Detail 2.5.9.

- The associated, relative velocity and acceleration vectors are

$$\boldsymbol{\beta}(t') = \frac{\boldsymbol{v}(t')}{c_{\text{med}}} \quad \text{and} \quad \dot{\boldsymbol{\beta}}(t') = \frac{\dot{\boldsymbol{v}}(t')}{c_{\text{med}}} \ .$$

- The derivative of the function f, which has already been obtained in (2.87), is

$$\frac{df(t')}{dt'} \equiv g(t') = 1 - \boldsymbol{n}(t') \cdot \boldsymbol{\beta}(t') \ . \tag{2.95}$$

The result for the electric field of the moving point charge, using this notation, is

$$\boldsymbol{E}(\boldsymbol{r},t) = \frac{q}{\varepsilon} \left[\frac{1}{g^3(t') R^2(t')} \left(\boldsymbol{n}(t') - \boldsymbol{\beta}(t') \right) \left(1 - \beta^2(t') \right) \right. \tag{2.96}$$

$$+ \frac{1}{c_{\text{med}}\, g^3(t')\, R(t')} \left(\boldsymbol{n}(t') \times \right.$$

$$\left. \left. \left[\left(\boldsymbol{n}(t') - \boldsymbol{\beta}(t') \right) \times \dot{\boldsymbol{\beta}}(t') \right] \right) \right]_{t'=t_{\text{ret}}(t)} .$$

The triple vector product in the second term can be given in an explicit form with the relation

$$\boldsymbol{a} \times (\boldsymbol{b} \times \boldsymbol{c}) = (\boldsymbol{a} \cdot \boldsymbol{c})\,\boldsymbol{b} - (\boldsymbol{a} \cdot \boldsymbol{b})\,\boldsymbol{c}$$

as

$$\boldsymbol{n} \times (\boldsymbol{n} - \boldsymbol{\beta}) \times \dot{\boldsymbol{\beta}} = (\boldsymbol{n} \cdot \dot{\boldsymbol{\beta}})\,\boldsymbol{n} - \dot{\boldsymbol{\beta}} - (\boldsymbol{n} \cdot \dot{\boldsymbol{\beta}})\,\boldsymbol{\beta} + (\boldsymbol{n} \cdot \boldsymbol{\beta})\,\dot{\boldsymbol{\beta}} \ .$$

For the magnetic induction one finds in the same manner (with an additional simplification of the notation)

$$\boldsymbol{B}(\boldsymbol{r},t) = \frac{q}{\varepsilon} \left[\frac{(1-\beta^2)}{g^3 R^2} (\boldsymbol{\beta} \times \boldsymbol{n}) + \frac{1}{c_{\text{med}} R} \left\{ \frac{(\boldsymbol{n} \cdot \dot{\boldsymbol{\beta}})}{g^3} (\boldsymbol{\beta} \times \boldsymbol{n}) \right. \right. \tag{2.97}$$

$$\left. \left. + \frac{1}{g^2} \left(\dot{\boldsymbol{\beta}} \times \boldsymbol{n} \right) \right\} \right]_{t'=t_{\text{ret}}(t)} .$$

One can check directly that the two vector fields are related by

$$\boldsymbol{B}(\boldsymbol{r},t) = \boldsymbol{n}(t_{\text{ret}}) \times \boldsymbol{E}(\boldsymbol{r},t) \ . \tag{2.98}$$

2.4 Generation of Waves: The Radiation of Moving Point Charges

Fig. 2.31 The Geometry of the Liénard-Wiechert fields

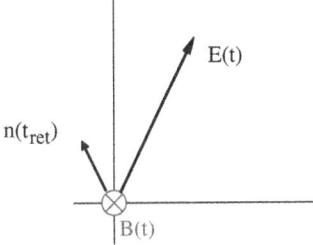

The geometry of the fields is indicated in Fig. 2.31. The vector of the magnetic induction is perpendicular to the plane spanned by the vectors $n(t_{ret})$ and $E(r, t)$. Each of the electromagnetic fields contains one contribution, which is determined by the velocity of the point charge at the retarded time, and a contribution, which depends linearly on the acceleration. The part of the electric field due the acceleration is, as a consequence of the triple product orthogonal to n. For the velocity part of the electric field one obtains by contrast

$$n \cdot E = \left[\frac{q}{\varepsilon g^2(t') R^2(t')} \left(1 - \beta(t')^2\right) \right]_{t'=t_{ret}(t)}.$$

The electric field is not transverse with respect to $n(t_{ret})$.

The term depending on the acceleration is responsible for the experimentally observed radiation in the far zone, as the contribution due to the acceleration decreases more slowly with the separation from the point charge in comparison to the term depending on the velocity ($1/R$ in comparison with $1/R^2$).

2.4.2 The Classical Bremsstrahlung

The electromagnetic fields, which are emitted by a moving charge, are rather complicated in the near zone. The discussion becomes, however simpler, for points which are sufficiently far away from the charge. If one considers the Poynting vector in a medium with a simple material equation

$$S(r, t) = \frac{c}{4\pi \mu} (E(r, t) \times B(r, t))$$

$$= \frac{c}{4\pi \mu} (E(r, t) \times [n(t_{ret}) \times E(r, t)])$$

and resolves the triple vector product, one finds

$$S(r, t) = \frac{c}{4\pi \mu} \left[E^2(r, t) n(t_{ret}) - (n(t_{ret}) \cdot E(r, t)) E(r, t) \right]. \quad (2.99)$$

The energy, which is radiated through a large sphere around the position of the point charge, is determined by the acceleration term. For terms starting with the order $R^{-3}(t_{\text{ret}})$ the relation

$$\oint \mathbf{S} \cdot d\mathbf{f} \xrightarrow{S \propto O(R^{-3})} \oint \frac{R^2}{R^3} d\Omega \longrightarrow \frac{1}{R} \xrightarrow{R \to \infty} 0$$

is valid. It is therefore sufficient, to insert the acceleration term of the $\mathbf{E}$-field in (2.99), in order to determine the energy loss or the radiation pattern of a moving point charge. Only accelerated point charges loose energy, from the point of view of the far zone. There exists an exchange of energy between the different components of the fields in the near zone. The details are more complicated, but in most cases not of interest.

As the acceleration term of the electric field is orthogonal to the vector $\mathbf{n}$, one can write for the Poynting vector of an accelerated point charge in sufficiently large distance from the point charge in good approximation

$$\mathbf{S}(\mathbf{r}, t) = \frac{c}{4\pi\mu} \mathbf{E}(\mathbf{r}, t)^2 \mathbf{n}(t_{\text{ret}}) \tag{2.100}$$

$$= \frac{q^2}{4\pi c \varepsilon} \left[\left(\frac{\mathbf{n}(t') \times \left((\mathbf{n}(t') - \boldsymbol{\beta}(t')) \times \dot{\boldsymbol{\beta}}(t') \right)}{g(t')^3 R(t')} \right)^2 \mathbf{n}(t') \right]_{t'=t_{\text{ret}}(t)}.$$

The direction of the flow of energy is identical with the direction of the vector $\mathbf{n}(t_{\text{ret}})$, that is starting at the position of the particle at the time t_{ret} to the point $\mathbf{r}$, where the radiation is registered. The angular distribution of the radiation is given by (compare p. 107)

$$\frac{dP}{d\Omega} = (\mathbf{S}(\mathbf{r}, t) \cdot \mathbf{n}(t_{\text{ret}})) \, R(t_{\text{ret}})^2 \tag{2.101}$$

$$= \frac{q^2}{4\pi c \varepsilon} \left(\frac{\mathbf{n}(t') \times \left((\mathbf{n}(t') - \boldsymbol{\beta}(t')) \times \dot{\boldsymbol{\beta}}(t') \right)}{g(t')^3} \right)^2_{t'=t_{\text{ret}}}.$$

One can neglect the terms depending on β, if the point charge moves slowly, so that $\beta \ll 1$. The quantity g can then be set equal to 1. There remains

$$\frac{dP}{d\Omega} = \frac{\mu q^2}{4\pi c^3} \left[\left(\mathbf{n}(t') \times (\mathbf{n}(t') \times \dot{\mathbf{v}}(t')) \right)^2 \right]_{t'=t_{\text{ret}}(t)},$$

2.4 Generation of Waves: The Radiation of Moving Point Charges

respectively after explicit resolution of the vector product

$$\frac{dP}{d\Omega} = \frac{\mu q^2}{4\pi c^3} \dot{v}^2(t_{\text{ret}}(t)) \sin^2 \theta(t_{\text{ret}}(t)) \,. \tag{2.102}$$

The angle $\beta \ll 1$ is the angle between the acceleration vector and the direction of the radiation at the time t_{ret}. The $\sin^2$-dependence is the typical pattern of the non-relativistic bremsstrahlung. The formula for the total radiated energy per unit of time

$$P = \iint \left(\frac{dP}{d\Omega}\right) d\varphi \sin\alpha \, d\alpha = \frac{2\mu q^2 \dot{v}(t_{\text{ret}})^2}{3c^3} \tag{2.103}$$

has been given for the first time by J. Larmor.

The evaluation of the complete intermediate result (2.101) yields a relativistic extension of the Larmor formula. Instead of the quantity, which has been considered above

$$\frac{dP(\mathbf{r},t)}{d\Omega} = (\mathbf{S}(\mathbf{r},t) \cdot \mathbf{n}(t_{\text{ret}}(t))) \, R(t_{\text{ret}}(t))^2 \,,$$

which gives the energy per time and surface at a point $\mathbf{r}$ at the time t (as a consequence of the emission at the time $t' = t - R(t')/c_{\text{med}}$), one usually considers the quantity

$$\frac{dP(\mathbf{r},t')}{d\Omega} = \frac{dP(\mathbf{r},t)}{d\Omega} \frac{\partial t}{\partial t'} \,,$$

which represents the emitted power per solid angle at the time t'. The partial derivative, that one needs, is according to (2.87) and (2.95)

$$\frac{\partial t}{\partial t'} = \frac{\partial f(t')}{\partial t'} = 1 - \mathbf{n}(\mathbf{r},t') \cdot \boldsymbol{\beta}(t') = g(\mathbf{r},t') \,,$$

so that one finds, in the notation of p. 123 for the angular distribution of the radiation

$$\frac{dP(\mathbf{r},t')}{d\Omega} = \frac{q^2}{4\pi c \varepsilon} \frac{\left(\mathbf{n}(t') \times \left((\mathbf{n}(t') - \boldsymbol{\beta}(t')) \times \dot{\boldsymbol{\beta}}(t')\right)\right)^2}{(1 - \mathbf{n}(\mathbf{r},t') \cdot \boldsymbol{\beta}(t'))^5} \,. \tag{2.104}$$

The relation (2.104) is used for the calculation of the emission of radiation by particle accelerators, with particular interest for linear motion of the particles (linear accelerators) and a circular motion of the point charges (circular accelerator as the betatron, the cyclotron or the synchrotron).

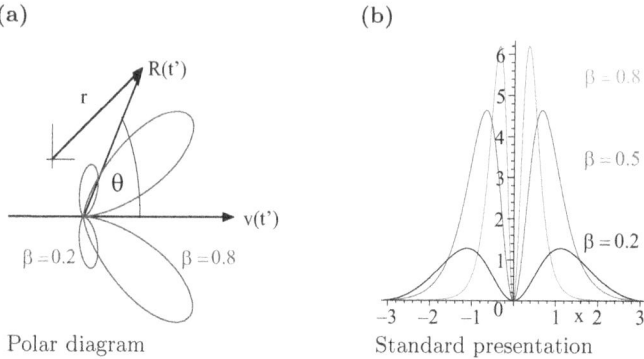

Fig. 2.32 Pattern of the bremsstrahlung of a straight line motion as a function of $x = \theta$ for two respectively three values of the parameter $\beta = v/c$. (**a**) Polar diagram. (**b**) Standard presentation

- The velocity- and the acceleration vectors are parallel in a linear accelerator (Fig. 2.32). The evaluation of the vector products gives in this case the radiation pattern

$$\frac{dP(r,t')}{d\Omega} = \frac{\mu q^2}{4\pi c^3} \dot{v}^2(t') \frac{\sin^2\theta(t')}{(1-\beta(t')\cos\theta(t'))^5} . \qquad (2.105)$$

The angle θ is the angle between the vector from the point charge to the point of observation at the time t' (the vector $n(t')$) and the direction of the beam at the time t' (compare Fig. 2.31, p. 125). This result turns into the formula of Larmor, which describes a maximal emission perpendicular to the direction of the beam, for small velocities. The direction of the maximal emission turns, as shown in Fig. 2.31a and b (without the prefactors), more and more in the beam direction with increasing value of β. At the same time the intensity of the radiation in this direction increases (The values for $\beta = 0.8$ are scaled with a factor of $1/20$). The total power emitted per unit time is obtained by integration of (2.105) over the complete range of the solid angle (Detail 2.5.10.1)

$$P(t') = \frac{\mu q^2}{2c^3}\dot{v}^2(t') \int_0^\pi d\theta \frac{\sin^3\theta}{(1-\beta\cos\theta)^5} = \frac{2\mu q^2}{3c^3}\dot{v}^2(t')\left[1-\beta^2(t')\right]^{-3} .$$

- The evaluation of (2.104) is more involved for the circular motion. If the *momentary* direction of the three vectors $\boldsymbol{\beta}(t')$, $\dot{\boldsymbol{\beta}}(t')$ and $\boldsymbol{n}(t')$ is specified as

$$\boldsymbol{\beta} = (0,0,\beta) \quad \dot{\boldsymbol{\beta}} = (\dot{\beta}, 0, 0) \quad \boldsymbol{n} = (\cos\varphi\sin\theta, \sin\varphi\sin\theta, \cos\theta) ,$$

2.4 Generation of Waves: The Radiation of Moving Point Charges

one finds for the angular distribution of the radiation the expression (see Detail 2.5.10.2)

$$\frac{dP(r,t')}{d\Omega} = \frac{\mu q^2}{4\pi c^3} \dot{v}^2 \frac{1}{(1-\beta\cos\theta)^3} \left[1 - \frac{(1-\beta^2)\sin^2\theta\cos^2\varphi}{(1-\beta\cos\theta)^2}\right].$$

(2.106)

In order to compare this result with the one of the linear motion, one integrates over the azimuthal angle and considers

$$\frac{1}{2\pi}\frac{dP(r,t')}{d\cos\theta} = \frac{\mu q^2}{4\pi c^3} \dot{v}^2 \frac{1}{(1-\beta\cos\theta)^3} \left[1 - \frac{(1-\beta^2)\sin^2\theta}{2(1-\beta\cos\theta)^2}\right].$$

This function of θ is shown in Fig. 2.33 (without the prefactors) for two values of the parameter β. The radiation pattern is nearly isotropic for small velocities, but one has to keep in mind, that the pattern has to be scaled with

$$\frac{\mu q^2 \dot{v}^2}{4\pi c^3}.$$

In the case of a uniform circular motion with radius R one finds that the scaling factor is proportional to β^2/R^2. For increasing velocity one observes also a preferred emission in the forward direction, that is a direction tangential to the circle, for the circular motion.

- If one compares the result for the linear motion (see Fig. 2.32) with that of the circular motion, one can note, that the denominator with the function $(1-\beta\cos\theta)$ affects the pattern considerably. This is a marked consequence of relativistic effects besides the influence on the radiation pattern caused by the relative orientation of the velocity- and the acceleration vectors. The total emitted power

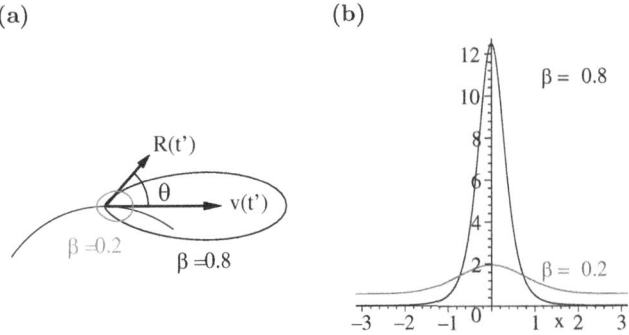

Fig. 2.33 Pattern of the bremsstrahlung for a circular motion as a function of $x = \theta$ for $\beta = 0.2, 0.8$. (**a**) Polar diagram. (**b**) Standard representation

is also quite similar in the two cases (Detail 2.5.10.2)

$$P_{\text{circle}}(t') = \frac{2\mu q^2}{3c^3} \dot{v}^2(t') \left[1 - \beta^2(t')\right]^{-2},$$

even if the increase of the power with growing β is weaker for the circular motion.

2.4.3 Remarks on the Čerenkov Radiation

The function

$$f(t') = t' + |\mathbf{r} - \mathbf{s}(t')|/c_{\text{med}},$$

which has been discussed in (2.86), is a monotonous function, if the radiation of a moving charge propagates in the vacuum. This leads to *exactly* one retarded time in physically relevant situations. The statement $df(t')/dt' > 0$ is, however, not necessarily correct, if the radiation propagates in a medium, as the particle velocity v_0 can be larger than the velocity of light in the medium c_{med}. The function $t - f(t')$ can have multiple zeros. A point charge, which is moving uniformly with

$$\mathbf{s}(t) = \mathbf{r}_0 + \mathbf{v}_0 t$$

can in this case emit a radiation, which can be observed in the far zone. This radiation is the **Čerenkov radiation**.

The structure of this radiation can be explained by the following argument: Instead of the electromagnetic potentials (2.88) and (2.89) there exists a sum of retarded contributions, so that the scalar potential can e.g. be produced by the sum of contributions of two retarded times

$$V(\mathbf{r},t) = \frac{q}{\varepsilon} \left\{ \left| \frac{1}{R(t')g(t')} \right|_{t'=t_{\text{ret1}}(t)} + \left| \frac{1}{R(t')g(t')} \right|_{t'=t_{\text{ret2}}(t)} \right\}$$

$$= \frac{q}{\varepsilon} \left\{ \left| \frac{1}{|\mathbf{r}-\mathbf{s}(t')| - (\mathbf{r}-\mathbf{s}(t')) \cdot \frac{\mathbf{v}(t')}{c_{\text{med}}}} \right|_{t'=t_{\text{ret1}}(t)} \right.$$

$$\left. + \left| \frac{1}{|\mathbf{r}-\mathbf{s}(t')| - (\mathbf{r}-\mathbf{s}(t')) \cdot \frac{\mathbf{v}(t')}{c_{\text{med}}}} \right|_{t'=t_{\text{ret2}}(t)} \right\}. \quad (2.107)$$

The contributions of the waves, which arrive at the same time t at the position $\mathbf{r}$, as a consequence of the δ-function in Eq. (2.84) have to be superimposed. The character of the radiation can be changed by the superposition.

2.4 Generation of Waves: The Radiation of Moving Point Charges

The zeros of the function $t - f(t')$ are given in the case of a uniform motion of a point charge by (see (2.92))

$$(t - t') = \frac{\left\{-X(t) \cdot v_0 \pm \left[(X(t) \cdot v_0)^2 - (v_0^2 - c_{med}^2)X(t) \cdot X(t)\right]^{1/2}\right\}}{(v_0^2 - c_{med}^2)}.$$
(2.108)

For $v_0 > c_{med}$ the following situation arises: The square root in (2.108) is real, if the angle γ between the vectors $X(t)$ and v_0 satisfies the relation

$$\cos^2 \gamma - \left(1 - \frac{c_{med}^2}{v_0^2}\right) \geq 0.$$

It is then smaller than the scalar product $X(t) \cdot v_0$. Causality requires the relation

$$-\cos \gamma \pm \left[\cos^2 \gamma - \left(1 - \frac{c_{med}^2}{v_0^2}\right)\right]^{1/2} \geq 0.$$

This condition is, as shown in Fig. 2.34, satisfied, independent of the sign of the square root for backward angles with

$$-\pi \leq \gamma \leq -\arccos\left(\left[1 - \frac{c_{med}^2}{v_0^2}\right]^{1/2}\right).$$
(2.109)

In the figure one sees the radicand of the square root featuring in (2.108) as a function of $\cos \theta$. If this is positive, there exist two causal solutions for the time difference, indicated by $f_\pm$ in the figure. The electromagnetic potentials are only defined within the cone (2.109), about the axis of the radiation.

The function in the denominator $R(t')g(t')$ in (2.107) differs for the two retarded times (2.108) only by a sign

$$(R(t')g(t'))_{t_{ret1}(t), t_{ret2}(t)} = \pm X(t)\left[1 - (v/c_{med})^2 \sin^2 \gamma\right]^{1/2}.$$

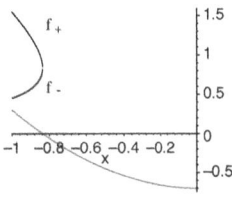

Fig. 2.34 Conditions for Čerenkov radiation ($x = \cos \theta$)

As only the absolute value of the quantities enters in (2.107), one finds for the scalar potential

$$V(\mathbf{r},t) = \frac{2q}{\varepsilon} \frac{1}{X(t)\left[1 - (v/c_{\text{med}})^2 \sin^2 \gamma\right]^{1/2}}. \qquad (2.110)$$

The vector potential for a point charge with a constant velocity is

$$\mathbf{A}(\mathbf{r},t) = \frac{\sqrt{\varepsilon\mu}}{c_{\text{med}}} \mathbf{v}\, V(\mathbf{r},t).$$

The potentials are within a cone, the Čerenkov cone, different from zero, they are singular on the cone and vanish outside of the cone. The singularity is an artefact of the implicit assumptions, that the dielectric constant is independent of the frequency. Without this assumption a regular distribution of the border of the Čerenkov cone would have been obtained.

A similar phenomenon can be observed for the propagation of sound, if an object is moving with supersonic velocity. This is a sound shock wave. The formation of a corresponding Čerenkov shock wave is explained in Fig. 2.35. A point charge is at the position $x = 0$ at the time $t = 0$ and is emitting a elementary wave. The charge moves by the distance $x = v_0 t$ in a time interval $[0, t]$. The radius of the wave front of an elementary spherical wave, which was generated at time $t = 0$, is $R = c_{\text{med}} t$. The tangent starting from the point $x = v_0 t$ on this spherical wave (and all spherical waves created in the intermediate time) marks the front (three dimensional in realty) of the shock wave respectively of the Čerenkov cone. The (electromagnetic) potentials describe in this fashion a wave front, which propagates with the velocity c_{med} in a direction, which is, as indicated in the figure, characterised by the angle θ_C. This angle is determined by $\cos \theta_C = c_{\text{med}}/v_0$. A measurement of this angle makes it possible to determine the velocity of highly energetic elementary particles moving through matter.

Fig. 2.35 Construction of the Čerenkov cone with elementary waves

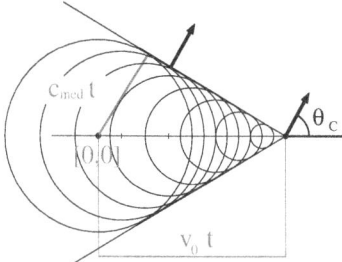

2.5 Details

2.5.1 The Real and the Ideal Transformer

The equivalent circuit diagram (Fig. 2.36b) of the transformer (Fig. 2.36a) shows the two circuits with the resistors (R_i) and the coils (L_{ii}) (L_{ii} with N_i windings) as well as the coupling by mutual induction ($L_{12} = L_{21}$). The primary circuit is connected to an alternating voltage $U = U_0 \cos \omega t$, which will be used in the form

$$U = U_0 \, e^{i\omega t}$$

in the calculation to follow. On the secondary side the voltage U_2 is picked off. If this voltage is connected to a consumer with a resistor R_2, this resistor is to be added to the value of the resistor in series. The coupling is effected by a laminated iron core in order to avoid losses by eddy currents. On the theoretical side one distinguishes the options:

(a) Real Transformer: Complete solution of the equations (2.2) and (2.3) (p. 67)

$$U_1(t) = U(t) + U_1^s(t) + U_1^m(t)$$

$$= U(t) - L_{11}\frac{d}{dt} i_1(t) - L_{12}\frac{d}{dt} i_2(t) = R_1 i_1(t)$$

$$U_2(t) = U_2^s(t) + U_2^m(t)$$

$$= -L_{22}\frac{d}{dt} i_2(t) - L_{12}\frac{d}{dt} i_1(t) = R_2 i_2(t) \,.$$

(b) Ideal Transformer: Solution with the condition

$$R_1 \ll \omega L_{11} \,,$$

that is the resistor in the primary circuit can be neglected.

Fig. 2.36 The transformer problem. (**a**) Representation. (**b**) Equivalent circuit diagram

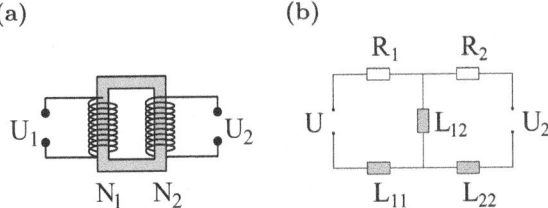

2.5.1.1 The Real Transformer

The discussion of the real transformer is based on the input

$$U(t) = U_0 \, e^{i\omega t} \quad (U_0, \, \omega \text{ specified})$$

$$i_1(t) = i_{10} \, e^{i(\omega t - \varphi_1)} \quad \to \quad i_{10} \cos(\omega t - \varphi_1)$$

$$i_2(t) = i_{20} \, e^{i(\omega t - \varphi_2)} \quad \to \quad i_{20} \cos(\omega t - \varphi_2).$$

The task is the determination of the real amplitudes and phases of the currents in the primary and the secondary circuits with

$$\frac{d}{dt} i_1(t) = i\omega i_1(t) \qquad \frac{d}{dt} i_2(t) = i\omega i_2(t),$$

on the basis of the relations

$$U(t) = (R_1 + i\omega L_{11}) i_1(t) + i\omega L_{12} i_2(t)$$

$$0 = (R_2 + i\omega L_{22}) i_2(t) + i\omega L_{12} i_1(t).$$

The second equation gives the relation between the two currents

$$i_1(t) = -\frac{(R_2 + i\omega L_{22})}{i\omega L_{12}} i_2(t).$$

If this relation is inserted into the first equation, one finds after sorting

$$i_2(t) = -\frac{i\omega L_{12}}{D} U(t), \tag{2.111}$$

where the denominator is given by

$$D = (R_1 + i\omega L_{11})(R_2 + i\omega L_{22}) + \omega^2 L_{12}^2 \tag{2.112}$$

$$= \left(R_1 R_2 + \omega^2 \left(L_{12}^2 - L_{11} L_{22}\right)\right) + i\omega \left(L_{11} R_2 + L_{22} R_1\right).$$

The primary current is connected with the applied voltage by

$$i_1(t) = \frac{R_2 + i\omega L_{22}}{D} U(t). \tag{2.113}$$

For the further evaluation two options offer themselves:

2.5 Details

(i) Determine the ratio of the current strength and the phase differences of the two currents. The ration can be obtained from

$$\frac{i_1(t)}{i_2(t)} = \frac{i_{10}}{i_{20}} e^{i(\varphi_2 - \varphi_1)} = -\frac{R_2 + i\omega L_{22}}{i\omega L_{12}} = \frac{1}{\omega L_{12}}(-\omega L_{22} + iR_2) \ .$$

The phase difference can be found from the ration of imaginary part to real part, explicitly

$$\tan(\varphi_2 - \varphi_1) = -\frac{R_2}{\omega L_{22}} \ .$$

The ratio of the strength of the currents is given by the absolute value of the relation above

$$\frac{|i_{10}|^2}{|i_{20}|^2} = \frac{1}{\omega^2 L_{12}^2}\left(R_2^2 + \omega^2 L_{22}^2\right)$$

or

$$i_{20} = \pm \frac{\omega L_{12}}{\left[R_2^2 + \omega^2 L_{22}^2\right]^{1/2}} i_{10} \ .$$

If the resistance in the secondary circuit is small compared to the self-induction (times frequency)

$$\omega L_{22} \gg R_2 \ ,$$

the ratio of the current strengths is essentially given by the ratio of the mutual induction and the self-induction

$$i_{20} \xrightarrow{R_2 \ll \omega L_{22}} \pm \frac{L_{12}}{L_{22}} i_{10} \ .$$

The amplitude of the induced current is the larger the stronger the coupling of the two circuits is. For the phases one finds in this limit

$$\tan(\varphi_2 - \varphi_1) \longrightarrow 0^- \ .$$

The phase difference goes to zero if one is coming from the left hand side

$$\varphi_2 - \varphi_1 \approx 0 \quad \text{resp.} \quad \varphi_2 - \varphi_1 \approx \pi \ .$$

Fig. 2.37 Current flow in a transformer

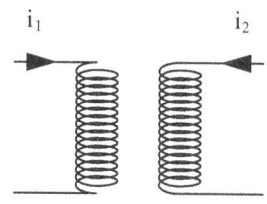

In the second case

$$\cos \varphi_2 = \cos(\pi + \varphi_1) = -\cos \varphi_1 ,$$

the two currents flow in opposite directions (Fig. 2.37).

(ii) Somewhat more intricate is the solution of the transformer problem by the separation of imaginary and real parts in (2.111) and (2.113). For the time $t = 0$ one has

$$i_{10} (\cos \varphi_1 - i \sin \varphi_1) = \frac{R_2 + i\omega L_{22}}{D} U_0$$

$$i_{20} (\cos \varphi_2 - i \sin \varphi_2) = -\frac{i\omega L_{12}}{D} U_0 .$$

If the denominator is written with the abbreviation (see (2.112))

$$D = A + iB \quad \text{with} \quad A = \left(R_1 R_2 + \omega^2 \left(L_{12}^2 - L_{11} L_{22}\right)\right)$$

$$B = \omega (L_{11} R_2 + L_{22} R_1) ,$$

one first finds

$$i_{10} (\cos \varphi_1 - i \sin \varphi_1) = \frac{(R_2 + i\omega L_{22})(A - iB)}{A^2 + B^2} U_0$$

$$= \frac{(AR_2 + \omega B L_{22}) - i(BR_2 - \omega A L_{22})}{A^2 + B^2} U_0$$

$$i_{20} (\cos \varphi_2 - i \sin \varphi_2) = -\frac{i\omega L_{12} (A - iB)}{A^2 + B^2} U_0$$

$$= \frac{\omega L_{12} (B + iA)}{A^2 + B^2} U_0 .$$

2.5 Details

For the individual phases one can win the relations

$$\tan \varphi_2 = -\frac{A}{B}$$

$$\tan \varphi_1 = \frac{BR_2 - \omega A L_{22}}{AR_2 + \omega B L_{22}},$$

which gives

$$\tan(\varphi_2 - \varphi_1) = \frac{\tan \varphi_2 - \tan \varphi_1}{1 + \tan \varphi_1 \tan \varphi_2}$$

$$= \frac{-A(AR_2 + \omega B L_{22}) - B(BR_2 - \omega A L_{22})}{B(AR_2 + \omega B L_{22}) - A(BR_2 - \omega A L_{22})}$$

$$= -\frac{R_2}{\omega L_{22}}.$$

The current strengths can be obtained by looking at the absolute value of the two relations

$$|i_{10}|^2 = \frac{U_0^2}{(A^2 + B^2)^2} \left\{ (AR_2 + \omega B L_{22})^2 + (\omega A L_{22} - BR_2)^2 \right\}$$

$$= \frac{(R_2^2 + \omega^2 L_{22}^2)}{(A^2 + B^2)} U_0^2$$

$$|i_{20}|^2 = \frac{U_0^2}{(A^2 + B^2)^2} \left\{ (-\omega B L_{12})^2 + (A\omega L_{12})^2 \right\}$$

$$= \frac{\omega^2 L_{12}^2}{(A^2 + B^2)} U_0^2.$$

2.5.1.2 The Ideal Transformer

Here one assumes that the resistance in the primary circuit is small

$$R_1 \ll \omega L_{11}.$$

In this limit the transformer equations reduce to

$$U(t) = i\omega \left(L_{11} i_1(t) + L_{12} i_2(t) \right)$$

$$-U_2(t) = i\omega \left(L_{12} i_1(t) + L_{22} i_2(t) \right) \ .$$

One uses two times t_1 and t_2, for which the current in the primary and the secondary circuit have the value zero

$$t_1 \longrightarrow \quad i_1(t_1) = i_{10} \cos(\omega t_1 - \varphi_1) = 0$$

$$\longrightarrow \omega t_1 = \varphi_1 + \frac{\pi}{2}$$

$$t_2 \longrightarrow \quad i_2(t_2) = i_{20} \cos(\omega t_2 - \varphi_2) = 0$$

$$\longrightarrow \omega t_2 = \varphi_2 + \frac{\pi}{2} \ .$$

For the first (second) time is

$$-\frac{U(t_1)}{U_2(t_1)} = \frac{L_{12}}{L_{22}} \quad \text{resp.} \quad -\frac{U(t_2)}{U_2(t_2)} = \frac{L_{11}}{L_{12}} \ .$$

In both relations is the right hand side independent of time. This implies that the left hand side must also be independent of time. There follows with

$$U(t) = U_0 \exp\{i\omega t\}$$

the relation

$$U_2(t) = U_2 \exp\{i\omega t\}$$

and

$$\frac{L_{12}}{L_{22}} = \frac{L_{11}}{L_{12}} \iff L_{11} L_{22} = L_{12}^2 \ .$$

With the formula of the coefficient of the self-induction of a solenoid

$$L_{11} = N_1^2 L_0 \qquad L_{22} = N_2^2 L_0 \ ,$$

one finds

$$L_{12} = N_1 N_2 L_0 \ .$$

2.5 Details

If one neglects the terms in R_1 in the denominator D and uses the above relation for the mutual induction one obtains for quantities defined in part 2.5.1.1

$$A = 0 \qquad B = \omega L_{11} R_2 \qquad D = i\omega L_{11} R_2 \,.$$

It then follows

1. for the phase in the second circuit

$$\tan \varphi_2 = 0$$

(in accordance with the result in part 2.5.1.1.ii),

2. for the phase in the primary circuit

$$\tan \varphi_1 = \frac{R_2}{\omega L_{22}},$$

3. for the primary current

$$i_{10}(\cos \varphi_1 - i \sin \varphi_1) = \frac{R_2 + i\omega L_{22}}{i\omega L_{11} R_2} U_0$$

with the resolution

$$i_{10} = \frac{\left[R_2^2 + \omega^2 L_{22}^2\right]^{1/2}}{\omega L_{11} R_2} U_0 \,,$$

as well as

$$\sin \varphi_1 = \frac{R_2}{\left[R_2^2 + \omega^2 L_{22}^2\right]^{1/2}} \qquad \cos \varphi_1 = \frac{\omega L_{22}}{\left[R_2^2 + \omega^2 L_{22}^2\right]^{1/2}},$$

4. for the secondary current

$$i_{20} = -\frac{L_{12}}{L_{11} R_2} U_0 \,.$$

With the results above one find the well known relation between applied voltage and secondary voltage (multiply by $\cos \omega t$)

$$R_2 i_2(t) = U_2(t) = -\frac{L_{12}}{L_{11}} U(t) = -\frac{N_2}{N_1} U(t) \,.$$

2.5.2 Refraction: Fresnel Formulae and Energy Flow

2.5.2.1 The Fresnel Formulae

In the discussion of reflection and refraction one addresses kinematical aspects, which are due to geometrical considerations, and dynamical aspects, which are concerned with the properties of electromagnetic waves. The first class deals with the laws of reflection and refraction, the second to the Fresnel formulae, which are concerned with the ratio of the field amplitudes of the refracted and the reflected waves in comparison to the incoming wave as well as the energy flow through boundary layers in terms of refection- and transmission coefficients.

The derivation of the Fresnel formulae for linearly polarised plane waves is different for the cases

- The vectors of the electric field are perpendicular to the plane of the wave vectors.
- The vectors of the electric field lie in the plane of the wave vectors.

The case of an arbitrary polarisation of the incoming wave can be reduced to these two cases.

The situation of the fields in the first (and simpler) case has been discussed in Chap. 2.2.1 in a more graphic manner. The two possibilities mentioned will be treated more formally here. The first step is the introduction of a complete coordinate trihedron with the unit vectors (see Fig. 2.38a,b)

- The vector e_n is perpendicular to the plane boundary. It points into the medium 2.
- The vector e_{t1} lies in the boundary layer and is perpendicular to the wave vector plane.
- The vector e_{t2} lies in the boundary layer and marks the wave vector plane.

In reference to this coordinate system the components of the wave number vectors k_1 (for the incoming wave), k'_1 (for the reflected wave) and k_2 (for the refracted wave) have the coordinates

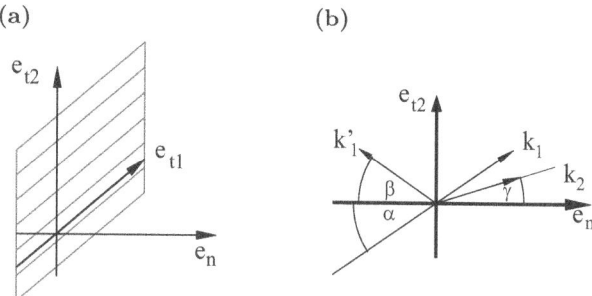

Fig. 2.38 Reflection and refraction: preparation. (**a**) Coordinate trihedron. (**b**) Decomposition of the wave Vectors

2.5 Details

$$k_1 = (k_1 \cos\alpha)e_n + (k_1 \sin\alpha)e_{t2} \equiv (k_1 \cos\alpha, 0, k_1 \sin\alpha)$$

$$k'_1 = (-k_1 \cos\alpha, 0, k_1 \sin\alpha)$$

$$k_2 = (k_2 \cos\gamma, 0, k_2 \sin\gamma) .$$

The magnetic fields for the case, that the electric field vectors are perpendicular to the plane of wave number vectors, can be obtained from

$$E_1 = (0, E_{10}, 0), \quad E'_1 = (0, E'_{10}, 0), \quad E_2 = (0, E_{20}, 0)$$

with the help of the relation (see (1.47), Chap. 1.3.3)

$$B_i = \frac{c}{\omega_i}(k_i \times E_i) = \frac{n_i}{k_i}(k_i \times E_i) . \qquad (2.114)$$

The three B-vectors are

$$B_1 = \begin{vmatrix} e_n & e_{t1} & e_{t2} \\ n_1 \cos\alpha & 0 & n_1 \sin\alpha \\ 0 & E_{10} & 0 \end{vmatrix} = n_1 E_{10}(-\sin\alpha, 0, \cos\alpha)$$

$$B'_1 = n_1 E'_{10}(-\sin\alpha, 0, -\cos\alpha)$$

$$B_2 = n_2 E'_{20}(-\sin\gamma, 0, \cos\gamma) .$$

The application of the conditions of continuity will not be repeated.

The decomposition of the E-vectors in the wave number plane are decomposed as (see Figs. 2.39 a, b and 2.40)

$$E_1 = E_{10}(-\sin\alpha, 0, \cos\alpha)$$

$$E'_1 = E'_{10}(-\sin\alpha, 0, -\cos\alpha)$$

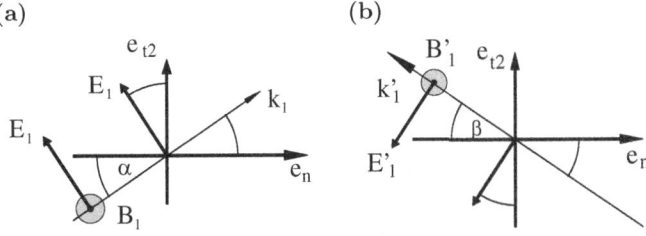

Fig. 2.39 Decomposition of the electric field vectors. (a) E_1 and (b) E'_1

Fig. 2.40 Decomposition of the electric field vector E_2

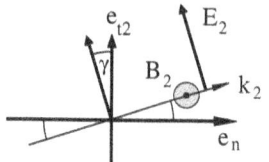

$$E_2 = E_{20}(-\sin\gamma, 0, \cos\gamma).$$

For the magnetic induction one finds for the vector product (2.114) in this case

$$B_1 = \begin{vmatrix} e_n & e_{t1} & e_{t2} \\ n_1\cos\alpha & 0 & n_1\sin\alpha \\ -E_{10}\cos\alpha & 0 & E_{10}\sin\alpha \end{vmatrix} = (0, -n_1 E_{10}, 0)$$

$$B_1' = (0, -n_1 E_{10}', 0)$$

$$B_2 = (0, -n_2 E_{20}, 0).$$

All three B-vectors point in the e_{t1}-direction.

The relevant continuity condition for tangential components (the e_{t1}-direction) of the magnetic fields

$$\frac{1}{\mu_1}(B_{1,t1} + B_{1,t1}') = \frac{1}{\mu_2} B_{2,t1}$$

yields then the following relations among the amplitudes of the electric fields

$$\frac{n_1}{\mu_1}(E_{10} + E_{10}') = \frac{n_2}{\mu_2} E_{20}.$$

This result can be rewritten with $n_i = \sqrt{\varepsilon_i \mu_i}$, in

$$\sqrt{\frac{\varepsilon_1}{\mu_1}}(E_{10} + E_{10}') = \sqrt{\frac{\varepsilon_2}{\mu_2}} E_{20}. \qquad (2.115)$$

The continuity conditions for the components of the electric fields lead to the following results: With the relation

$$\varepsilon_1(E_{1,n} + E_{1,n}') = \varepsilon_2 E_{2,n}$$

2.5 Details

one finds

$$\varepsilon_1 \sin\alpha (E_{10} + E'_{10}) = \varepsilon_2 \sin\gamma\, E_{20},$$

respectively using the law of refraction

$$\sin\alpha = \frac{n_2}{n_1}\sin\gamma = \frac{\sqrt{\varepsilon_2\mu_2}}{\sqrt{\varepsilon_1\mu_1}}\sin\gamma$$

once more the result (2.115). The connection condition for the tangential components (in the $\boldsymbol{e}_{t2}$-direction) yields directly

$$(E_{10} - E'_{10})\cos\alpha = E_{20}\cos\gamma. \qquad (2.116)$$

Again two linear equations for the amplitude ratios have to be solved. The system of equations

$$-\left(\frac{E'_{10}}{E_{10}}\right) + \frac{\sqrt{\varepsilon_2\mu_1}}{\sqrt{\varepsilon_1\mu_2}}\left(\frac{E_{20}}{E_{10}}\right) = 1$$

$$\left(\frac{E'_{10}}{E_{10}}\right) + \frac{\cos\gamma}{\cos\alpha}\left(\frac{E_{20}}{E_{10}}\right) = 1$$

has the solution

$$\frac{E'_{10}}{E_{10}} = \frac{\cos\alpha\sqrt{\dfrac{\varepsilon_2\mu_1}{\varepsilon_1\mu_2}} - \cos\gamma}{\cos\alpha\sqrt{\dfrac{\varepsilon_2\mu_1}{\varepsilon_1\mu_2}} + \cos\gamma} \qquad (2.117)$$

$$\frac{E_{20}}{E_{10}} = \frac{2\cos\alpha}{\cos\alpha\sqrt{\dfrac{\varepsilon_2\mu_1}{\varepsilon_1\mu_2}} + \cos\gamma}. \qquad (2.118)$$

These results are mostly rewritten with the law of refraction. If one uses

$$\sqrt{\frac{\varepsilon_2\mu_1}{\varepsilon_1\mu_2}} = \frac{\mu_1}{\mu_2}\sqrt{\frac{\varepsilon_2\mu_2}{\varepsilon_1\mu_1}} = \frac{\mu_1}{\mu_2}\frac{n_2}{n_1} = \frac{\mu_1}{\mu_2}\frac{\sin\alpha}{\sin\gamma},$$

one obtains after a few steps the standard result

$$\frac{E'_{10}}{E_{10}} = \frac{\left(\frac{\mu_1 \sin\alpha}{\mu_2 \sin\gamma}\right)\cos\alpha - \cos\gamma}{\left(\frac{\mu_1 \sin\alpha}{\mu_2 \sin\gamma}\right)\cos\alpha + \cos\gamma} = \frac{\left(\frac{\mu_1}{\mu_2}\right)\sin 2\alpha - \sin 2\gamma}{\left(\frac{\mu_1}{\mu_2}\right)\sin 2\alpha + \sin 2\gamma}, \quad (2.119)$$

as well as

$$\frac{E_{20}}{E_{10}} = \frac{2\cos\alpha}{\left(\frac{\mu_1 \sin\alpha}{\mu_2 \sin\gamma}\right)\cos\alpha + \cos\gamma} = \frac{2\cos\alpha \sin\gamma}{\left(\frac{\mu_1}{\mu_2}\right)\cos\alpha \sin\alpha + \cos\gamma \sin\gamma}$$

$$= \left(\sqrt{\frac{\varepsilon_1\mu_1}{\varepsilon_2\mu_2}}\right)\frac{2\cos\alpha \sin\alpha}{\left(\left(\frac{\mu_1}{\mu_2}\right)\cos\alpha \sin\alpha + \cos\gamma \sin\gamma\right)}$$

$$= \left(\sqrt{\frac{\varepsilon_1\mu_1}{\varepsilon_2\mu_2}}\right)\frac{2\sin 2\alpha}{\left(\left(\frac{\mu_1}{\mu_2}\right)\sin 2\alpha + \sin 2\gamma\right)}. \quad (2.120)$$

Generally the difference in the permeabilities is negligible ($\mu_1 = \mu_2$). The Fresnel formulae (2.119) and (2.120) are then simplified to

$$\frac{E'_{10}}{E_{10}} = \frac{\tan(\alpha - \gamma)}{\tan(\alpha + \gamma)} \quad (2.121)$$

$$\frac{E_{20}}{E_{10}} = \frac{2\cos\alpha \sin\gamma}{\sin(\alpha + \gamma)\cos(\alpha - \gamma)}. \quad (2.122)$$

The dielectric constants have been eliminated with the law of refraction in order to obtain these results and the relations

$$\sin(\alpha \pm \gamma)\cos(\alpha \mp \gamma) = (\sin\alpha \cos\gamma \pm \cos\alpha \sin\gamma)$$

$$\cdot(\cos\alpha \cos\gamma \mp \sin\alpha \sin\gamma)$$

$$= \sin\alpha \cos\alpha \pm \sin\gamma \cos\gamma$$

between the trigonometric functions have been used.

2.5 Details

For a vertical incidence of the radiation on the plane interface ($\alpha = \beta = \gamma = 0$) one obtains for $\mu_1 = \mu_2$ and

$$\sqrt{\frac{\varepsilon_2}{\varepsilon_1}} \approx \frac{n_2}{n_1}$$

starting with the direct solutions (2.117) and (2.118) the relations

$$\frac{E'_{10}}{E_{10}} \longrightarrow \frac{n_2 - n_1}{n_2 + n_1} \tag{2.123}$$

$$\frac{E_{20}}{E_{10}} \longrightarrow \frac{2n_1}{n_2 + n_1}. \tag{2.124}$$

One realises, that the difference between the two possible orientations of the electric fields does not apply in this limit.

2.5.2.2 Energy Flow

The result for the energy flow (see Chap. 2.2.1, p. 78)

$$\left(\frac{E'_{10}}{E_{10}}\right)^2 + \frac{n_2 \mu_1 \cos\gamma}{n_1 \mu_2 \cos\alpha} \left(\frac{E_{20}}{E_{10}}\right)^2 = 1 \tag{2.125}$$

can be verified directly with the two formulae of Fresnel for the linearly polarisation of plane electromagnetic waves.

- One finds with insertion for the case that the electric vector is perpendicular to the wave number plane

$$\frac{E'_{10}}{E_{10}} = 1 - \frac{E_{20}}{E_{10}} \qquad \frac{E_{20}}{E_{10}} = \frac{2\tan\gamma}{\left(\frac{\mu_1}{\mu_2}\tan\alpha + \tan\gamma\right)}$$

indeed

$$1 - \frac{4\tan\gamma}{\left(\frac{\mu_1}{\mu_2}\tan\alpha + \tan\gamma\right)} + \frac{4\tan^2\gamma}{\left(\frac{\mu_1}{\mu_2}\tan\alpha + \tan\gamma\right)^2} \left\{1 + \frac{\mu_1 \tan\alpha}{\mu_2 \tan\gamma}\right\}$$

$$= 1.$$

- The statement on energy conservation (2.117) and (2.118) is also satisfied, if the electric vector lies in the wave number plane

$$\frac{1}{\left(\sqrt{\frac{\varepsilon_2\mu_1}{\varepsilon_1\mu_2}}\cos\alpha + \cos\gamma\right)^2}\left\{\left(\sqrt{\frac{\varepsilon_2\mu_1}{\varepsilon_1\mu_2}}\cos\alpha\right)^2\right.$$
$$\left. - 2\sqrt{\frac{\varepsilon_2\mu_1}{\varepsilon_1\mu_2}}\cos\alpha\cos\gamma + \cos^2\gamma + 4\sqrt{\frac{\varepsilon_2\mu_2}{\varepsilon_1\mu_1}}\frac{\mu_1\cos\gamma}{\mu_2\cos\alpha}\cos^2\alpha\right\} = 1.$$

2.5.3 The Telegraph Equation

If an electromagnetic wave strikes a metallic surface, there is, as a consequence of the conductivity, a difference with respect to a dielectric material. The discussion in Chap. 2.2.2 (p. 80) is, however, restricted to the part of the electromagnetic wave, which penetrates into the metal. The following presents the derivation of the appropriate wave equation, which is discussed in this chapter.

The differential equation, which governs the propagation of waves in a metal is, provided the material is neutral on the average,

$$\nabla \times \boldsymbol{H}(\boldsymbol{r},t) = \frac{1}{c}\frac{\partial}{\partial t}\boldsymbol{D}(\boldsymbol{r},t) + \frac{4\pi}{c}\sigma\,\boldsymbol{E}(\boldsymbol{r},t)$$

(or with incorporation of Ohm's law $\boldsymbol{j}_{tr}(\boldsymbol{r},t) = \sigma \boldsymbol{E}(\boldsymbol{r},t)$)

$$\nabla \cdot \boldsymbol{B} = 0 \qquad \nabla \times \boldsymbol{E} = -\frac{1}{c}\frac{\partial \boldsymbol{B}}{\partial t} \qquad \nabla \cdot \boldsymbol{E} = 0.$$

The material equation

$$\boldsymbol{D} = \varepsilon \boldsymbol{E} \qquad \boldsymbol{B} = \mu \boldsymbol{H}$$

has been used. The wave equations for the $\boldsymbol{E}$- and the $\boldsymbol{B}$-field can be won in analogy to the case of the free Maxwell equations.

- The $\boldsymbol{E}$-field:
  The rotation of the law of induction (interchange the sequence of the differentiation) gives

$$\nabla \times (\nabla \times \boldsymbol{E}) = -\frac{1}{c}\frac{\partial}{\partial t}(\nabla \times \boldsymbol{B}). \qquad (2.126)$$

2.5 Details

On the left hand side one obtains with $\nabla \cdot \mathbf{E} = 0$

$$\nabla \times (\nabla \times \mathbf{E}) = -\Delta \mathbf{E} .$$

On the right hand side of (2.126) one inserts the extended Ampère law

$$-\frac{1}{c}\frac{\partial}{\partial t} (\nabla \times \mathbf{B}) = -\frac{1}{c}\frac{\partial}{\partial t} \left(\frac{\varepsilon\mu}{c}\frac{\partial \mathbf{E}}{\partial t} + \frac{4\pi}{c}\sigma\mu\mathbf{E} \right)$$

$$= -\frac{\varepsilon\mu}{c^2}\frac{\partial^2}{\partial t^2}\mathbf{E} - \frac{4\pi}{c^2}\sigma\mu\frac{\partial \mathbf{E}}{\partial t} .$$

The resulting wave equation, the telegraph equation for the electric field is therefore

$$\Delta \mathbf{E} - \frac{\varepsilon\mu}{c^2}\frac{\partial^2}{\partial t^2}\mathbf{E} - \frac{4\pi}{c^2}\sigma\mu\frac{\partial \mathbf{E}}{\partial t} = 0 .$$

- the **B**-field:
One begins with the rotation of the Ampère law

$$\nabla \times (\nabla \times \mathbf{B}) = \frac{\varepsilon\mu}{c}\frac{\partial}{\partial t} (\nabla \times \mathbf{E}) + \frac{4\pi}{c}\sigma\mu (\nabla \times \mathbf{E})$$

and inserts on the right hand side the law of induction

$$= -\frac{\varepsilon\mu}{c^2}\frac{\partial^2}{\partial t^2}\mathbf{B} - \frac{4\pi}{c^2}\sigma\mu\frac{\partial \mathbf{B}}{\partial t} .$$

On the left hand side one obtains because of $\nabla \cdot \mathbf{B} = 0$ again

$$\nabla \times (\nabla \times \mathbf{B}) = -\Delta \mathbf{B}$$

and hence

$$\Delta \mathbf{B} - \frac{\varepsilon\mu}{c^2}\frac{\partial^2 \mathbf{B}}{\partial t^2} - \frac{4\pi}{c^2}\sigma\mu\frac{\partial \mathbf{B}}{\partial t} = 0 .$$

2.5.4 Diffraction

The formulae for the calculation of patterns of diffraction requires the overcoming of a few smaller mathematical cliffs. This task is tackled in two sections. The diffraction pattern of a circular opening is discussed and illustrated in the last section.

2.5.4.1 Calculation of Kirchhoff's Integral Representation

Starting point is Eq. (2.45) of Chap. 2.2.4 (p. 94)

$$\psi(\mathbf{r},t) = \frac{1}{4\pi} \int_{t_i}^{t_f} dt' \oiint_S \left[G^{(+)}(\mathbf{r},t;\mathbf{r}',t') \frac{\partial \psi(\mathbf{r}',t')}{\partial n'} \right.$$

$$\left. - \psi(\mathbf{r}',t') \frac{\partial G^{(+)}(\mathbf{r},t;\mathbf{r}',t')}{\partial n'} \right] df' , \qquad (2.127)$$

in which the wave function of a scalar wave (as a place holder of an electromagnetic wave) is represented by a surface integral over the area of the diffractive object. The additional time integral is restricted by the retarded Green's function

$$G^{(+)}(\mathbf{r},t;\mathbf{r}',t') = \frac{\delta(|\mathbf{r}-\mathbf{r}'|/c + t' - t)}{|\mathbf{r}-\mathbf{r}'|} .$$

The geometry of the problem is illustrated again in Fig. 2.41.

The explicit evaluation of the time integral requires some preparation. First one has to calculate the normal derivative of the Green's function. This derivative can be connected with the gradient (see Dreizler and Lüdde, 2010, Theoretical Mechanics (Springer Berlin Heidelberg), Math. Chap. 5.2)

$$\frac{\partial}{\partial n'} = \mathbf{e}_n \cdot \nabla' ,$$

where, in the present situation, $\mathbf{e}_n$ represents the normal vector in all points of the lateral face $S = S_1 + S_2$ and the gradient is defined as

$$\nabla' = \mathbf{e}_x \partial_{x'} + \mathbf{e}_y \partial_{y'} + \mathbf{e}_n \partial_{z'} .$$

There follows in explicit notation

$$\frac{\partial}{\partial n'} G^{(+)}(\mathbf{r},t;\mathbf{r}',t') = \mathbf{e}_n \cdot \nabla' \left\{ \frac{\delta(|\mathbf{r}-\mathbf{r}'|/c + t' - t)}{|\mathbf{r}-\mathbf{r}'|} \right\} .$$

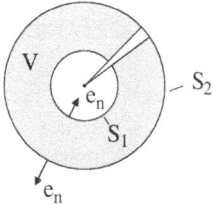

Fig. 2.41 Geometry of the Kirchhoff representation of diffraction

2.5 Details

The following derivatives have to be calculated:

- The gradient of the separation

$$\partial_{x'}|\mathbf{r}-\mathbf{r}'| = \partial_{x'}[(x-x')^2 + (y-y')^2 + (z-z')^2]^{1/2}$$

$$= -\frac{(x-x')}{[(x-x')^2+(y-y')^2+(z-z')^2]^{1/2}} = -\frac{(x-x')}{|\mathbf{r}-\mathbf{r}'|}$$

and corresponding expressions for the partial derivatives with respect to the other coordinates

$$\nabla'|\mathbf{r}-\mathbf{r}'| = -\frac{(\mathbf{r}-\mathbf{r}')}{|\mathbf{r}-\mathbf{r}'|} .$$

- The gradient of the inverse separation is

$$\nabla' \frac{1}{|\mathbf{r}-\mathbf{r}'|} = \frac{(\mathbf{r}-\mathbf{r}')}{|\mathbf{r}-\mathbf{r}'|^3} .$$

- With the abbreviation $\mathbf{R} = \mathbf{r} - \mathbf{r}'$ one obtains then

$$\nabla' \left\{ \frac{\delta(R/c+t'-t)}{R} \right\} = \frac{\mathbf{R}}{R^3} \delta(R/c+t'-t)$$

$$+ \frac{1}{R} \delta^{(1)}(R/c+t'-t) \nabla'(R/c+t'-t) ,$$

where the second term originates by application of the chain rule ($\delta^{(1)}$ stands for the derivative of the δ-function with respect to the complete argument). The final result is

$$\nabla' \left\{ \frac{\delta(R/c+t'-t)}{R} \right\} = \frac{\mathbf{R}}{R^3} \delta(R/c+t'-t)$$

$$- \frac{\mathbf{R}}{cR^2} \delta^{(1)}(R/c+t'-t) .$$

The time integral in (2.127) contains two terms, in which the δ-function appears. The time argument t' of the related factor of the integrand is the retarded time

$$t' = t - \frac{R}{c} .$$

The term, in which one finds the derivative of the δ-function, can be treated with the following steps:

$$\int_{t_i}^{t_f} dt'\, \delta'(R/c + t' - t)\psi(r', t') =$$

- substitute $s = R/c + t' - t$ with $ds = dt'$

$$= \int_{s_i}^{s_f} ds\, \frac{d\delta^{(1)}(s)}{ds} \psi(r', s - R/c + t) = -\left.\frac{d\psi(r', s - R/c + t)}{ds}\right|_{s=0}.$$

- the still pending derivative is transformed with the chain rule

$$= -\left.\frac{\partial \psi(r', t')}{\partial t'} \frac{\partial t'}{\partial s}\right|_{t'=t-R/c}$$

- respectively as the partial derivative of t' with respect to s equals 1

$$= -\left.\frac{\partial \psi(r', t')}{\partial t'}\right|_{t'=t-R/c}.$$

One obtains the desired Kirchhoff integral representation, if the normal derivative (2.127) of the wave function is rewritten and all terms are put together

$$\psi(r, t) = \frac{1}{4\pi} \oiint_S df'\, e_n \cdot \left[\frac{\nabla' \psi(r', t')}{R} \right. \tag{2.128}$$

$$\left. - \frac{R\psi(r', t')}{R^3} - \frac{R}{cR^2} \frac{\partial \psi(r', t')}{\partial t'} \right]_{t'=t-R/c}.$$

2.5.4.2 Angular Integrals for Airy's Diffraction Problem

The evaluation of the angular integral (p. 98)

$$WI = \int_0^{2\pi} d\varphi'\, e^{ik\rho'(\sin\alpha\cos\varphi' - \sin\beta\cos(\varphi'-\gamma))}$$

requires some manipulations, which will be presented and executed here in a more explicit fashion. The angular part of the exponent (WA) is explicitly displayed and sorted

$$WA = \sin\alpha \cos\varphi' - \sin\beta \cos(\varphi' - \gamma)$$
$$= \sin\alpha \cos\varphi' - \sin\beta(\cos\varphi' \cos\gamma + \sin\varphi' \sin\gamma)$$

2.5 Details

$$= \cos\varphi'(\sin\alpha - \sin\beta\cos\gamma) - \sin\varphi'(\sin\beta\sin\gamma).$$

With the definitions

$$A = \sin\alpha - \sin\beta\cos\gamma \quad \text{and} \quad B = -\sin\beta\sin\gamma,$$

as well as

$$\cos\varphi_0 = -\frac{A}{\sqrt{(A^2+B^2)}} \quad \text{and} \quad \sin\varphi_0 = -\frac{B}{\sqrt{(A^2+B^2)}},$$

the conversion given in Chap. 2.2.4 can be executed

$$A\cos\varphi' + B\sin\varphi' = -[A^2+B^2]^{1/2}\cos(\varphi' - \varphi_0).$$

The prefactor, which will be denoted by σ, is

$$\sigma = [A^2 + B^2]^{1/2} = \left[\sin^2\alpha + \sin^2\beta - 2\sin\alpha\sin\beta\cos\gamma\right]^{1/2}.$$

The integral resulting via these arguments

$$\mathrm{WI} = \int_0^{2\pi} d\varphi' \, e^{-ik\sigma\rho'\cos(\varphi'-\varphi_0)}$$

can be simplified due to the periodicity of the cosine function in the exponent. One substitutes $\varphi'' = \varphi' - \varphi_0$ in

$$\mathrm{WI} = \int_0^{2\pi} d\varphi' \, e^{-iC\cos(\varphi'-\varphi_0)} \qquad (C = k\sigma\rho'),$$

and finds

$$\mathrm{WI} = \int_{-\varphi_0}^{2\pi-\varphi_0} d\varphi'' \, e^{-iC\cos\varphi''} = \int_{-\varphi_0}^{0} d\varphi'' \, e^{-iC\cos\varphi''}$$

$$+ \int_0^{2\pi} d\varphi'' \, e^{-iC\cos\varphi''} - \int_{2\pi-\varphi_0}^{2\pi} d\varphi'' \, e^{-iC\cos\varphi''}. \qquad (2.129)$$

In the last integral one can now substitute $\varphi''' = \varphi'' - 2\pi$ and obtain

$$\int_{2\pi-\varphi_0}^{2\pi} d\varphi'' \, e^{-iC\cos\varphi''} = -\int_{-\varphi_0}^{0} d\varphi''' \, e^{-iC\cos\varphi'''}.$$

The first and the last contribution in (2.129) cancel each other and there remains

$$\text{WI} = \int_0^{2\pi} d\varphi' \, e^{-ik\sigma\rho'\cos\varphi'} \, ,$$

the integral representation of the (up to a factor 2π) Bessel function J_0.

2.5.4.3 The Diffraction Pattern of the Airy Problem

The diffraction pattern of a circular opening (radius a) is given (for an appropriate set of values of the parameters (see below)) by the intensity distribution (see Chap. 2.2.4 (2.50), p. 99)

$$\frac{dP}{d\Omega} = \frac{c(ka^2)^2|\psi_0|^2}{32\pi r^2} (\cos\alpha + \cos\beta)^2 \left| \frac{J_1(ak\sigma)}{ak\sigma} \right|^2 . \tag{2.130}$$

The angular coordinates β, γ of the observation point with the radius r from the middle of the opening are given in Fig. 2.42 again. The magnitude σ depends only on the polar angle β, if one restricts the observation to the vertical incidence of the light ($\alpha = 0$)

$$\sigma = \sin\beta \qquad 0 \leq \beta \leq \pi$$

and (2.130) is reduced to

$$\frac{dP}{d\Omega} = \frac{c(ka^2)^2|\psi_0|^2}{32\pi r^2} (1 + \cos\beta)^2 \left| \frac{J_1(\eta)}{\eta} \right|^2 \tag{2.131}$$

with $\eta = ak\sin\beta$. The distribution has cylinder symmetry with respect to the z-axis marked in Fig. 2.42. The distribution (2.131) is determined, besides the intensity $|\psi_0|^2$ of the incident light, by three essential parameters:

- the wave length of the light $\lambda = 2\pi/k$,
- the radius of the opening a and
- the polar angle β.

Fig. 2.42 Geometry for the diffraction by a circular opening

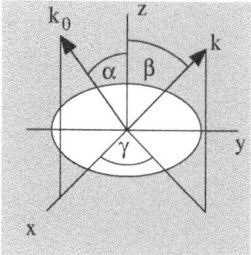

2.5 Details

The strength of the distribution decreases according to the spherical symmetry with the distance r as $1/r^2$.

The discussion of the dependence of the distribution on the three parameters is formed by the Bessel function $J_1(\eta)$ respectively the function $g_1(\eta) = J_1(\eta)/\eta$. The function can be defined by a power series

$$g_1(\eta) = \frac{1}{2} \sum_n \frac{(-)^n}{n!(n+1)!} \left(\frac{\eta}{2}\right)^{2n}$$

$$= \frac{1}{2} - \frac{1}{16}\eta^2 + \frac{1}{384}\eta^4 - \ldots .$$

This alternating function does not entirely justice to one property of special interest. Figure 2.43 shows, that it is an oscillating function. The zeros of $g_1(\eta)$ can only be calculated numerically. The values, rounded to four digits after the comma for the first five zeros and their difference are shown in Table 2.1. One can recognise, that the function is not entirely periodic. The asymptotic behaviour of the function $J_n(\eta)$

$$J_n(\eta) \xrightarrow{\eta \to \infty} \sqrt{\frac{2}{\pi\eta}} \cos\left(\eta - \frac{\pi}{4} - \frac{n\pi}{2}\right) \qquad (2.132)$$

shows, that the difference of neighbouring zeros approaches slowly the number π. The extrema of the function $g_1(\eta)$ are determined via the derivative

$$g_2(\eta) \equiv \frac{dg_1(\eta)}{d\eta} = \frac{d}{d\eta}\left(\frac{J_1(\eta)}{\eta}\right) = \frac{J_2(\eta)}{\eta} .$$

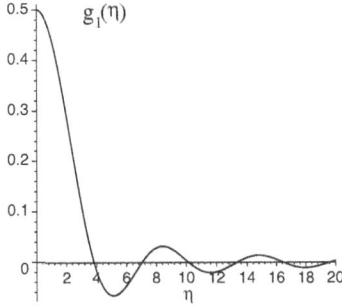

Fig. 2.43 The function $g_1(\eta) = J_1(\eta)/\eta$

Table 2.1 Zeros of the function $g_1(\eta)$

i	1	2	3	4	5
$\eta(i)$	3.8317	7.0156	10.1735	13.3237	16.4706
$\eta(i+1) - \eta(i)$	3.1839	3.1579	3.1502	3.1469	

Fig. 2.44 The function
$g_2(\eta) = J_2(\eta)/\eta$

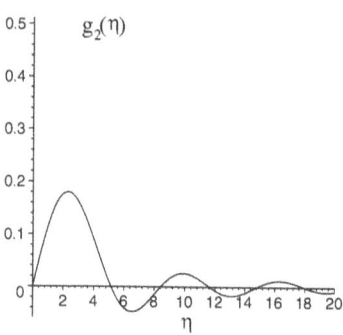

Table 2.2 Zeros of the function $g_2(\eta)$

i	1	2	3	4	5	6
$\eta(i)$	0.0	5.1356	8.4172	11.6198	14.7960	17.9598
$\eta(i+1) - \eta(i)$	5.1356	3.2816	3.2026	3.1762	3.1638	

The function $g_2(\eta)$ can, as $g_1(\eta)$, defined by a power series

$$g_2(\eta) = \frac{\eta}{4} \sum_n^\infty \frac{(-)^n}{n!\,(n+2)!} \left(\frac{\eta}{2}\right)^{2n}$$

$$= \frac{\eta}{8} - \frac{1}{96}\eta^2 + \frac{1}{1536}\eta^4 - \cdots,$$

the function itself is plotted in Fig. 2.44. The zeros of the function $g_2(\eta)$ or of $J_2(\eta)$ can also be obtained numerically. The first six zeros are collected in Table 2.2. The difference of neighbouring zeros of the function $g_2(\eta)$ also approaches the number π in the asymptotic regime (see 2.132).

The zeros of the function $g_1(\eta)$ are the points with completely destructive interference. They mark the dark rings of the pattern of interference in a plane perpendicular to the z-axis (or a corresponding observation screen). The extrema of the function, the zeros of the function $g_2(\eta)$, are the points with the maximal constructive interference, which can be observed as bright rings. The difference values given in the Tables 2.1 and 2.2 represent the space between the bright respectively the dark rings. The pattern of interference, which is given by

$$\frac{dP}{d\Omega} = \left\{\frac{a^2 c\pi |\psi_0|^2}{8d^2}\right\} \left\{\left(\frac{a}{\lambda}\right)^2 (1+\cos\beta)^2 \cos^2\beta \left|\frac{J_1(2\pi(a/\lambda)\sin\beta)}{2\pi(a/\lambda)\sin\beta}\right|^2\right\}$$

can be observed, if one uses a screen in a distance d from the plane of diffraction as one has $r = d/\cos\beta$ and $k = 2\pi/\lambda$. The maxima respectively the zeros describe the bright and the dark rings (with a gradual transition) of the pattern of interference

2.5 Details

Fig. 2.45 Intensity distribution for the diffraction by a circular hole for $a/\lambda = 8$

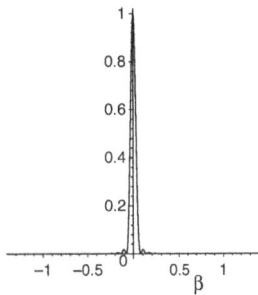

about the axis of symmetry, the z-axis. The factor

$$(1 + \cos \beta)^2 \cos^2 \beta \left| \frac{J_1(2\pi (a/\lambda) \sin \beta)}{2\pi (a/\lambda) \sin \beta} \right|^2$$

is shown in Figs. 2.45, 2.46, 2.47, and 2.48.

These figures show, that the ratio a/λ is the relevant parameter for the pattern of diffraction of the system. If the opening is large in comparison to the wave length ($a/\lambda = 8$, Fig. 2.45), essentially only the image of the circle and a very compressed pattern of the structures of the diffraction are seen. The first dark ring appears (Fig. 2.46) at about 4° (0.07 rad), the first bright ring under ≈5.8° (0.102 rad).

If the two parameters (a and λ) are more or less equal ($a/\lambda = 2$, Fig. 2.47), the central spot is broader and the pattern more extended. The position of the first dark respectively the first bright ring is found at angles of 17.75° and 23.5°.

This trend continues, if the opening is smaller than the wave length ($a/\lambda = 0.8$, Fig. 2.48), the angles are here 47.25° and 62.5°.

However one is soon in a range of the parameters, for which the conditions for the derivation of Eq. (2.130) are not given, so that the results do not reproduce the

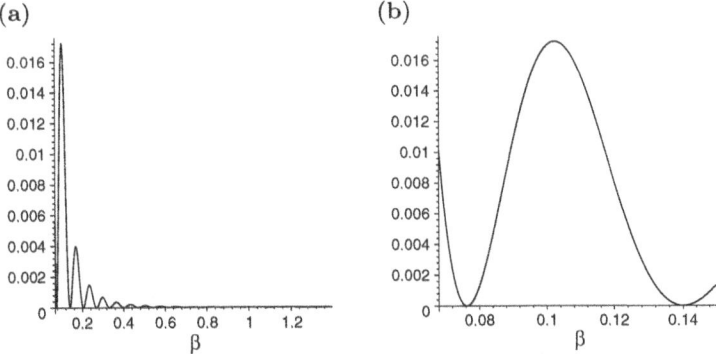

Fig. 2.46 Details for a range of angles. (a) $3.8° \leq \beta \leq 80°$. (b) $3.8° \leq \beta \leq 8.6°$

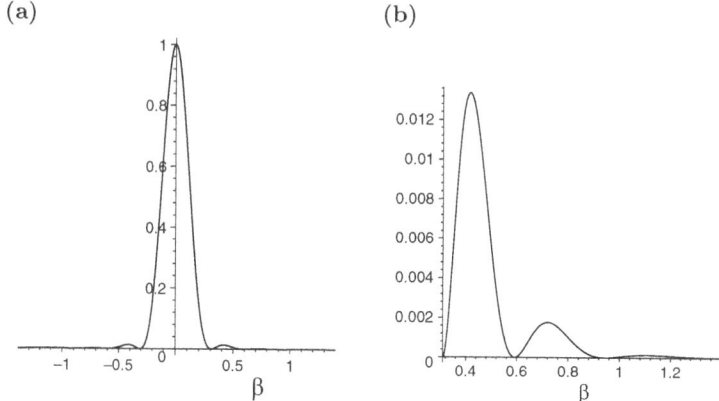

Fig. 2.47 Intensity distribution for $a/\lambda = 2$. (**a**) Total distribution. (**b**) Details for the range of angles $17.75° \leq \beta \leq 80°$

observed pattern. The angles quoted are obtained from the relations

$$\sin\beta(\text{dark}) \approx \frac{0.61\lambda}{a} \qquad \text{resp.} \qquad \sin\beta(\text{bright}) \approx \frac{0.82\lambda}{a}.$$

The distance to the next and between further neighbouring rings of the same type is given in good approximation by

$$\Delta\sin\beta \approx \frac{0.5\,\lambda}{a}.$$

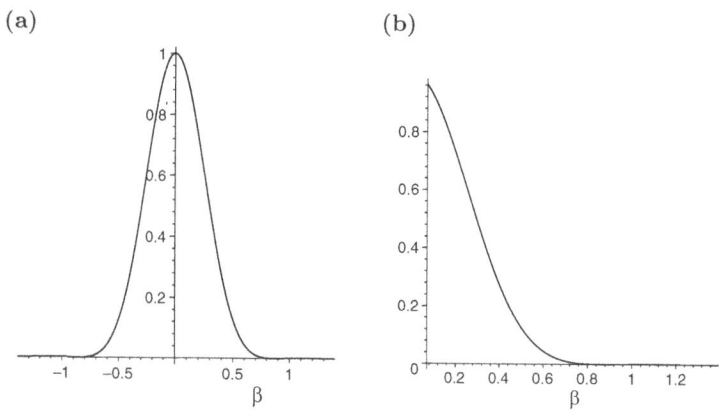

Fig. 2.48 Intensity distribution for $a/\lambda = 0.8$. (**a**) Total distribution. (**b**) Details for the range of angles $47.25° \leq \beta \leq 80°$

2.5 Details

2.5.5 The Hertz Dipole

The calculation of the radiant emission of electromagnetic waves is rather complicated. The emitted radiation acts back on the current distribution in the antenna. The calculation of the exact pattern of the radiation is even for idealised antennae not simple, even if this effect is not included. However, the discussion of the far field of electromagnetic waves in a larger distance from the sender plays a special role, as one can assume, that the receiver is not found in the vicinity of the sender. The far field is dominated by the dipole radiation, the Hertz dipole (provided the oscillating charge distribution in the antenna is overall neutral, so that no monopole radiation is present). The discussion of the dipole in Chap. 2.3.2 did not include two points:

(a) The verification of the relation (p. 105)

$$\iiint dV' \, \boldsymbol{j}(\boldsymbol{r}') = - \iiint dV' \, \boldsymbol{r}' \left(\boldsymbol{\nabla}' \cdot \boldsymbol{j}(\boldsymbol{r}') \right) . \tag{2.133}$$

(b) The calculation of the contribution of the scalar potential (p. 106) and hence the calculation of the electric field according to the formula

$$\boldsymbol{E}(\boldsymbol{r},t) = - \left(\boldsymbol{\nabla} V(\boldsymbol{r},t) + \frac{1}{c} \frac{\partial}{\partial t} \boldsymbol{A}(\boldsymbol{r},t) \right) . \tag{2.134}$$

2.5.5.1 An Integral of the Current Density

The verification of the formula (2.133) starts with the discussion of the terms

$$\boldsymbol{\nabla} \cdot (x \, \boldsymbol{j}(\boldsymbol{r})), \quad \boldsymbol{\nabla} \cdot (y \, \boldsymbol{j}(\boldsymbol{r})), \quad \boldsymbol{\nabla} \cdot (z \, \boldsymbol{j}(\boldsymbol{r})) .$$

As the evaluation of each of these expressions follows the same pattern, it is sufficient to look at one of them

$$\boldsymbol{\nabla} \cdot (x \, \boldsymbol{j}(\boldsymbol{r})) = \frac{\partial}{\partial x}(x \, j_x) + \frac{\partial}{\partial y}(x \, j_y) + \frac{\partial}{\partial z}(x \, j_z)$$

$$= j_x + x \left(\frac{\partial}{\partial x} j_x + \frac{\partial}{\partial y} j_y + \frac{\partial}{\partial z} j_z \right)$$

$$= j_x + x (\boldsymbol{\nabla} \cdot \boldsymbol{j}(\boldsymbol{r})) .$$

The next task is the integration over the complete space

$$\iiint dV' \, \boldsymbol{\nabla}' \cdot \left(x' \boldsymbol{j}(\boldsymbol{r}') \right) = \iiint dV' \, j_x(\boldsymbol{r}') + \iiint dV' \, x' \left(\boldsymbol{\nabla}' \cdot \boldsymbol{j}(\boldsymbol{r}') \right) .$$

The integral on the right hand side can be changed into a surface integral over an infinite sphere with the aid of the divergence theorem

$$\iiint dV' \, \nabla' \cdot (x' \boldsymbol{j}(\boldsymbol{r}')) = \oiint_{O(\infty)} d\boldsymbol{S}' \cdot (x' \boldsymbol{j}(\boldsymbol{r}')) \ .$$

The value of this integral is, however, zero, as it was assumed that the source of the radiation is restricted to the surrounding of the origin.

The argumentation for the terms with $y\,\boldsymbol{j}$ and $z\,\boldsymbol{j}$ is the same, so that one can write in summary

$$\iiint dV' \, \boldsymbol{j}(\boldsymbol{r}') = - \iiint dV' \, \boldsymbol{r}' \left(\nabla' \cdot \boldsymbol{j}(\boldsymbol{r}') \right) \ .$$

2.5.5.2 Alternative Calculation of the Dipole Field

For the scalar potential

$$V(\boldsymbol{r}, t) = V(\boldsymbol{r}) \, e^{-i\omega t}$$

$$V(\boldsymbol{r}) = \iiint dV' \, \frac{\rho_{tr}(\boldsymbol{r}')}{|\boldsymbol{r} - \boldsymbol{r}'|} \, e^{ik|\boldsymbol{r}-\boldsymbol{r}'|}$$

one can (as in the discussion of the vector potential, Eqs. (2.56)/(2.57) in Chap. 2.3.1) use the approximation

$$V_{RZ}(\boldsymbol{r}, t) = V_{RZ}(\boldsymbol{k}) \, \frac{1}{r} \, e^{i(kr-\omega t)}$$

$$V_{RZ}(\boldsymbol{k}) = \iiint dV' \, \rho(\boldsymbol{r}') \, e^{i \boldsymbol{k} \cdot \boldsymbol{r}'}$$

in the radiation zone (RZ). The expansion of the exponential function yields there

$$V_{RZ}(\boldsymbol{k}) = V^{(0)}(\boldsymbol{k}) + V^{(1)}(\boldsymbol{k}) + \ldots$$

$$= \iiint dV' \, \rho(\boldsymbol{r}') - i\boldsymbol{k} \cdot \iiint dV' \, \boldsymbol{r}' \rho(\boldsymbol{r}') + \ldots \ .$$

For a dipole antenna one assumes, that

$$\iiint dV' \, \rho(\boldsymbol{r}') = 0$$

is. The total charge of the antenna is zero, but the charge distribution changes locally with time.

2.5 Details

In the second term of the expansion one can again recognise the dipole moment p of the charge distribution. The scalar potential is thus in lowest order

$$V^{(1)}(r, t) = V^{(1)}(k)\frac{1}{r} e^{i(kr-\omega t)} = -i(k \cdot p)\frac{1}{r} e^{i(kr-\omega t)} .$$

The gradient of the total potential in the order $1/r$

$$\nabla V^{(1)}(r, t) = -i(k \cdot p)\nabla \left(\frac{1}{r} e^{ikr}\right) e^{-i\omega t}$$

with the definition $k = k e_r$ is

$$\nabla \left(\frac{1}{r} e^{ikr}\right) \approx \frac{1}{r}\left(\nabla e^{ikr}\right) = \frac{1}{r} ik \left(\frac{x}{r}, \frac{y}{r}, \frac{z}{r}\right) e^{ikr}$$

$$= \frac{ik}{r} e_r e^{ikr} = \frac{ik}{r} e^{ikr} .$$

The long wave limit in the radiation zone is found to be

$$\nabla V^{(1)}(r, t) = (k \cdot p) k \frac{e^{i(kr-\omega t)}}{r} .$$

For the vector potential (Chap. 2.3.2, (2.59), p. 105)

$$A^{(0)}(r, t) = -p \left(\frac{ik}{r}\right) e^{i(kr-\omega t)}$$

one has with $\omega/c = k$ the time derivative

$$\frac{1}{c}\frac{\partial}{\partial t} A^{(0)}(r, t) = -p k^2 \frac{e^{i(kr-\omega t)}}{r} ,$$

so that the electric field with (2.134) can be written in the dipole approximation

$$E_{(E1)}(r, t) = \left(-(k \cdot p)k + k^2 p\right) \frac{e^{i(kr-\omega t)}}{r} . \tag{2.135}$$

The factor in the wave number can be changed with the formula

$$k \times (k \times p) = k(k \cdot p) - p k^2$$

into

$$-(k \cdot p)k + k^2 p = (k \times p) \times k .\quad (2.136)$$

The result agrees with (2.61) in Chap. 2.3.2.

2.5.6 Multipole Radiation (Higher Multipole Moments)

The discussion of the next contributions to the radiation of an (ideal) antenna also needs the preparation of a number of intermediate steps.

2.5.6.1 A Segmentation
The segmentation of the product $(k \cdot r)j$ (Chap. 2.3.3, (2.67), p. 108) into a part, which is symmetric with respect to the permutation of the vectors r and j, plus an antisymmetric part, can be achieved in the following fashion. One writes in a first step

$$(k \cdot r)j = \frac{1}{2}[(k \cdot r)j + (k \cdot r)j]$$

and adds and subtracts the symmetric complement

$$(k \cdot r)j = \frac{1}{2}[(k \cdot r)j + (k \cdot j)r] + \frac{1}{2}[(k \cdot r)j - (k \cdot j)r] .$$

The second term is the standard double vector product

$$a \times (b \times c) = b(a \cdot c) - c(a \cdot b) .$$

This allows to write

$$\frac{1}{2}[(k \cdot r)j - (k \cdot j)r] = \frac{1}{2}[k \times (j \times r)] = -\frac{1}{2}[(j \times r) \times k]$$

$$= \frac{1}{2}[(r \times j) \times k] .$$

2.5.6.2 The Quadrupole Contribution
If one inserts $f(r) = (k \cdot r)$ and $b(r) = j(r)$ into

$$\nabla \cdot (f b) = b \cdot \nabla f + f(\nabla \cdot b) ,$$

2.5 Details

multiplies the expression with the vector r and integrates over the complete space, one finds

$$\iiint dV' \; r' \nabla' \cdot [(k \cdot r')j(r')] = \iiint dV' \; r'(j(r') \cdot k)$$

$$+ \iiint dV' \; r'(k \cdot r')(\nabla' \cdot j(r')) \, .$$

The integral on the left hand side can be treated with partial integration, thus one obtains e.g. (insert $(k \cdot r')j(r') = F(r')$)

$$\iiint dV' \; x'(\partial_{x'} F_{x'}) = \iint dy' dz' \; x' F_{x'} \Big|_{-\infty}^{\infty} - \iiint dV' \; F_{x'}$$

$$= - \iiint dV' \; F_{x'}$$

and

$$\iiint dV' \; x'(\partial_{y'} F_{y'}) = \iint dx' dz' \; x' F_{y'} \Big|_{-\infty}^{\infty} = 0 \, ,$$

as the quantities F_i contain a current distribution, which is restricted to the origin. If these statements (the ones given plus seven additional ones) are combined, one arrives at

$$\iiint dV' \; r' \nabla' F(r') = - \iiint dV' \; F(r') \, ,$$

or after sorting (and reinsertion of F)

$$\iiint dV' \; \{(k \cdot r')j(r') + (k \cdot j(r'))r'\} =$$

$$- \iiint dV' \; r'(k \cdot r')(\nabla' \cdot j(r')) \, .$$

2.5.6.3 The Angular Distribution of the Quadrupole Radiation

The task is the calculation of the angular distribution

$$\frac{dP}{d\Omega} = \frac{ck^2}{288\pi} |k \times Q(k)|^2$$

with the components of the quadrupole vector $\boldsymbol{Q} = (Q_1, Q_2, Q_3)$, which are constructed with the elements Q_{il} of the quadrupole tensor

$$Q_i(\boldsymbol{k}) = \sum_{l=1}^{3} Q_{il} k_l .$$

One inserts here the given tensor elements

$$Q_{il} = \delta_{il} Q_{ll} \quad \text{with} \quad Q_{11} = Q_{22} = \frac{1}{2} Q_0 \quad Q_{33} = -Q_0$$

and obtains

$$Q_1 = \frac{Q_0}{2} k_1 \quad Q_2 = \frac{Q_0}{2} k_2 \quad Q_3 = -Q_0 k_3 .$$

If the representation of components of the wave number vector in terms of spherical coordinates is used here

$$k_1 = k \cos\varphi \sin\theta \quad k_2 = k \sin\varphi \sin\theta \quad k_3 = k \cos\theta ,$$

one finds for the individual contributions to the square of the absolute value of the vector product

$$|\boldsymbol{k} \times \boldsymbol{Q}(\boldsymbol{k})|^2 = (k_2 Q_3 - k_3 Q_2)^2 + (k_3 Q_1 - k_1 Q_3)^2 + (k_1 Q_2 - k_2 Q_1)^2$$

the results

$$(k_2 Q_3 - k_3 Q_2)^2 = \frac{Q_0^2}{4} (-2k_2 k_3 - k_3 k_2)^2$$

$$= \frac{9}{4} Q_0^2 k^4 \sin^2\varphi \sin^2\theta \cos^2\theta$$

$$(k_3 Q_1 - k_1 Q_3)^2 = \frac{Q_0^2}{4} k^4 (k_3 k_1 + 2k_1 k_3)^2$$

$$= \frac{9}{4} Q_0^2 k^4 \cos^2\varphi \sin^2\theta \cos^2\theta$$

$$(k_1 Q_2 - k_2 Q_1)^2 = \frac{Q_0^2}{4} (k_1 k_2 - k_1 k_2)^2 = 0 .$$

2.5 Details

The sum of these terms is

$$|k \times Q(k)|^2 = \frac{9}{4} Q_0^2 k^4 \sin^2\theta \cos^2\theta ,$$

so that the angular distribution of the emitted power is given by

$$\frac{dP}{d\Omega} = \frac{ck^6 Q_0^2}{128 \pi} \sin^2\theta \cos^2\theta .$$

The function $f(\theta) = \sin^2\theta \cos^2\theta$ is zero at the locations

$$\theta = 0, \frac{\pi}{2}, \pi .$$

The extrema of this function are given by

$$\frac{df}{d\theta} = 2 \left(\sin\theta \cos^3\theta - \cos\theta \sin^3\theta \right) = 0$$

and correspond to

$$\sin\theta \cos\theta = 0$$

(the zeroes indicated above) and

$$\cos^2\theta - \sin^2\theta = \cos 2\theta = 0 ,$$

that is

$$2\theta = \frac{\pi}{2}, \frac{3}{2}\pi \quad \text{resp.} \quad \theta = \frac{\pi}{4}, \frac{3}{4}\pi .$$

This characterises (check!) the maxima. The standard representation of this function in the region 0 to π is plotted in Fig. 2.49. This pattern is rotationally symmetric with respect to the z-axis.

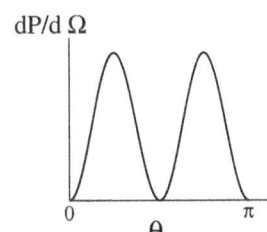

Fig. 2.49 Quadrupole radiation: standard plot of the pattern of emission

The total emitted radiation is

$$P = \iint \left(\frac{dP}{d\Omega}\right) d\Omega = \iint \left(\frac{dP}{d\Omega}\right) \sin\theta \, d\theta \, d\varphi$$

$$= \iint \left(\frac{ck^6 Q_0^2}{128\pi} \sin^2\theta \cos^2\theta\right) \sin\theta \, d\theta \, d\varphi$$

calculated by the integral (substitute $x = \cos\theta$)

$$\int_0^\pi \sin^3\theta \cos^2\theta \, d\theta = \int_{-1}^1 x^2[1-x^2] \, dx = \left[\frac{1}{3}x^3 - \frac{1}{5}x^5\right]_{-1}^1 = \frac{4}{15}.$$

The result is

$$P = \frac{ck^6 Q_0^2}{128\pi} \frac{4}{15} 2\pi = \frac{ck^6 Q_0^2}{240}.$$

2.5.7 The Complete Multipole Expansion

The complete multipole expansion of the sender problem allows a systematic access to the calculation of the electrodynamic potentials. The calculation of the corresponding potentials is relatively intricate due to the complexity of the finally emerging angular functions. The treatment of this topic is only broached in Chap. 2.3.4. In order to deal definitely with the problem one needs the multipole expansion of the Green's function of the sender problem

$$G^{(+)}(r, r') = \frac{e^{ik|r-r'|}}{|r-r'|}. \tag{2.137}$$

The discussion of this function differs only in the details of the discussion of the spherical Dirichlet problem for the calculation of electric potentials in Dreizler and Lüdde, 2024, Electrostatics and Magnetostatics (Springer Berlin Heidelberg), Detail 4.7.3. If one inserts the multipole expansion of this Green's function

$$G^{(+)}(r, r') = \sum_{l,m} g_l(r, r') Y_{l,m}^\star(\Omega') Y_{l,m}(\Omega) \tag{2.138}$$

into the Helmholtz equation (see (2.55), Chap. 2.3.1, p. 102), one obtains a differential equation for the radial Green's function $g_l(r, r')$

$$\left(\frac{d^2}{dr^2} + \frac{2}{r}\frac{d}{dr} + k^2 - \frac{l(l+1)}{r^2}\right) g_l(r, r') = -\frac{1}{r^2}\delta(r-r'). \tag{2.139}$$

2.5 Details

The function (2.137) satisfies the symmetry condition

$$G^{(+)}(r, r') = G^{(+)}(r', r)$$

and the boundary conditions:

- The function has to be finite at the coordinate origin

$$G^{(+)}(r, r') \xrightarrow{r \to 0} \text{finite} \qquad (r < r') \,.$$

- The function is an outgoing wave in the asymptotic region

$$G^{(+)}(r, r') \xrightarrow{r \to \infty} \text{const.} \frac{e^{ikr}}{r} \qquad (r > r') \,.$$

The inhomogeneous term in (2.139) causes, as shown in Dreizler and Lüdde, 2024, Electrostatics and Magnetostatics (Springer Berlin Heidelberg), Detail 4.7.3, the discontinuity

$$\lim_{\varepsilon \to 0} \left\{ \frac{d}{dr} \left(r g_l(r, r') \right)_{r=r'+\varepsilon} - \frac{d}{dr} \left(r g_l(r, r') \right)_{r=r'-\varepsilon} \right\} = -\frac{4\pi k_e}{r'} \,. \qquad (2.140)$$

The homogeneous differential equation, which corresponds to (2.139), for a function of one variable

$$\left(\frac{d^2}{dr^2} + \frac{2}{r} \frac{d}{dr} + k^2 - \frac{l(l+1)}{r^2} \right) f_l(r) = 0 \qquad (2.141)$$

is the differential equation of a spherical Bessel function. Some properties of this function, which will be needed, are collected in the next subsection.

2.5.7.1 Spherical Bessel Functions

As fundamental solutions of the differential equation (2.141) one can use either

- the spherical Bessel functions $j_l(kr)$ and
- the spherical Neumann functions $n_l(kr)$,

or alternatively the spherical Hankel functions

- $h_l^{(+)}(kr) = j_l(kr) + i n_l(kr)$
- $h_l^{(-)}(kr) = j_l(kr) - i n_l(kr) \,.$

Properties of these functions, which are needed for the present discussion, are the behaviour of these functions for $r \to 0$ and $r \to \infty$ as well as a statement on the first derivative.

At the origin (more precisely for $kr \to 0$) one has (remember: $(2l+1)!! = 1 \cdot 3 \cdot 5 \cdots (2l+1)$)

$$j_l(kr) = \frac{(kr)^l}{(2l+1)!!} \qquad n_l(kr) = -\frac{(2l-1)!!}{(kr)^{l+1}} . \qquad (2.142)$$

The Bessel functions are regular at the origin, the Neumann functions are divergent. The asymptotic behaviour of these functions is determined by the decaying trigonometric functions

$$j_l(kr) \xrightarrow{r \to \infty} \frac{1}{kr} \sin\left(kr - \frac{l\pi}{2}\right)$$

$$n_l(kr) \xrightarrow{r \to \infty} -\frac{1}{kr} \cos\left(kr - \frac{l\pi}{2}\right) . \qquad (2.143)$$

For the Hankel functions is then

$$h_l^{(+)}(kr) \xrightarrow{r \to \infty} \frac{e^{ikr}}{kr}$$

$$h_l^{(-)}(kr) \xrightarrow{r \to \infty} \frac{e^{-ikr}}{kr} . \qquad (2.144)$$

For the discussion of the singularity of the derivative of the function $g_l(r, r')$ one uses a recursion formula, which involves the first derivative of the solution of the Bessel differential equation. The derivative of all four functions satisfies

$$\frac{d}{dr} f_l(kr) = k f_{l-1}(kr) - \frac{(l+1)}{r} f_l(kr) \qquad f_l = j_l, n_l, h_l^{(\pm)} . \qquad (2.145)$$

An additional exploitation of the discontinuity can be achieved with a relation, in which the difference of products of the Bessel- or the von Neumann functions occur with crossed indices

$$j_l(kr) n_{l-1}(kr) - j_{l-1}(kr) n_l(kr) = \frac{1}{(kr)^2} . \qquad (2.146)$$

2.5.7.2 Determination of the Green's Function $g_l(r, r')$

The solutions (2.138) of the homogeneous differential equation in the regions $r < r'$ and $r' < r$ are constructed with the Bessel- and von Neumann functions

$$g_{1l}(r, r') = a_{1l}(r') j_l(kr) + b_{1l}(r') n_l(kr) \qquad r < r'$$

2.5 Details

$$g_{2l}(r, r') = a_{2l}(r') j_l(kr) + b_{2l}(r') n_l(kr) \qquad r' < r \qquad (2.147)$$

The boundary condition, that g_l has to be finite at the origin, requires as a consequence of the property (2.142)

$$b_{1l}(r') = 0 .$$

The condition for the asymptotic region

$$g_l(r, r') \xrightarrow{r' \to \infty} \text{const. } e^{ikr}$$

can only be satisfied for

$$b_{2l}(r') = i \, a_{2l}(r') ,$$

as (2.143)

$$g_{2l}(r, r') = a_{2l}(r')\{j_l(kr) + i n_l(kr)\} = a_{2l}(r') h_l^{(+)}(kr)$$

$$\xrightarrow{r' \to \infty} a_{2l}(r') \left(\frac{e^{ikr}}{kr} \right) .$$

The remaining expression for the solution in both regions

$$g_{1l}(r, r') = a_{1l}(r') j_l(kr) \qquad g_{2l}(r, r') = a_{2l}(r') h_l^{(+)}(kr)$$

is restricted further by the symmetry condition $g_l(r, r') = g_l(r', r)$. This condition demands

$$a_{1l}(r') j_l(kr) = a_{2l}(r) h_l^{(+)}(kr')$$

and thus

$$a_{1l}(r) = c_l \, h_l^{(+)}(kr) \qquad a_{2l}(r) = c_l \, j_l(kr) ,$$

respectively in explicit form

$$g_{1l}(r, r') = c_l \, h_l^{(+)}(r') \, j_l(kr) \qquad g_{2l}(r, r') = c_l \, j_l(r') \, h_l^{(+)}(kr) .$$

The constant c_l can be determined by the discontinuity (2.140) at the position $r = r'$. One uses the recursion (2.145) for the derivative and finds

$$\lim_{\varepsilon \to 0} c_l \left\{ j_l(kr') \left[h_l^{(+)}(kr) + r \left(k h_{l-1}^{(+)}(kr) - \frac{(l+1)}{r} h_l^{(+)}(kr) \right) \right]_{r=r'-\varepsilon} \right.$$
$$\left. - h_l^{(+)}(kr') \left[j_l(kr) + r \left(k j_{l-1}(kr) - \frac{(l+1)}{r} j_l(kr) \right) \right]_{r=r'+\varepsilon} \right\}$$
$$= -\frac{4\pi}{r'}.$$

The terms, in which the two functions of a product have the same index, cancel each other in taking the limit. There remains

$$k c_l r' \left(h_l^{(+)}(kr') j_{l-1}(kr') - j_l(kr') h_{l-1}^{(+)}(kr') \right) = \frac{4\pi}{r'}.$$

If one inserts the Hankel functions, one finds that the products with two Bessel functions cancel. The remaining expression

$$i k c_l r' \left(j_l(kr') n_{l-1}(kr') - j_{l-1}(kr') n_l(kr') \right) = -\frac{4\pi}{r'}$$

can be simplified with (2.146) to

$$\frac{i k c_l}{k^2 r'^2} = -\frac{4\pi}{r'^2},$$

so that one obtains

$$c_l = 4\pi i k .$$

The multipole expansion (2.138) of the Green's function of the sender problem is thus

$$G^{(+)}(\boldsymbol{r}, \boldsymbol{r}') = \frac{e^{ik|\boldsymbol{r}-\boldsymbol{r}'|}}{|\boldsymbol{r} - \boldsymbol{r}'|}$$

$$= 4\pi i k \sum_{l=0}^{\infty} j_l(kr_<) h_l^{(+)}(kr_>) \sum_{m=-l}^{l} Y_{l,m}^*(\Omega') Y_{l,m}(\Omega) .$$

2.5.8 The Centrally Fed, Thin Antenna

The exact evaluation of the relation (2.57) (p. 103)

$$\boldsymbol{A}_{RZ}(\boldsymbol{k}) = \frac{1}{c} \iiint dV' \, \boldsymbol{j}(\boldsymbol{r}') e^{-i\boldsymbol{k}\cdot\boldsymbol{r}'}$$

2.5 Details

in the radiation zone RZ is possible for the model of a centrally fed, thin antenna. With the specification

$$j(r,t) = \begin{cases} i_0 \sin\left(\frac{kd}{2} - k|z|\right) e^{-i\omega t} \delta(x)\delta(y) e_z & |z| < \frac{d}{2} \\ 0 & |z| \geq \frac{d}{2} \end{cases} \quad (2.148)$$

one can reduce the integral over space to an integral in one dimension. Besides the individual steps for the calculation of the vector potential one can in this case consider a simple and direct expansion of the angular distribution of the radiated power as well as the multipole expansion presented in Chap. 2.3.4.

2.5.8.1 The Vector Potential

The space part of the vector potential in the far zone (compare (2.56)) is given by

$$A(r) = \frac{i_0 e^{ikr}}{cr} \int_{-d/2}^{d/2} dz' \sin\left(\frac{kd}{2} - k|z'|\right) e^{-ikz'\cos\theta} e_z .$$

The evaluation of the integral

$$I = \int_{-d/2}^{d/2} dz' \sin\left(\frac{kd}{2} - k|z'|\right) e^{-ikz'\cos\theta}$$

is best done by the division of the range of integration into the intervals $[-d/2, 0]$ and $[0, d/2]$

$$I = I_1 + I_2 = \int_{-d/2}^{0} dz' \sin\left(\frac{kd}{2} + kz'\right) e^{-ikz'\cos\theta}$$
$$+ \int_{0}^{d/2} dz' \sin\left(\frac{kd}{2} - kz'\right) e^{-ikz'\cos\theta} .$$

If the sine function in the integrand is resolved with the addition theorem

$$\sin\left(\frac{kd}{2} \pm kz'\right) = \left\{\sin\frac{kd}{2}\cos kz' \pm \cos\frac{kd}{2}\sin kz'\right\},$$

one can use the integrals (see Table of Integrals or use partial integration)

$$\int dz\, e^{az} \sin bz = \frac{e^{az}}{(a^2 + b^2)}(a \sin bz - b \cos bz)$$

$$\int dz\, e^{az} \cos bz = \frac{e^{az}}{(a^2+b^2)} (a\cos bz + b\sin bz) .$$

Insertion of the limits and sorting yields the intermediate result in the form

$$I = \frac{1}{k^2 \sin^2 \theta} \left\{ k e^{-i(kd/2)\cos\theta} + k e^{+i(kd/2)\cos\theta} - 2k \cos\frac{kd}{2} \right\},$$

so that one finds for the vector potential

$$\mathbf{A}(\mathbf{r}) = \frac{2i_0\, e^{ikr}}{ckr} \frac{\left(\cos\left[\frac{kd}{2}\cos\theta\right] - \cos\frac{kd}{2}\right)}{\sin^2\theta} \mathbf{e}_z \qquad (2.149)$$

and thus, as shown in Chap. 2.3.5, for the power emitted by the antenna per solid angle and unit of time

$$\frac{dP}{d\Omega} = \frac{i_0^2}{2\pi c} \frac{\left(\cos\left(\frac{kd}{2}\cos\theta\right) - \cos\left(\frac{kd}{2}\right)\right)^2}{\sin^2\theta} . \qquad (2.150)$$

2.5.8.2 A Simple Expansion for the Power of the Radiation

The Poynting vector or the power is up to a numerical factor given by the function

$$f(\theta) = \frac{\left(\cos\left(\frac{kd}{2}\cos\theta\right) - \cos\left(\frac{kd}{2}\right)\right)^2}{\sin^2\theta} . \qquad (2.151)$$

If one expands, with the assumption $kd/2 = a < 1$ the function in the numerator with respect to powers of a, one arrives at a first summary

$$(\cos(a\cos\theta) - \cos a)^2 = \left(\frac{a^2}{2}(1-\cos^2\theta) - \frac{a^4}{24}(1-\cos^4\theta) \right.$$

$$\left. + \frac{a^6}{720}(1-\cos^6\theta) - \ldots \right)^2 .$$

A factor $1 - \cos^2\theta$ can be extracted from each of the terms with $(1 - \cos^{2n}\theta)$, so that one obtains after squaring and sorting up to a^8 the result

$$f(\theta) = \frac{a^4}{4}\sin^2\theta \left(1 - \frac{a^2}{6}(1+\cos^2\theta) \right.$$

$$\left. + \frac{a^4}{720}(9 + 14\cos^2\theta + 9\cos^4\theta)\ldots \right) .$$

2.5 Details

Fig. 2.50 Emission pattern for the centrally fed antenna in the far zone: comparison of the exact solution (2.150, full line) with the simple approximation (broken line). (**a**) Up to order a^6. (**b**) Up to order a^8

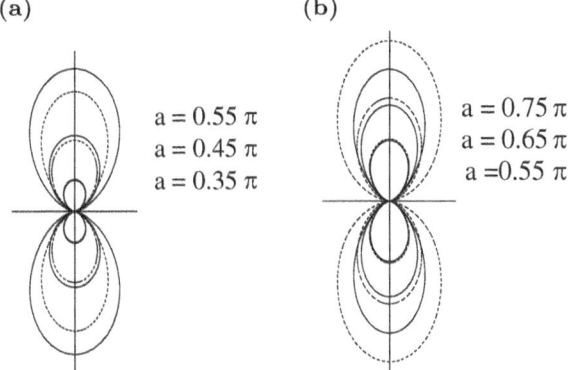

This expansion should be adequate for small values of the parameter a. A comparison of the result of this expansion up to order $(a)^6$ with the result (2.150) in Fig. 2.50a for a sequence of a-values $a = 0.35\pi$, 0.45π, 0.55π and up to order $(a)^8$ in Fig. 2.50b for $kd = 0.55\pi$, 0.65π, 0.75π demonstrates, that this is indeed the case.

One notices, that the error induced by the truncation of the expansion is acceptable for values of $a = 0.35\pi$ respectively $a = 0.55\pi$, but that it becomes quickly larger with growing values of a. One can also notice, that the expansion up to order a^6 underestimates the exact value in each case and the expansion up to order a^8 overestimates. A comparison of the function $f(\theta)$, expanded to order a^6 respectively a^8 with the exact result is shown in Fig. 2.51 for values of a between 0.25π and 0.65π.

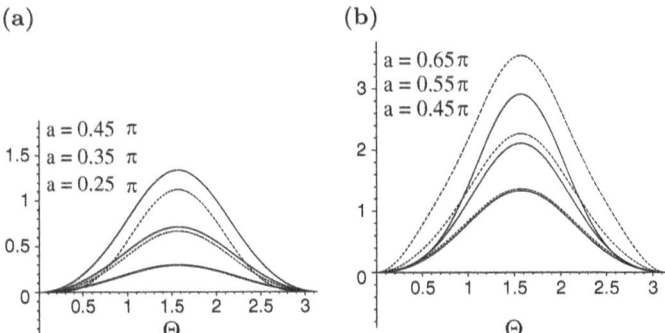

Fig. 2.51 Emission pattern: Comparison of the expansion in powers of a with exact values (full line). (**a**) Expansion up to a^6. (**b**) Expansion up to a^8

2.5.8.3 The Multipole Expansion

The starting point for the discussion of the emission problem is here the multipole expansion (2.78) of the vector potential in Chap. 2.3.4 (p. 114) in the radiation zone

$$A_{RZ}(r) = \frac{4\pi}{c} \frac{e^{ikr}}{r} \sum_{l,m} (-i)^l Y_{l,m}(\Omega) A_{l,m}(k) \qquad (2.152)$$

with the coefficients (2.77)

$$A_{l,m}(k) = \iiint dV' \, j_l(kr') Y^*_{l,m}(\Omega') \, \boldsymbol{j}(r') \,. \qquad (2.153)$$

Necessary properties of the spherical harmonics and the Bessel functions, as well as additional technical details are collected in Chap. 2.5.8.4.

It is necessary to express the current density (2.148) in terms of spherical coordinates, as the problem is now formulated in these coordinates. See 2.5.8.4.1, that the result is

$$\boldsymbol{j}(r) = \frac{i_0}{2\pi r^2} \sin\left(\frac{kd}{2} - k|z|\right) (\delta(\cos\theta - 1) - \delta(\cos\theta - 1)) \, \boldsymbol{e}_r$$

$$\left(|z| < \frac{d}{2}\right) \,.$$

The δ-functions restrict the charge density to the interval $[-d/2, d/2]$ on the z-axis. It vanishes on the outside of this interval. One has to calculate therefore

$$A_{l,m}(k) = \frac{i_0}{2\pi} \int_0^{d/2} dr' \int_0^{2\pi} d\varphi' \int_{-1}^1 d\cos\theta' \sin\left(\frac{kd}{2} - k|r'\cos\theta'|\right)$$

$$(\delta(\cos\theta' - 1) - \delta(\cos\theta' - 1))$$

$$\left(\sin\theta' \cos\varphi' \boldsymbol{e}_x + \sin\theta' \sin\varphi' \boldsymbol{e}_y + \cos\theta' \boldsymbol{e}_z\right)$$

$$Y^*_{l,m}(\Omega') j_l(kr') \,.$$

Integration over the angle θ should be executed first, so that one has

$$A_{l,m}(k) = \frac{i_0}{2\pi} \int_0^{d/2} dr' \int_0^{2\pi} d\varphi' \sin\left(\frac{kd}{2} - kr'\right)$$

$$\left(Y^*_{l,m}(0, \varphi') + Y^*_{l,m}(\pi, \varphi')\right) j_l(kr') \boldsymbol{e}_z \,.$$

2.5 Details

With the formulae given in 2.5.8.4.2 for the spherical harmonics and integration over φ' follows

$$A_{l,m}(k) = \begin{cases} i_0 \sqrt{\dfrac{(2l+1)}{\pi}} \delta_{m0} \left[\displaystyle\int_0^{d/2} dr' \sin\left(\dfrac{kd}{2} - kr'\right) j_l(kr') \right] e_z \\ 0 \qquad (l \text{ odd}) \end{cases}$$

for even (upper line) respectively for odd values of l. The remaining integral over the radial coordinate is treated with the substitution $t = kr'$. It depends only on the parameter $a = kd/2$

$$k\, I_l(a) = k \int_0^{d/2} dr' \sin\left(\dfrac{kd}{2} - kr'\right) j_l(kr')$$

$$= \int_0^a dt\, \sin(a-t) j_l(t)\,.$$

As the spherical Bessel functions have the structure

$$j_l(t) \longrightarrow \sin(t) \cdot \text{Poly}_1 + \cos(t) \cdot \text{Poly}_2\,,$$

where Poly_i are polynomials in $1/t$, which are balanced with respect to each other, one can show (see example in 2.5.8.4.2), that the radial integrals can be represented by the sine integral and the cosine integral

$$Si(x) = \int_0^x \dfrac{\sin t}{t}\, dt$$

$$Ci(x) = \int_0^x \dfrac{(\cos t - 1)}{t}\, dt + \gamma + \ln x$$

(γ is Euler's constant $\gamma = 0.5772156649\ldots$). As the values of these functions have to be extracted from appropriate Tables or one might evaluate the integrals I_l numerically. The corresponding functions $I_l(a)$ are illustrated in Fig. 2.52 for $l = 0, 2$. The final result for the expansion coefficients

$$A_{l,m}(k) = \dfrac{i_0}{k} \sqrt{\dfrac{(2l+1)}{\pi}} \delta_{m0} I_l(a)\, e_z$$

yields the multipole expansion of the vector potential

$$A_{RZ}(r) = \dfrac{2i_0}{c} \dfrac{e^{ikr}}{kr} \sum_{l=\text{even}} (-i)^l (2l+1)\, I_l(a)\, P_l(\cos\theta)\, e_z\,.$$

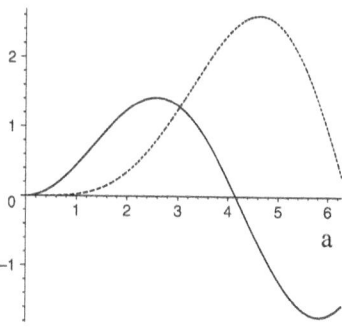

Fig. 2.52 The radial integral I_0 (full line) and I_2 (broken line) as functions of a

The fact, that a spherical harmonic with the index 0 corresponds to a Legendre polynomial is used in order to write this result. Comparison with (2.149) shows, that the statement

$$\frac{(\cos(a\cos\theta) - \cos a)}{\sin^2\theta} = \sum_{l=\text{even}} (-\mathrm{i})^l (2l+1) I_l(a) P_l(\cos\theta) \qquad (2.154)$$

is valid.

2.5.8.4 Technical Aspects Concerning the Multipole Expansion

This section contains some details of the necessary calculations and the properties of the special functions involved.

1. The equivalence of the representations of the charge density in Cartesian and spherical coordinates can be obtained by comparing the integrals

$$I_1 = \int_{-d/2}^{d/2} dz \iint dx\, dy\, f(x, y, |z|) \delta(x)\delta(y)$$

$$= \int_{-d/2}^{d/2} dz\, f(0, 0, |z|) = 2 \int_0^{d/2} dz\, f(0, 0, z)$$

and

$$I_2 = \frac{1}{2\pi} \int_0^{d/2} dr \int_0^{2\pi} d\varphi \int_{-1}^{1} d\cos\theta\, f(r\cos\varphi\sin\theta, r\sin\varphi\sin\theta, |r\cos\theta|)$$

$$\cos\theta(\delta(\cos\theta - 1) - \delta(\cos\theta - 1))$$

$$= \frac{1}{\pi} \int_0^{d/2} dr \int_0^{2\pi} d\varphi\, f(0, 0, r) = 2 \int_0^{d/2} dr\, f(0, 0, r)\ .$$

2.5 Details

2. The definition of the spherical harmonics is

$$Y_{l,m}(\theta, \varphi) = \left[\frac{(2l+1)(l-m)!}{4\pi(l+m)!} \right]^{1/2} P_{l,m}(\cos\theta)\, e^{im\varphi} ,$$

of the associated Legendre polynomials

$$P_{l,m}(z) = (-1)^m \left[1 - z^2\right]^{m/2} \frac{d^m}{dz^m} P_l(z) .$$

There follows

$$P_{l,m}(z = \pm 1) = \delta_{m0} P_l((z = \pm 1) = \delta_{m0}(\pm 1)^l$$

$$Y_{l,m}(\theta = 0, \varphi) = \left[\frac{2l+1}{4\pi}\right]^{1/2} \delta_{m0}$$

$$Y_{l,m}(\theta = \pi, \varphi) = \left[\frac{2l+1}{4\pi}\right]^{1/2} \delta_{m0}(-1)^l$$

$$Y_{l0}(\theta, \varphi) = \left[\frac{2l+1}{4\pi}\right]^{1/2} P_l(\cos\theta) .$$

The simplest Legendre polynomials are

$$P_0(\cos\theta) = 1 \qquad P_2(\cos\theta) = \frac{1}{2}\left(3\cos^2\theta - 1\right)$$

$$P_4(\cos\theta) = \frac{1}{8}\left(35\cos^4\theta - 30\cos^2\theta + 3\right) .$$

3. The definitions of the simplest spherical Bessel functions, needed here, are

$$j_0(t) = \frac{\sin t}{t} \qquad j_2(t) = \left(\frac{3}{t^3} - \frac{1}{t}\right)\sin t - \frac{3}{t^2}\cos t$$

$$j_4(t) = \left(\frac{105}{t^5} - \frac{45}{t^3} + \frac{1}{t}\right)\sin t - \left(\frac{105}{t^4} - \frac{10}{t^2}\right)\cos t .$$

4. The analytic evaluation of the integral

$$I_0(a) = \int_0^a dt\, \sin(a-t)\frac{\sin t}{t}$$

calls for the steps: The first sine function is resolved with the addition theorem

$$I_0(a) = \int_0^a dt \left\{ \sin a \left(\frac{\sin^2 t}{t} \right) - \cos a \left(\frac{\sin t \cos t}{t} \right) \right\},$$

the trigonometric functions of t are combined and the integration variable is renamed ($s = 2t$)

$$I_0(a) = \frac{1}{2} \int_0^{2a} ds \left\{ \sin a \left(\frac{\sin s}{s} \right) - \cos a \left(\frac{\cos s}{s} \right) \right\}.$$

One recognises the definition of the sine integral (Si) and the cosine integral (Ci) and writes

$$I_0(a) = \frac{1}{2} \left\{ \sin a \ Si(2a) - \cos a \ (Ci(2a) - \ln(2a) - \gamma) \right\}.$$

2.5.9 Calculation of the Liénard-Wiechert Fields

The solution of the inhomogeneous wave equation for a moving point charge is used for a wide range of applications, as e.g. for the discussion of brems- and synchroton radiation as well as for collision problems. In order to avoid calculational details in Chap. 2.4.1 the basic formulae for these potentials are treated here.

The situation at hand can be characterised in the following fashion: A point charge is moving on a trajectory $s(t')$. The electromagnetic fields generated by this charge are registered at the position r (Fig. 2.53). The finite time, that the electromagnetic radiation needs to travel from the point s to the point r causes the fact, that the radiation emitted at the time t' arrives only at a later time $t = t_{\text{ret}}(t')$.

Fig. 2.53 Vectors for the calculation of the electromagnetic fields of a moving point charge

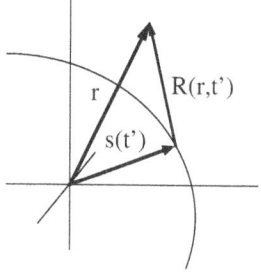

2.5 Details

For the calculation of the electromagnetic fields $E(r,t)$ and $B(r,t)$ exist two options

(a) Use the integral representations (2.84), (2.85) for the electromagnetic potentials (p. 121)

$$V^{(+)}(r,t) = \frac{q}{\varepsilon} \int dt' \frac{1}{R(r,t')} \delta\left(t' + \frac{R(r,t')}{c_{med}} - t\right)$$

$$A^{(+)}(r,t') = \frac{q\mu}{c} \int dt' \frac{v(t')}{R(r,t')} \delta\left(t' + \frac{R(r,t')}{c_{med}} - t\right)$$

(2.155)

with the difference vector (see Fig. 2.53)

$$R(r,t') = r - s(t')$$

and determine the fields with the relations

$$E(r,t) = -\nabla V(r,t) - \frac{1}{c} \frac{\partial A^{(+)}(r,t)}{\partial t}$$

$$B(r,t) = \nabla \times A^{(+)}(r,t).$$

(2.156)

(b) Start with the Liénard-Wiechert potentials (2.88), (2.89)

$$V(r,t) = \frac{q}{\varepsilon} \left[\frac{1}{R(r,t') - R(r,t') \cdot v(t')/c_{med}} \right]_{t'=t_{ret}(t)}$$

$$A^{(+)}(r,t) = \frac{q\mu}{c} \left[\frac{v(t')}{R(r,t') - R(r,t') \cdot v(t')/c_{med}} \right]_{t'=t_{ret}(t)}$$

(2.157)

and calculate the fields with the relations (2.156) as in option (a). The times t and t' are related by

$$t' + \frac{R(r,t')}{c_{med}} - t = 0.$$

(2.158)

Resolution leads to $t' = t_{ret}(t)$.

The two options must give the same result. As, however, each of the calculations has its peculiarities, it is worthwhile to consider both of them. In both cases it is worthwhile to introduce the abbreviations

$$n(r, t') \equiv n(t') = \frac{R(r, t')}{R((r, t')}$$

$$\beta(t') = \frac{v(t')}{c_{med}} \equiv \frac{\dot{s}(t')}{c_{med}} \quad (2.159)$$

$$g(r, t') \equiv g(t') = 1 - n(r, t') \cdot \beta(t') .$$

The notation in the functions n and $g(n)$ implies the time argument t' instead of r, t'. At a later stage all arguments will be suppressed in order to keep the equations sufficiently simple.

2.5.9.1 The Electric Field with Method (a)

Application of the relation (2.156) for the calculation of the electric field requires the gradients and the time derivatives of the potentials. For this step one assumes, that differentiation and integration can be interchanged. The gradient operator acts on the denominator of the integrand of the representation of the scalar potential and on the δ-function. One finds

$$\nabla V^{(+)}(r, t) = \frac{q}{\varepsilon} \int dt' \left\{ -\frac{n(t')}{R(t')^2} \delta \left(t' + \frac{R(t')}{c_{med}} - t \right) \right.$$

$$\left. + \frac{n(t')}{c_{med} R(t')} \delta^{(1)} \left(t' + \frac{R(t')}{c_{med}} - t \right) \right\} ,$$

where $\delta^{(1)}$ is the derivative of the δ-function with respect to the complete argument

$$\delta^{(1)}(x) = \frac{\partial}{\partial x} \delta(x)$$

and where the gradient of the function $|R(t')|$

$$|R(t')| = [(x - s_x(t'))^2 + (y - s_y(t'))^2 + (z - s_z(t'))^2]^{1/2}$$

corresponds to the n

$$\nabla R(t') = n(t') . \quad (2.160)$$

The time derivative of the vector potential is

$$\frac{\partial}{\partial t} A^{(+)}(r, t') = -\frac{q\mu}{c} \int dt' \frac{v(t')}{R(t')} \delta^{(1)} \left(t' + \frac{R(t')}{c_{med}} - t \right) .$$

2.5 Details

The two derivatives can be combined to give the expression for the electric field

$$E(r, t) = -\nabla V(r, t) - \frac{1}{c}\frac{\partial A^{(+)}(r, t)}{\partial t}$$

$$= \frac{q}{\varepsilon}\int dt' \left\{ \frac{n(t')}{R(t')^2}\delta\left(t' + \frac{R(t')}{c_{med}} - t\right) \right.$$

$$\left. + \frac{(\boldsymbol{\beta}(t') - \boldsymbol{n}(t'))}{c_{med}R(t')}\delta^{(1)}\left(t' + \frac{R}{c_{med}} - t\right)\right\} .$$

In order to evaluate the integrals involved, the substitution

$$f(t') = t' + \frac{R(t')}{c_{med}} \tag{2.161}$$

can be used. The derivative of this function of t' (the coordinates r in the function $R(r, t)$ have to be considered as a parameter)

$$\frac{df}{dt'} = \left(1 - \frac{\boldsymbol{n}(t') \cdot \boldsymbol{v}(t')}{c_{med}}\right) = (1 - \boldsymbol{n}(t') \cdot \boldsymbol{\beta}(t'))$$

can be used to provide the derivative of the inverse function, respectively

$$dt' = \frac{df}{g(t')} .$$

The resulting expression for the electric field

$$E(r, t) = \frac{q}{\varepsilon}\int \frac{df}{g(t')} \left\{ \frac{n(t')}{R(t')^2}\delta(f - t) + \frac{(\boldsymbol{\beta}(t') - \boldsymbol{n}(t'))}{c_{med}R(t')}\delta^{(1)}(f - t)\right\}$$

can be integrated directly

$$= \frac{q}{\varepsilon}\left\{\frac{n(t')}{g(t')R(t')^2} - \frac{1}{c_{med}}\frac{d}{df}\left(\frac{\boldsymbol{\beta}(t') - \boldsymbol{n}(t')}{g(t')R(t')}\right)\right\}\bigg|_{t'=t_{ret}(t)} .$$

In order to represent the time t' by the time t, it is necessary to resolve the equation $f(t') - t = t' - R(t')/c_{med} - t = 0$ with respect to t'.

For the calculation of the derivative with respect to the 'variable' f one relies on the chain rule

$$\frac{d}{df} = \frac{d}{dt'}\frac{dt'}{df} = \frac{1}{g(t')}\frac{d}{dt'}$$

and obtains

$$E(r,t) = \frac{q}{\varepsilon}\left\{\frac{n(t')}{g(t')R(t')^2} - \frac{1}{c_{med}g(t')}\frac{d}{dt'}\left(\frac{\beta(t')-n(t')}{g(t')R(t')}\right)\right\}\Bigg|_{t'=t_{ret}(t)}.$$

This compact result has to be processed further. As the individual results become rather lengthy, it is useful to suppress the variable t'. One needs the derivatives

- of the unit vector n

$$\frac{1}{c_{med}}\frac{d}{dt'}n = \frac{1}{c_{med}}\frac{d}{dt'}\frac{R}{R} = \frac{1}{c_{med}}\left\{-\frac{v}{R} + \frac{1}{R^2}(n\cdot v)R\right\}$$

$$= -\frac{\beta}{R} + \frac{1}{R}(n\cdot\beta)n = \frac{1}{R}(n\times(n\times\beta)). \quad (2.162)$$

This result arises from the decomposition of the double vector product

$$(n\times(n\times\beta)) = (n\cdot\beta)n - (n\cdot n)\beta = (n\cdot\beta)n - \beta,$$

- and of the product gR

$$\frac{1}{c_{med}}\frac{d}{dt'}(gR) = \frac{1}{c_{med}}\frac{d}{dt'}(R - R\cdot\beta) = \frac{1}{c_{med}}\left(-n\cdot v + v\cdot\beta - R\cdot\dot\beta\right)$$

$$= -n\cdot\beta + \beta^2 - \frac{R}{c_{med}}(n\cdot\dot\beta). \quad (2.163)$$

In the expression for E, in which the derivatives are spread out

$$E(r,t) = \frac{q}{\varepsilon}\left\{\frac{n}{gR^2} + \frac{n}{c_{med}g}\frac{d}{dt'}\left(\frac{1}{gR}\right) + \frac{1}{c_{med}g^2R}\frac{d}{dt'}n\right.$$

$$\left. - \frac{1}{c_{med}g}\frac{d}{dt'}\left(\frac{\beta}{gR}\right)\right\}\Bigg|_{t'=t_{ret}(t)},$$

one has to insert the derivatives of n and obtains

$$E(r,t) = \frac{q}{\varepsilon}\left\{\frac{n}{gR^2} + \frac{n}{c_{med}g}\frac{d}{dt'}\left(\frac{1}{gR}\right) + \frac{1}{g^2R}\left(-\frac{\beta}{R} + \frac{1}{R}(n\cdot\beta)n\right)\right.$$

$$\left. - \frac{1}{c_{med}g}\frac{d}{dt'}\left(\frac{\beta}{gR}\right)\right\}\Bigg|_{t'=t_{ret}(t)}. \quad (2.164)$$

2.5 Details

The first term and part of the third term in the bracket can be combined

$$\frac{n}{g^2 R^2}(g + n \cdot \beta) = \frac{n}{g^2 R^2}(1 - n \cdot \beta + n \cdot \beta) = \frac{n}{g^2 R^2}.$$

In the next step one inserts the derivative (2.163) of gR

$$E(r,t) = \frac{q}{\varepsilon} \left\{ \frac{(n-\beta)}{g^2 R^2} - \frac{n}{g}\left(\beta^2 - (n \cdot \beta) - \frac{R}{c_{med}}(n \cdot \dot{\beta})\right) \frac{1}{(gR)^2} \right.$$

$$-\frac{1}{g}\frac{\dot{\beta}}{c_{med} g R}$$

$$\left. + \frac{\beta}{g}\left(\beta^2 - (n \cdot \beta) - \frac{R}{c_{med}}(n \cdot \dot{\beta})\right)\frac{1}{(gR)^2} \right\}\Bigg|_{t'=t_{ret}(t)}$$

and sorts

$$E(r,t) = \frac{q}{\varepsilon}\left\{(n-\beta)\left(\frac{1}{g^2 R^2} - \frac{\beta^2}{g^3 R^2} + \frac{(n \cdot \beta)}{g^3 R^2}\right)\right.$$

$$\left. + \frac{(n-\beta)}{c_{med} g^3 R}(n \cdot \dot{\beta}) - \frac{\dot{\beta}}{c_{med} g^2 R}\right\}\Bigg|_{t'=t_{ret}(t)}.$$

Once more one can combine some of the terms, as two expressions each of the first and the second term

$$\frac{1}{g^2 R^2} + \frac{(n \cdot \beta)}{g^3 R^2} = \frac{1}{g^3 R^2}(1 - n \cdot \beta + n \cdot \beta) = \frac{1}{g^3 R^2}.$$

One finds then

$$E(r,t) = \frac{q}{\varepsilon}\left\{(n-\beta)(1-\beta^2)\frac{1}{g^3 R^2}\right. \tag{2.165}$$

$$\left. + \frac{1}{c_{med} g^3 R}\left((n \cdot \dot{\beta})(n-\beta) - \dot{\beta}(1 - n \cdot \beta)\right)\right\}\Bigg|_{t'=t_{ret}(t)}.$$

The factor in the second term can be written in the form of a double vector product

$$n \times \left[(n-\beta) \times \dot{\beta}\right] = (n \cdot \dot{\beta})n - \dot{\beta} - (n \cdot \dot{\beta})\beta + (n \cdot \beta)\dot{\beta}.$$

All the quantities (β, n, R, g) in the final result (2.165) are after the replacement of t' by t_{ret} functions of r and t.

2.5.9.2 The Electric Field via Method (b)

Starting point is in this case the relations (2.157), abbreviated in the form

$$V^{(+)}(r,t) = \frac{q}{\varepsilon} \left(\frac{1}{R(t')g(t')} \right) \Big|_{t'=t_{\text{ret}}(t)}$$

$$A^{(+)}(r,t') = \frac{q\mu}{c} \left(\frac{v(t')}{R(t')g(t')} \right) \Big|_{t'=t_{\text{ret}}(t)} .$$

The calculation of the electromagnetic fields can be done directly with application of (2.156). For the calculation of the gradient of the scalar potential $\nabla V^{(+)}(r,t)$ one has to keep in mind, that the retarded time is a function of t and of r

$$t' = t_{\text{ret}}(t, r) .$$

This dependence is not applied for the calculation of the electric field with the Kirchhoff representation, as t' is a variable, with respect to which one has to differentiate before one substitutes $t' = t_{\text{ret}}$. As the retarded time depends on the position, one has in this case

$$\nabla V^{(+)}(r,t) = \frac{q}{\varepsilon} \left\{ \nabla_r \left(\frac{1}{Rg} \right) + \frac{\partial}{\partial t'} \left(\frac{1}{Rg} \right) \nabla_r t'(t,r) \right\} .$$

The derivatives needed here can be calculated directly:

- The gradient of $(Rg)^{-1}$ (compare (2.156))

$$\nabla_r \left(\frac{1}{Rg} \right) = \nabla_r \frac{1}{(R - R \cdot \beta)} = -\frac{1}{R^2 g^2} (\nabla_r R - \nabla_r (R \cdot \beta))$$

$$= -\frac{1}{R^2 g^2} (n - \beta) . \tag{2.166}$$

- The time derivative (with respect to t') of $(Rg)^{-1}$

$$\frac{\partial}{\partial t'} \left(\frac{1}{Rg} \right) = -\frac{1}{R^2 g^2} \left\{ \frac{\partial}{\partial t'} R - \frac{\partial}{\partial t'} (R \cdot \beta) \right\}$$

$$= \frac{1}{R^2 g^2} \left\{ n \cdot v - v \cdot \beta + R \cdot \dot{\beta} \right\} . \tag{2.167}$$

- In order to calculate the gradient of t'

$$\nabla_r (t') = \left(\partial_x t', \partial_y t', \partial_z t' \right)$$

2.5 Details

one has to start with the total differential of the implicit function

$$F(t, t', \boldsymbol{r}) = t' + \frac{R}{c_{med}} - t = 0 .$$

As one needs the *partial* derivatives of t' with respect to the coordinates

$$dt' + \frac{1}{c_{med}} \left(\nabla R \cdot d\boldsymbol{r} + \frac{\partial R}{\partial t'} dt' \right) - dt = 0 \qquad (2.168)$$

one sets $dt = 0$ and obtains e.g. for the derivative with respect to the x-coordinate (n_x is the x-component of $\boldsymbol{n}$)

$$\frac{\partial t'}{\partial x} = -\frac{1}{c_{med} g} \frac{\partial R}{\partial x} = -\frac{n_x}{c_{med} g} .$$

Corresponding results can be obtained for the derivatives with respect to the other coordinates, so that

$$\nabla_r t' = -\frac{1}{c_{med} g} \boldsymbol{n} \qquad (2.169)$$

follows.

The combination of these separate statements leads to the gradient of the scalar potential

$$-\nabla V^{(+)}(\boldsymbol{r}, t) = \frac{q}{\varepsilon} \left\{ \frac{1}{R^2 g^2} (\boldsymbol{n} - \boldsymbol{\beta}) \right.$$

$$\left. + \frac{1}{R^2 g^3} \left\{ (\boldsymbol{n} \cdot \boldsymbol{\beta}) \boldsymbol{n} - \beta^2 \boldsymbol{n} + \frac{R}{c_{med} g} (\boldsymbol{n} \cdot \dot{\boldsymbol{\beta}}) \boldsymbol{n} \right\} \right\}_{t' = t_{ret}(t)} ,$$

respectively after an extension of the first term with g and sorting

$$-\nabla V^{(+)}(\boldsymbol{r}, t) = \frac{q}{\varepsilon} \frac{1}{R^2 g^3} \left\{ ((1 - \beta^2) \boldsymbol{n} \right.$$

$$\left. -(1 - (\boldsymbol{n} \cdot \boldsymbol{\beta})) \boldsymbol{\beta} + \frac{R}{c_{med} g} (\boldsymbol{n} \cdot \dot{\boldsymbol{\beta}}) \boldsymbol{n} \right\}_{t' = t_{ret}(t)} . \qquad (2.170)$$

For the calculation of the time derivative (with respect to t) of the vector potential with the chain rule

$$-\frac{1}{c} \frac{\partial}{\partial t} \boldsymbol{A} = -\frac{1}{c} \frac{\partial \boldsymbol{A}}{\partial t'} \frac{\partial t'}{\partial t}$$

one needs the partial derivative of t' with respect to t. The total differential (2.168) gives in this case (set $d\mathbf{r} = 0$)

$$\frac{\partial t'}{\partial t} = \frac{1}{g}.$$

The second factor, the partial derivative of the vector potential with respect to t' can be determined with (2.167) and the relation $\mathbf{R} = R\,\mathbf{n}$

$$-\frac{1}{c}\frac{\partial}{\partial t'}\mathbf{A} = \frac{q}{\varepsilon c_{\text{med}}^2}\left\{\frac{\dot{\mathbf{v}}}{Rg} + \mathbf{v}\frac{1}{R^2 g^2}\right.$$

$$\left.\Big((\mathbf{n}\cdot\mathbf{v}) - (\mathbf{v}\cdot\boldsymbol{\beta}) + R(\mathbf{n}\cdot\dot{\boldsymbol{\beta}})\Big)\right\}\bigg|_{t'=t_{\text{ret}}(t)}. \qquad (2.171)$$

The prefactor in (2.171) is composed from the individual factors according to

$$-\frac{1}{c}\frac{q\mu}{c} = \frac{q}{\varepsilon c_{\text{med}}^2}.$$

With the intermediates results (2.170) and (2.171) one obtains for the electric field

$$\mathbf{E}(\mathbf{r},t) = \frac{q}{\varepsilon}\left\{\frac{1}{g^3 R^2}\left\{(1-\beta^2)\mathbf{n} - (1-\mathbf{n}\cdot\boldsymbol{\beta})\boldsymbol{\beta} + \frac{1}{c_{\text{med}}Rg^3}(\mathbf{n}\cdot\dot{\boldsymbol{\beta}})\mathbf{n}\right.\right.$$

$$-\frac{1}{c_{\text{med}}}\frac{\dot{\boldsymbol{\beta}}}{Rg^2} - \frac{1}{g^3 R^2}(\mathbf{n}\cdot\boldsymbol{\beta})\dot{\boldsymbol{\beta}} + \frac{1}{g^3 R^2}\beta^2\boldsymbol{\beta}$$

$$\left.\left.-\frac{1}{c_{\text{med}}}\frac{1}{g^2 R}(\mathbf{n}\cdot\dot{\boldsymbol{\beta}})\boldsymbol{\beta}\right\}\right\}\bigg|_{t'=t_{\text{ret}}(t)},$$

which can be sorted in the form

$$\mathbf{E}(\mathbf{r},t) = \frac{q}{\varepsilon}\left\{\frac{1}{g^3 R^2}(\mathbf{n}-\boldsymbol{\beta})(1-\beta^2) + \frac{1}{c_{\text{med}}g^3 R}\left\{(\mathbf{n}\cdot\dot{\boldsymbol{\beta}})\mathbf{n}\right.\right.$$

$$\left.\left.-(\mathbf{n}\cdot\dot{\boldsymbol{\beta}})\boldsymbol{\beta} - \dot{\boldsymbol{\beta}}(1-(\mathbf{n}\cdot\boldsymbol{\beta}))\right\}\right\}\bigg|_{t'=t_{\text{ret}}(t)}. \qquad (2.172)$$

The result (2.172) agrees with the previous result (2.165).

2.5.9.3 Calculation of the Magnet Field

For the calculation of the magnet field only the implementation of the integral representation will be considered. Starting point is the relation (2.156)

$$\begin{aligned}
\boldsymbol{B}(\boldsymbol{r},t) &= \nabla \times \boldsymbol{A}^{(+)}(\boldsymbol{r},t) \\
&= \frac{q\mu}{c} \nabla \times \int dt' \frac{\boldsymbol{v}(t')}{R(\boldsymbol{r},t')} \delta\left(t' + \frac{R(\boldsymbol{r},t')}{c_{med}} - t\right) \\
&= \frac{q}{\varepsilon} \nabla \times \int dt' \frac{\boldsymbol{\beta}(t')}{R(\boldsymbol{r},t')} \delta\left(t' + \frac{R(\boldsymbol{r},t')}{c_{med}} - t\right).
\end{aligned}$$

Then the nabla operator acts only on the function $R(\boldsymbol{r},t')$. With the partial derivative (R_x is the x-component of $\boldsymbol{R}$, compare (2.160))

$$\frac{\partial R}{\partial x} = \frac{R_x}{R} = n_x \quad \text{etc.}$$

one finds with the first step

$$\boldsymbol{B}(\boldsymbol{r},t) = \frac{q}{\varepsilon} \int dt' (\boldsymbol{n} \times \boldsymbol{\beta}) \left\{ -\frac{1}{R^2} \delta\left(t' + \frac{R(\boldsymbol{r},t')}{c_{med}} - t\right) \right.$$

$$\left. + \frac{1}{c_{med} R} \delta^{(1)}\left(t' + \frac{R(\boldsymbol{r},t')}{c_{med}} - t\right) \right\},$$

a result, that can also be treated with the substitution (2.161)

$$f(t') = t' + \frac{R(t')}{c_{med}}$$

$$\boldsymbol{B}(\boldsymbol{r},t) = \frac{q}{\varepsilon} \int \frac{df}{g} (\boldsymbol{n} \times \boldsymbol{\beta}) \left\{ -\frac{1}{R^2} \delta(f-t) + \frac{1}{c_{med} R} \delta^{(1)}(f-t) \right\}.$$

The integration can be executed

$$\boldsymbol{B}(\boldsymbol{r},t) = \frac{q}{\varepsilon} \left[\frac{1}{gR^2}(\boldsymbol{n} \times \boldsymbol{\beta}) - \frac{1}{c_{med}g} \frac{d}{dt'}\left(\frac{(\boldsymbol{n} \times \boldsymbol{\beta})}{gR}\right) \right]_{t'=t_{ret}(t)}$$

and the consecutive differentiation can be prepared

$$B(r,t) = \frac{q}{\varepsilon}\left[-\frac{(n\times\beta)}{gR^2} - \frac{1}{c_{\text{med}}g}n\times\frac{d}{dt'}\left(\frac{\beta}{gR}\right)\right.$$
$$\left.-\frac{1}{g}\left(\frac{1}{c_{\text{med}}}\frac{d}{dt'}n\right)\times\left(\frac{\beta}{gR}\right)\right]_{t'=t_{\text{ret}}(t)}.$$

First one differentiates n in the last term using (2.162) and resolves the triple vector product

$$(n \times (n \times \beta)) \times \beta = \{(n \cdot \beta) - \beta\} \times \beta = (n \cdot \beta)(n \times \beta),$$

sorts the intermediate result

$$B(r,t) = \frac{q}{\varepsilon}\left[-\frac{(n\times\beta)}{gR^2}\right.$$
$$\left. -\frac{1}{c_{\text{med}}g}n\times\frac{d}{dt'}\left(\frac{\beta}{gR}\right) - \frac{1}{g^2R^2}(n\cdot\beta)(n\times\beta)\right]_{t'=t_{\text{ret}}(t)}$$

by extension of the first term

$$B(r,t) = \frac{q}{\varepsilon}\left[\frac{1}{g^2R^2}(n\times\beta)[-1+(n\cdot\beta)-(n\times\beta)]\right.$$
$$\left.-\frac{1}{c_{\text{med}}g}n\times\frac{d}{dt'}\left(\frac{\beta}{gR}\right)\right]_{t'=t_{\text{ret}}(t)}$$

in the form

$$B(r,t) = \frac{q}{\varepsilon}\left[\frac{1}{g^2R^2}(\beta\times n) + \frac{1}{c_{\text{med}}g}\frac{d}{dt'}\left(\frac{\beta}{gR}\right)\times n\right]_{t'=t_{\text{ret}}(t)}. \quad (2.173)$$

A comparison with the intermediate result (2.164) for the electric field shows, that the electric and the magnetic Liénard-Wiechert field are related by

$$B(r,t) = n \times E(r,t). \quad (2.174)$$

2.5 Details

In order to show the magnetic induction explicitly one evaluates the remaining differentiation using the derivative (2.166) of $(gR)^{-1}$ and finds

$$B(r,t) = \frac{q}{\varepsilon}\left[-\frac{(\boldsymbol{\beta}\times\boldsymbol{n})(1-\beta^2)}{g^3 R^2}\right.$$

$$\left.+\frac{1}{c_{\text{med}}g^3 R}\left[(\boldsymbol{\beta}\times\boldsymbol{n})(\boldsymbol{n}\cdot\dot{\boldsymbol{\beta}})+(\dot{\boldsymbol{\beta}}\times\boldsymbol{n})(1-(\boldsymbol{n}\cdot\boldsymbol{\beta}))\right]\right]_{t'=t_{\text{ret}}(t)}$$

or, after a reformulation with $g = 1 - (\boldsymbol{n}\cdot\boldsymbol{\beta})$,

$$B(r,t) = \frac{q}{\varepsilon}\left[\frac{(1-\beta^2)}{g^3 R^2}(\boldsymbol{\beta}\times\boldsymbol{n})\right.$$

$$\left.+\frac{1}{c_{\text{med}} R}\left\{\frac{(\boldsymbol{n}\cdot\dot{\boldsymbol{\beta}})}{g^3}(\boldsymbol{\beta}\times\boldsymbol{n})+\frac{1}{g^2}(\dot{\boldsymbol{\beta}}\times\boldsymbol{n})\right\}\right]_{t'=t_{\text{ret}}(t)}. \quad (2.175)$$

2.5.9.4 Properties of the Electromagnetic Fields

Some properties of the fields, which have been calculated, are listed here

- The electric field is perpendicular to the vector $\boldsymbol{n}$. Explicit evaluation of the scalar product using the result (2.165) gives

$$\boldsymbol{n}\cdot\boldsymbol{E} = \frac{q}{\varepsilon}\left[\frac{(1-\beta^2)}{g^2 R^2}+\frac{1}{c_{\text{med}} R g^3}\left[(\boldsymbol{n}\cdot\dot{\boldsymbol{\beta}})-(\boldsymbol{\beta}\cdot\boldsymbol{n})(\dot{\boldsymbol{\beta}}\cdot\boldsymbol{n})\right.\right.$$

$$\left.\left.-(\boldsymbol{n}\cdot\dot{\boldsymbol{\beta}})+(\boldsymbol{\beta}\cdot\boldsymbol{n})(\dot{\boldsymbol{\beta}}\cdot\boldsymbol{n})\right]\right]_{t'=t_{\text{ret}}(t)}$$

$$= \frac{q}{\varepsilon}\frac{1}{R^2 g^2}(1-\beta^2).$$

- The magnetic induction is perpendicular to the vector $\boldsymbol{n}$. This statement follows from the cyclic properties of scalar triple product (Fig. 2.54)

$$(\boldsymbol{n}\cdot\boldsymbol{B}) = (\boldsymbol{n}\cdot(\boldsymbol{n}\times\boldsymbol{E})) = \boldsymbol{E}\cdot(\boldsymbol{n}\times\boldsymbol{n}) = 0.$$

2.5.10 The Bremsstrahlung

The Larmor formula, which describes the angular distribution of the bremsstrahlung in the non-relativistic regime has to be modified if a relativistic point charge is decelerated. The starting point for the discussion is in this case Eq. (2.104) in

Fig. 2.54 The Liénard-Wiechert fields

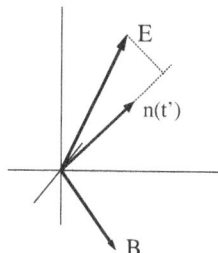

Chap. 2.4.2 (p. 127)

$$\frac{dP(r,t')}{d\Omega} = \frac{q^2}{4\pi c\,\varepsilon} \frac{\left(\boldsymbol{n}(t') \times \left[(\boldsymbol{n}(t') - \boldsymbol{\beta}(t')) \times \dot{\boldsymbol{\beta}}(t')\right]\right)^2}{(1 - \boldsymbol{n}(r,t') \cdot \boldsymbol{\beta}(t'))^5}, \qquad (2.176)$$

which will be evaluated for a linear and for a circular accelerator. The quantities involved are

$$\boldsymbol{\beta} = \frac{v}{c_{med}} = \sqrt{\frac{\varepsilon\mu}{c}}\,v \qquad \boldsymbol{n} = \frac{\boldsymbol{R}}{R}.$$

2.5.10.1 Linear Accelerator

The velocity vector and the vector for the acceleration are parallel in this case, so that one finds with a suitable choice of the coordinate systems e.g.

$$\boldsymbol{\beta} = (0, 0, \beta) \qquad \dot{\boldsymbol{\beta}} = (0, 0, \dot{\beta}).$$

If the components of the unit vector $\boldsymbol{n}$ are characterised by spherical coordinates

$$\boldsymbol{n} = (n_1, n_2, n_3) = (\cos\varphi\sin\theta,\ \sin\varphi\sin\theta,\ \cos\theta), \qquad (2.177)$$

one finds for the vector product in the numerator the equation (2.176), as for a linear accelerator $(\boldsymbol{\beta} \times \dot{\boldsymbol{\beta}}) = \mathbf{0}$ is valid

$$\boldsymbol{Z} = \boldsymbol{n} \times (\boldsymbol{n} \times \dot{\boldsymbol{\beta}}) = \dot{\beta}\left((n_1 n_3)\boldsymbol{e}_1 + (n_2 n_3)\boldsymbol{e}_2 - (n_1^2 + n_2^2)\boldsymbol{e}_3\right),$$

and for the square of this vector

$$(\boldsymbol{Z})^2 = \dot{\beta}^2(n_1^2 + n_2^2)(n_1^2 + n_2^2 + n_3^2) = \dot{\beta}^2 \sin^2\theta = \frac{\varepsilon\mu\dot{v}^2}{c^2}\sin^2\theta.$$

2.5 Details

Together with the obvious result for the denominator this result leads to Eq. (2.105)

$$\frac{dP(r,t')}{d\Omega} = \frac{\mu q^2}{4\pi c^3} \dot{v}^2(t') \frac{\sin^2\theta(t')}{(1-\beta(t')\cos\theta(t'))^5}$$

for the angular distribution with respect to the direction of the beam.

For the calculation of the total per unit of time emitted power one has to evaluate the integral in

$$P(t') = \frac{\mu q^2}{2c^3} \dot{v}^2 \int_0^\pi \frac{\sin^3\theta}{(1-\beta\cos\theta)^5} d\theta = \frac{\mu q^2}{2c^3} \dot{v}^2 \, I \qquad (2.178)$$

(the integration over φ gives the factor 2π).

The explicit evaluation involves the steps: A simpler denominator is obtained with the substitution

$$x = 1 - \beta\cos\theta \qquad dx = \beta\sin\theta \, d\theta \, .$$

It follows

$$\sin\theta = \sqrt{1-\cos^2\theta} = \left[1 - \left(\frac{(1-x)}{\beta}\right)^2\right]^{1/2} = \frac{[\beta^2 - (1-x)^2]^{1/2}}{\beta} \, ,$$

so that the integral I

$$I = \int_{1-\beta}^{1+\beta} \frac{\beta^2 - (1-x)^2}{\beta^3 x^5} \, dx \, .$$

Using partial integration or a Table of Integrals one obtains

$$\int \frac{R}{x^{m+1}} \, dx = -\frac{R^2}{a\,m\,x^m} + \frac{b(2-m)}{a\,m} \int \frac{R}{x^m} \, dx + \frac{c(4-m)}{a\,m} \int \frac{R}{x^{m-1}} \, dx$$

with $\quad R = a + bx + cx^2$

or the specific integral with

$$a = \beta^2 - 1 \qquad b = 2 \qquad c = -1$$

and the maximal m-value $m = 4$, so that one has to consider the individual integrals

$m = 4$

$$I = I_4 = \int \frac{R}{x^5} \, dx = -\frac{R^2}{4ax^4} - \frac{1}{a}\int \frac{R}{x^4} \, dx$$

$m = 3$

$$I_3 = \int \frac{R}{x^4}\,dx = -\frac{R^2}{3ax^3} - \frac{2}{3a}\int \frac{R}{x^3}\,dx - \frac{1}{3a}\int \frac{R}{x^2}\,dx$$

$m = 2$

$$I_2 = \int \frac{R}{x^3}\,dx = -\frac{R^2}{2ax^2} - \frac{1}{a}\int \frac{R}{x}\,dx$$

$m = 1$

$$I_1 = \int \frac{R}{x^2}\,dx = -\frac{R^2}{ax} + \frac{2}{a}\int \frac{R}{x}\,dx - \frac{3}{a}\int R\,dx\ .$$

If one inserts in the function $R(x)$ the values at the limits $x = 1 \pm \beta$, one finds $R(x = 1 \pm \beta) = 0$. The recursion for the integrals can be solved simply

$$I = -\frac{I_3}{a} = \frac{(2I_2 + I_1)}{3a^2} = \frac{1}{3a^3}(-2I_0 + 2I_0 - 3\int_{1-\beta}^{1+\beta} R(x)\,dx)$$

$$= -\frac{1}{a^3}\int_{1-\beta}^{1+\beta} R(x)\,dx\ .$$

The integral with $R(x)$ is elementary and gives

$$I = -\frac{1}{a^3}\left[ax + x^2 - \frac{1}{3}x^3\right]_{1-\beta}^{1+\beta}$$

$$= -\frac{1}{a^3}\left(2\beta^3 - \frac{2}{3}\beta^3\right) = -\frac{4}{3}\frac{\beta^3}{(\beta^2 - 1)^3}\ .$$

If this result is inserted into Eq. (2.178), one finds the result quoted in Chap. 2.4.2

$$P(t') = \frac{2}{3}\frac{\mu q^2}{c^3}\dot{v}^2(t')\frac{1}{(1-\beta^2)^3}\ .$$

2.5.10.2 Circular Accelerator

The velocity vector (tangential to the circle) and the acceleration vector are perpendicular, so that one can use an *instantaneous* coordinate system

$$\boldsymbol{\beta}(t') = (0, 0, \beta(t')) \qquad \dot{\boldsymbol{\beta}}(t') = (\dot{\beta}(t'), 0, 0)$$

(see Fig. 2.55). The vector $\boldsymbol{n}(t')$ is given, as before by Eq. (2.177). For the calculation of the numerator in (2.176) one evaluates once more a multiple vector

2.5 Details

Fig. 2.55 *Instantaneous coordinate system*

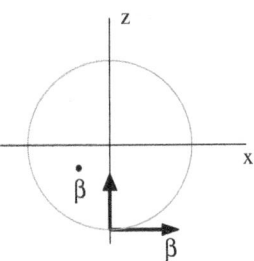

product

$$Z = n \times ((n - \beta) \times \dot{\beta})$$

$$= \left(-n_2^2 \dot{\beta} - n_3 \dot{\beta}(n_3 - \beta),\ \dot{\beta} n_1 n_2,\ n_1 \dot{\beta}(n_3 - \beta)\right)$$

and calculates the square

$$Z^2 = \dot{\beta}^2 \left\{ (n_2^4 + 2n_2^2 n_3^2 + n_3^4 + n_1^2 n_2^2 + n_1^2 n_3^2) \right.$$
$$\left. - \beta(2n_2^2 n_3 + 2n_3^3 + 2n_1^2 n_3) + \beta^2(n_3^2 + n_1^2) \right\}.$$

For the individual contributions one has

$$n_3^2 + n_1^2 = \cos^2 \theta + \cos^2 \varphi \sin^2 \theta,$$

$$(2n_2^2 n_3 + 2n_3^3 + 2n_1^2 n_3) = 2n_3(n_2^2 + n_3^2 + n_1^2) = 2\cos\theta,$$

$$n_2^4 + 2n_2^2 n_3^2 + n_3^4 + n_1^2 n_2^2 + n_1^2 n_3^2 = (n_2^2 + n_3^2)(n_2^2 + n_3^2 + n_1^2)$$
$$= \sin^2 \varphi \sin^2 \theta + \cos^2 \theta.$$

The total term is therefore

$$Z^2 = \dot{\beta}^2 \left\{ (1 - \beta \cos\theta)^2 + \sin^2 \theta \cos^2 \varphi (\beta^2 - 1) \right\},$$

so that one finds

$$\dot{\beta} = \frac{\dot{v}}{c_{\text{med}}} = \frac{\dot{v}\sqrt{\mu\varepsilon}}{c}$$

for the instantaneous angular distribution (compare Chap. 2.4.2, (2.106))

$$\frac{dP(r,t')}{d\Omega} = \frac{\mu q^2}{4\pi c^3}\dot{v}^2\frac{1}{(1-\beta\cos\theta)^3}\left[1 - \frac{(1-\beta^2)\sin^2\theta\cos^2\varphi}{(1-\beta\cos\theta)^2}\right].$$

The calculation of the per unit time emitted power requires two steps. After integration over the instantaneous angle ϕ with

$$\int_0^{2\pi}\cos^2\phi\,d\phi = \pi$$

there remains the integral

$$I = \int_{-1}^{1}\frac{1}{(1-\beta\cos\theta)^3}\left[1 - \frac{(1-\beta^2)\sin^2\theta}{2(1-\beta\cos\theta)^2}\right]d\cos\theta,$$

which can be decomposed into three individual integrals

$$I = I_1 - \frac{(1-\beta^2)}{2}I_2 + \frac{(1-\beta^2)}{2}I_3$$

with

$$I_1 = \int_{-1}^{1}\frac{1}{(1-\beta x)^3}\,dx \qquad I_2 = \int_{-1}^{1}\frac{1}{2(1-\beta x)^5}\,dx$$

$$I_3 = \int_{-1}^{1}\frac{x^2}{(1-\beta x)^5}\,dx \qquad (x = \cos\theta).$$

The integrals I_1 and I_2 can be calculated directly

$$I_1 = \int_{-1}^{1}\frac{1}{(1-\beta x)^3}\,dx = \frac{1}{2\beta}\frac{1}{(1-\beta x)^2}\bigg|_{-1}^{1} = \frac{2}{(1-\beta^2)^2}$$

$$I_2 = \int_{-1}^{1}\frac{1}{2(1-\beta x)^5}\,dx = \frac{1}{4\beta}\frac{1}{(1-\beta x)^4}\bigg|_{-1}^{1} = \frac{2(1+\beta^2)}{(1-\beta^2)^4}.$$

The last integral is evaluated with partial integration or found in a Table of Integrals as

$$I_3 = \int_{-1}^{1}\frac{x^2}{(1-\beta x)^5}\,dx = -\frac{1}{(1-\beta x)^4}\left(-\frac{x^2}{2\beta} + \frac{x}{3\beta^2} - \frac{1}{12\beta^3}\right)\bigg|_{-1}^{1}.$$

2.5 Details

$$= \frac{1}{(1-\beta)^4}\left(\frac{1}{2\beta} - \frac{1}{3\beta^2} + \frac{1}{12\beta^3}\right) - \frac{1}{(1+\beta)^4}\left(\frac{1}{2\beta} + \frac{1}{3\beta^2} + \frac{1}{12\beta^3}\right).$$

The terms in the two large brackets differ only in the sign. If one combines the individual terms with

$$\frac{1}{(1-\beta)^4} - \frac{1}{(1+\beta)^4} = \frac{8\beta(1+\beta^2)}{(1-\beta^2)^4}$$

and

$$\frac{1}{(1-\beta)^4} + \frac{1}{(1+\beta)^4} = 2\frac{(1+6\beta^2+\beta^4)}{(1-\beta^2)^4},$$

one obtains for the integral I_3

$$I_3 = \frac{1}{(1-\beta^2)^4}\left(\frac{10}{3}\beta^2 + \frac{2}{3}\right).$$

The total integral is then

$$I = I_1 - \frac{(1-\beta^2)}{2}(I_2 - I_3) = \frac{4}{3}\frac{1}{(1-\beta^2)^2},$$

so that one finds for the total at the time t' per unit time emitted power

$$P(t') = \frac{\mu q^2}{2c^3}\dot{v}^2 I = \frac{2\mu q^2}{3c^3}\dot{v}^2\frac{1}{(1-\beta^2)^2}.$$

Theory of Relativity and Electromagnetism 3

The incompatibility of the laws of mechanics and the laws of electrodynamics triggered the development of the (special) theory of relativity. At the core of the problem is the **theorem of relativity**, which states, that all inertial systems should be equivalent for the description of physical processes. This basic theorem is satisfied for the Galilei transformation

$$r'(t) = r(t) - v_{\text{rel}}\, t - r_0 , \tag{3.1}$$

which connects the description of mechanical processes from the point of view of two different systems of reference S and S', which are moving with respect to each other at a constant velocity (Fig. 3.1a).

The Galilei transformation implies the following statements

(1) The difference of two position vectors at a time t as measured by observers in the two systems S and S' are equal

$$r_2(t) - r_1(t) = r'_2(t) - r'_1(t) .$$

(2) The addition theorem for the velocities holds

$$v'(t) = v(t) - v_{\text{rel}} .$$

(3) The acceleration of an object measured by the two observers is the same

$$a'(t) = a(t) .$$

(4) These direct statements entail the consequence, that the equations of motion for a system of mass points/point charges, which are in accord with this theorem,

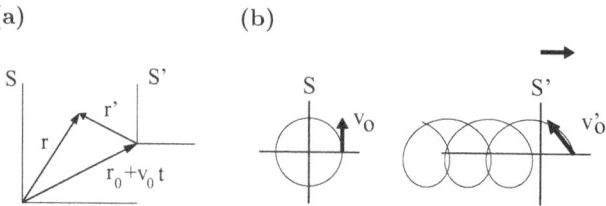

Fig. 3.1 The Galilei transformation. (**a**) The transformation law. (**b**) Trajectories

have the same form in the two inertial systems, provided the forces between the mass points/point charges depend only on the difference of the position vectors and/or the difference of the velocity vectors

$$m_i a_i = \sum_{k \neq i} f(r_i - r_k, v_i - v_k) \qquad (i = 1, \ldots N)$$

$$\overset{\text{G.transformation}}{\Longrightarrow} \qquad m_i a'_i = \sum_{k \neq i} f(r'_i - r'_k, v'_i - v'_k) \,.$$

The equations of motion are then called **form invariant** against Galilei transformations.

Trajectories of mass points/point charges are not only determined by the equations of motion, but also by initial conditions. Trajectories observed in S and S' differ in general (Fig. 3.1b), as they are influenced by the relative motion of the two inertial systems.

An observer in the system S' can e.g. observe a motion on a helix instead of a motion on a circle as observed by an observer in S. Statement (4) is satisfied for gravitation and for electrostatics, but not for magnetism. The magnetic interaction between two moving, charged mass points is determined by a force (see Dreizler and Lüdde, 2024, Electrostatics and Magnetostatics (Springer Berlin Heidelberg), Eq. (5.54), law of Lorentz), which depends on the velocities of the particles and not on the difference of the velocities

$$m\ddot{r}_i = \sum_k f_{\text{magn.}}(v_i, v_k) \,.$$

The equations of motion in a different inertial system are according to the addition theorem

$$m a'_i = \sum_k f_{\text{magn.}}(v'_i - v_{\text{rel}}, v'_k - v_{\text{rel}}) \neq \sum_k f_{\text{magn.}}(v'_i, v'_k) \,.$$

The equations of motion in the two inertial systems do not have the same form. One can also show directly, that the Maxwell equations are not form invariant against Galilei transformations (compare Chap. 3.5.2).

The following possibilities arise, if electrodynamics is not compatible with the principle of relativity:

(1) The principle of relativity is only valid in mechanics. It is not valid for the description of electrodynamic phenomena. It would then be necessary, that systems exist, which are special from the point of view of electrodynamics. It was the aim of the experiments[1] of A.A. Michelson and E.W. Morley (A.A. Michelson 1881, repeated: A.A. Michelson and E.W. Morley 1887) to verify or exclude the existence of such systems. The result of this experiment (and all experiments with the same objective) excluded the existence of such systems.
(2) The principle of relativity is generally valid. The Galilei transformation as well as the basic equations of mechanics (and/or electrodynamics) are not correct in this case. This was the suggestion of A. Einstein (1905). As the basic equations of mechanics proved their value quite well, the proposition of Einstein seemed to be too extreme (and is doubted on occasion even these days). The resolution of this apparent conflict of our experience is however quite simple: The difference between classical and relativistic mechanics presents itself only at very high velocities. Classical mechanics is a limiting case of relativistic mechanics, if all velocities involved are small in comparison with the velocity of light.

The question, which has to be answered, if one accepts the proposition of Einstein, is: Which transformation should be used instead of the Galilei transformation and/or what are the correct basic equations for the treatment of mechanical and electrodynamic phenomena? This question will be answered in the next section.

3.1 The Lorentz Transformation

The discussion of equations for a transformation between systems of reference is different for

(1) Uniform relative motion. The discussion of this case leads to the **special theory of relativity**, which is of interest here.
(2) Accelerated relative motion. The discussion of this case should establish in the end the **general theory of relativity**. This case will not be addressed here in detail.

---

[1] A.A. Michelson and E.W. Morley, Philosophical Magazine, Vol. 24 (1887), p. 449.

If one assumes, that the Galilei transformation is not correct, it is useful to look again at the assumptions, which have been made for the formation (in part implicit) of this transformation.

- The first one is: The distance between two points in space P_1 and P_2 is invariant under a Galilei transformation

$$\sum_{i=1}^{3} (x'_{i1} - x'_{i2})^2 = \sum_{i=1}^{3} (x_{i1} - x_{i2})^2 \ .$$

- With respect to the time coordinate, one assumes, that it is independent of the state of motion (or the system of reference)

$$t' = t \ .$$

The time is then called the absolute time.
- Another statement has to be added to the discussion of equations of motion: The mass of the object is the same for all inertial systems

$$m' = m \ .$$

It is astonishing, that none of these (apparently reasonable) conjectures are correct. The starting point for the formulation of the correct transformation between inertial systems is the experiment of Michelson and Morley.

3.1.1 The Michelson-Morley Experiment

The aim of the Michelson-Morley experiment was to find evidence for the existence or non-existence of an aether, in which electromagnetic waves propagate.[2] The notion at the end of the nineteenth century about the existence of an aether is based on the well understood analogy with the propagation of sound, which is based on oscillations of matter. It was thought, that all wave phenomena needed a medium, in which they propagate. On the basis of this picture one could ask, how the earth moves in the aether, which was thought to be absolutely at rest.

The tool, which was used by Michelson and Morley, was an interferometer with the following layout: A beam of light impinges on a semi-transparent plate, which decomposes the beam into two parts, which are perpendicular to each other. One part of the beam, which is perpendicular to the original direction of the beam is reflected by a mirror Sp_1. The other part passes through the plate and is reflected at a mirror Sp_2. Both beams return after a time t_i (i= 1,2) to the plate and interfere

---

[2] A short note on the historic development until the appearance of the theory of relativity is offered in Chap. 3.6.

3.1 The Lorentz Transformation

Fig. 3.2 The Michelson-Morley experiment. (a) Starting position. (b) After rotation

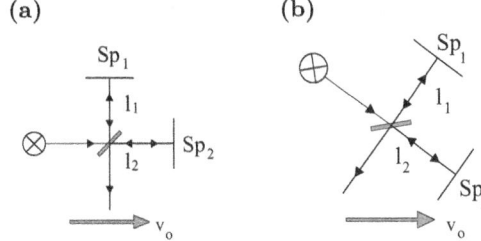

after passage (Sp_1) or reflection (Sp_2) in the direction indicated in Fig. 3.2a and b. The interference pattern depends on the difference of the run-time of the two beams. The following possibilities could be expected with respect to this difference.

(1) The apparatus is at rest in a specific system of reference, the aether system, in which the propagation of the light is isotropic, so that its speed is c in all directions. The difference for the run-time of the two beams is then

$$\Delta t = t_2 - t_1 = \frac{2(l_2 - l_1)}{c}. \tag{3.2}$$

The following remark is important for the discussion: If the total apparatus is rotated by an arbitrary angle (Fig. 3.2b, with $v_0 = 0$), there is no change of the difference of the run-time. The pattern of interference is the same for each final orientation of the apparatus.

(2) The apparatus moves with the velocity v_0 in the direction of the arm of the mirror Sp_2 (Fig. 3.3a with $v_0 \neq 0$) with respect to an inertial system, e.g. the aether system. The velocity of the light according to the addition theorem is $c - v_0$ on the way to the mirror Sp_2, on the way back it is $c + v_0$. The total run-time is therefore

$$t_2 = \frac{l_2}{c - v_0} + \frac{l_2}{c + v_0} = \frac{2cl_2}{c^2 - v_0^2}.$$

Fig. 3.3 Reflection at mirror 1. (a) Distances. (b) Velocities

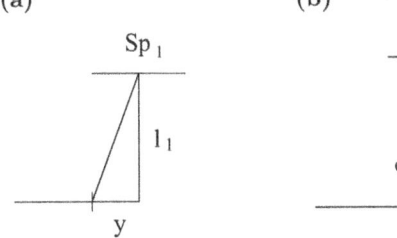
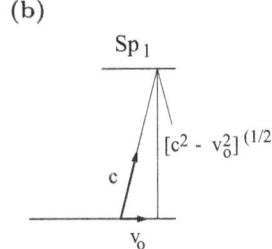

For the path to the mirror Sp_1 the effective run-time in both directions is $\sqrt{c^2 - v_0^2}$ also according to the addition theorem. The total run-time is therefore

$$t_1 = \frac{2l_1}{[c^2 - v_0^2]^{1/2}}.$$

The difference of the run-times of the two light beams is thus

$$\Delta t = \frac{2}{c}\left(\frac{l_2}{(1-(v_0/c)^2)} - \frac{l_1}{[1-(v_0/c)^2]^{1/2}}\right). \qquad (3.3)$$

If the apparatus is rotated[3] in this situation e.g. by 90°, the role of the two arms are interchanged. The difference of the run-times (also for $l_1 = l_2$) is changed and one expects a change of the interference pattern.

Both arms had equal length in the simplest variant of this experiment. The difference of the run-time is in this case

$$\Delta t(\text{rest}) = 0$$

$$\Delta t(v_0) = \frac{2l}{c}\left(\frac{1}{(1-(v_0/c)^2)} - \frac{1}{[1-(v_0/c)^2]^{1/2}}\right) \approx \frac{l}{c}(v_0/c)^2.$$

The experiment was performed at a specific time t on the trajectory of the earth around the sun and repeated six month later. A laboratory attached to the earth is a reasonable inertial system, if one ignores the rotation of the earth. The velocity of this system has the magnitude $v_{\text{earth}} \approx 30\,\text{km/s}$.

In the first experiment it was observed that the pattern of interference did not change after the apparatus was rotated. The difference of the run-times indicated, that the earth was at rest with respect to the aether system. The relative velocity was $v_0 = 2v_{\text{earth}} \approx 60\,\text{km/s}$ in the second experiment. Still one did not find any change in the interference pattern (Fig. 3.4). The experiment was repeated with light from the sun and fix stars, in order to exclude the motion of the light source. In another variant an interferometer with arms of different lengths was used in order to ascertain, that the additional path was traversed in every inertial system with the same velocity. The expected change of the interference pattern could not be found in any of these experiments. The different interferometers used, should have been able to respond to a relative velocity of $v_0 \approx 1\,\text{km/s}$. The experiments showed: A relative motion of the earth with respect to an aether system could not be demonstrated. The only other possibility, that the aether system moves with the earth, could be excluded by the observation of the stellar aberration (see Detail 3.7.1). The result, which had

---

[3] The complete interferometer was floated in a trough of mercury, so that it could be rotated easily.

3.1 The Lorentz Transformation

Fig. 3.4 Change of velocity in the first Michelson-Morley experiment

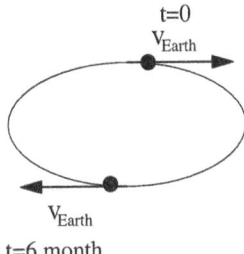

to be accepted in the end was: An absolute inertial system, the aether system, does not exist. No medium is necessary for the propagation of light. The velocity of the propagation is independent of the velocity of the source of the light, as demanded by Maxwell's equations.

This statement contradicts the Galilei transformation, which was used to predict the change of the pattern of interference and the relation

$$c' = c - v_0 \, .$$

All relevant experiments support the statement

$$c' = c \, , \qquad (3.4)$$

the velocity of light does not depend on the relative velocity of an inertial system of reference.

3.1.2 A Simple Derivation of Alternative Transformation Equations

The experimental result, that the velocity of light has the same magnitude in every inertial system, can be used to derive the equations for the transformation, which should replace the Galilei transformation. In order to present the derivation in a simple manner,[4] one uses two inertial systems S and S',

- which coincide for the time $t = 0$ (the coordinate axes are therefore parallel)
- and move both in the x-direction (Fig. 3.5), so that S' has the constant relative velocity $v_{\text{rel}} = v_0$ with respect to S.

The ansatz for the expected linear transformation is

$$x' = Ax + Bt \qquad y' = y \qquad z' = z \qquad t' = Cx + Dt \, .$$

---

[4] A general form of the transformation is given on p. 226.

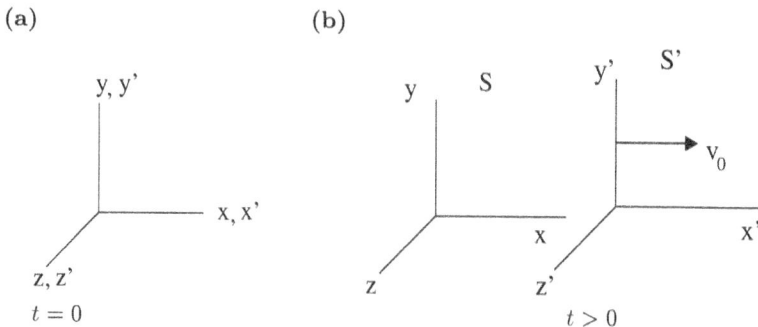

Fig. 3.5 Derivation of a simple form of the Lorentz transformation. (**a**) $t = 0$. (**b**) $t > 0$

The trivial statements for the y- and z-coordinates indicate, that one does not expect any transformation of the y- and z-coordinates for a relative motion in the x-direction. The equations for the x-direction and the time t have the only general form, which are compatible with the principle of relativity. The transformation has to be *linear* as otherwise a uniform motion from the point of view of one of the coordinate systems would be registered as an accelerated motion from the point of view of the other. The Galilei transformation (for a motion of the system S' with $+v_{\text{rel}}$ with respect to S) is obtained with the coefficients

$$A = 1 \qquad B = -v_{\text{rel}} \qquad C = 0 \qquad D = 1 \ .$$

The determination of the correct coefficients of the transformation is based on the result of the Michelson-Morley experiment together with the following argument: A pulse of light is emitted from the common origin at the time $t = 0$. It spreads out in all directions with the speed c. At the time t (>0) it reaches the point P (Fig. 3.6). The Michelson-Morley experiments allow the following statements

(S): An observer in S says: The distance from my origin to the point P is given by

$$r = ct \quad \text{or in three-dimensional space} \quad x^2 + y^2 + z^2 - c^2 t^2 = 0 \ .$$

(S'): An observer in S' notes according to his point of view: the distance from my origin to the point P is

$$r' = ct' \quad \text{or} \quad x'^2 + y'^2 + z'^2 - c^2 t'^2 = 0 \ .$$

Fig. 3.6 Illustration of the Michelson-Morley condition

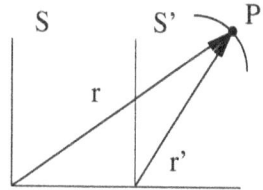

3.1 The Lorentz Transformation

These statements are possible, if the light propagates in **each** system with the speed c in all directions. The statements of the two observers can be combined in the **Michelson-Morley condition**

$$x^2 + y^2 + z^2 - c^2 t^2 = x'^2 + y'^2 + z'^2 - c^2 t'^2 . \tag{3.5}$$

This condition is not compatible with the Galilei transformation.

The determination of the coefficients A, B, C, D of the transformation via the condition (3.5) can be done in the following manner: The origin ($x' = 0$) of the system S', has, as observed by the system S, the coordinate $x = v_{\text{rel}}\, t$, so that one finds

$$0 = A v_{\text{rel}} t + B t \quad \text{or} \quad B = -A v_{\text{rel}} .$$

If the equations for the transformation are inserted now into the Michelson-Morley condition

$$x^2 - c^2 t^2 = x'^2 - c^2 t'^2 ,$$

one obtains

$$\left(A^2 - c^2 C^2\right) x^2 - 2\left(A^2 v_{\text{rel}} + c^2 C D\right) xt$$

$$-c^2 \left(D^2 - \frac{v_{\text{rel}}^2}{c^2} A^2\right) t^2 = x^2 - c^2 t^2 .$$

The coordinate x and the time t can be chosen freely, so that comparison of the coefficients yields a set of equations for the not yet determined coefficients A, C, D:

$$A^2 - c^2 C^2 = 1 \qquad A^2 v_{\text{rel}} + c^2 C D = 0 \qquad D^2 - \frac{v_{\text{rel}}^2}{c^2} A^2 = 1 .$$

The solution of this system of equations is simple (see Detail 3.7.2). One can choose the signs of the quantities, so that the Galilei transformation is obtained in the limit of small relative velocities, and finds

$$A = +\left[1 - \left(\frac{v_{\text{rel}}}{c}\right)^2\right]^{-1/2} \qquad C = -\left(\frac{v_{\text{rel}}}{c^2}\right)\left[1 - \left(\frac{v_{\text{rel}}}{c}\right)^2\right]^{-1/2}$$

$$D = +\left[1 - \left(\frac{v_{\text{rel}}}{c}\right)^2\right]^{-1/2} .$$

The ratio v_{rel}/c and the root are found in practically each equation, so it is convenient to use the abbreviations

$$\beta = \frac{v_{rel}}{c} \qquad \gamma = \frac{1}{[1-\beta^2]^{1/2}} . \qquad (3.6)$$

The equations of the transformation are then

$$x' = \frac{1}{\left[1-\left(\frac{v_{rel}}{c}\right)^2\right]^{1/2}} (x - v_{rel}t) \qquad y' = y \qquad z' = z$$

$$t' = \frac{1}{\left[1-\left(\frac{v_{rel}}{c}\right)^2\right]^{1/2}} \left(-\frac{v_{rel}}{c^2}x + t\right), \qquad (3.7)$$

or in compact form

$$x' = \gamma(x - v_{rel}t) \qquad t' = \gamma\left(-\frac{\beta}{c}x + t\right). \qquad (3.8)$$

This result for the **Lorentz transformation**, including the correct interpretation, was given by A. Einstein in his famous publication[5] in the year 1905. The equations had been obtained already in 1904 by H.A. Lorentz,[6] but Lorentz tried to save the concept of an absolute time rather than taking the step, that A. Einstein did.

The first comments on the Lorentz transformation are:

(i) The Lorentz transformation turns into the Galilei transformation in the limit of small relative velocities ($v_{rel}/c \to 0$) of the inertial systems

$$\left.\begin{array}{l} x' = x - v_{rel}t \\ t' = t \end{array}\right\} + \text{terms} \quad O\left(v_{rel}^2/c^2\right).$$

Classical physics is not in contradiction to relativistic physics. It is a limiting case of relativistic physics. The time plays a special role in the Galilei transformation, as it is not transformed. Space and time are treated on the same level in the Lorentz transformation. This property is the basis for the formal treatment of the special theory of relativity with the concept of the Minkowski space (Chap. 3.3).

---

[5] A. Einstein, Annalen der Physik, Vol. 17 (1905), p. 891.
[6] H.A. Lorentz, Proceedings of the Academy of Science, Amsterdam, Vol. 6 (1904), p. 809.

(ii) The special role of the velocity of light c as a limiting velocity is quite evident. The equations for the Lorentz transformation do not make any sense, if v_{rel} approaches c. This fact is interpreted as:
Every signal (also every object) can propagate (or move) only with a maximum speed, the speed of light.

(iii) The inverse transformation (can be found by sorting with respect to the quantities in the system S) is

$$x = \gamma \left(x' + v_{rel} t'\right) \qquad t = \gamma \left(\frac{\beta}{c} x' + t'\right). \qquad (3.9)$$

This is exactly the form one expects, if the role of the two inertial systems is interchanged. The system S moves with respect to the system S' with the velocity $-v_{rel}$.

It is advantageous to discuss the physical content of the Lorentz transformation on the basis of the present simple form before the assembly of the general transformation and the associated, mathematical language.

3.2 Implications of the Lorentz Transformation

The change of the equations for the transformation leads to a variant of the addition theorem for velocities. In this version one should recognise the velocity of light as a limiting velocity, which can not be surpassed. In addition, the new addition theorem should turn into the classical addition theorem for small relative velocities.

3.2.1 The Addition Theorem for Velocities

The addition theorem answers the following question: Two observers in the inertial systems S and S' monitor the motion of an object (Fig. 3.7). The observer in S registers the (momentary) velocity vector

$$\mathbf{v}(t) = \left(v_x(t),\ v_y(t),\ v_z(t)\right).$$

The question is: Which velocity components are measured by the observer in S', who is moving with the velocity v_{rel} in the common x-direction? The question can

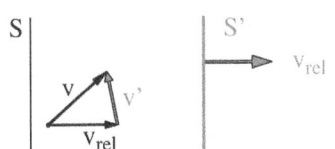

Fig. 3.7 Velocities as viewed from systems S and S'

be answered with the (relativistic) addition theorem, if one takes care of the fact that the velocity components are functions of all the coordinates *including the time*

$$v_x = \frac{dx}{dt}, \quad v_y = \frac{dy}{dt}, \quad v_z = \frac{dz}{dt}$$

$$v'_x = \frac{dx'}{dt'}, \quad v'_y = \frac{dy'}{dt'}, \quad v'_z = \frac{dz'}{dt'}.$$

The difference quotients and their limiting values, the differential quotients can be calculated with the total differentials of the equations of the Lorentz transformation

$$dx' = \gamma (dx - v_{rel} dt) \qquad dy' = dy \qquad dz' = dz$$

$$dt' = \gamma \left(-\frac{\beta}{c} dx + dt \right).$$

The explicit result is

$$v'_x = \frac{(v_x - v_{rel})}{\left(1 - \frac{v_x v_{rel}}{c^2}\right)} \qquad v'_y = \frac{v_y \left[1 - \left(\frac{v_{rel}}{c}\right)^2\right]^{1/2}}{\left(1 - \frac{v_x v_{rel}}{c^2}\right)}$$

$$v'_z = \frac{v_z \left[1 - \left(\frac{v_{rel}}{c}\right)^2\right]^{1/2}}{\left(1 - \frac{v_x v_{rel}}{c^2}\right)}. \tag{3.10}$$

These equations represent the relativistic addition theorem for velocities for the case of a uniform relative motion with the velocity $v_{rel} = (v_{rel}, 0, 0)$ of the system S' with respect to the system S with the same orientation. These equations imply:

- The limit for small relative velocities $v_{rel} \ll c$ is

$$v'_x = v_x - v_{rel} \qquad v'_y = v_y \qquad v'_z = v_z.$$

  One finds, as expected the classical limit, the addition theorem of the Galilei transformation.
- It is not possible to surpass the velocity of light. This statement is best illustrated by some direct examples.

3.2 Implications of the Lorentz Transformation

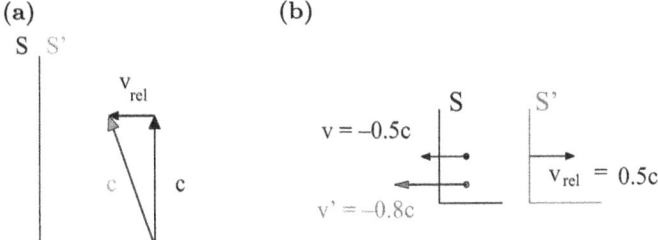

Fig. 3.8 Results of the relativistic addition theorem. (a) $c + v_{\text{rel}} = c!$. (b) $0.5c + 0.5c = 0.8c!$

(i) The observer S is looking at an electromagnetic wave, which propagates in the x-direction, so that $v = (c, 0, 0)$. The observer S' measures

$$v'_x = \frac{c - v_{\text{rel}}}{1 - v_{\text{rel}}/c} = c \qquad v'_y = v'_z = 0 \,.$$

This observer observes also an electromagnetic wave, which propagates with the velocity c in the x-direction (independent of the magnitude of the relative velocity v_{rel}).

(ii) A beam of light moves, observed by S, in the y-direction, so that its velocity is $v = (0, c, 0)$ (Fig. 3.8a). With the addition theorem one can calculate the velocity components as seen by S'

$$v'_x = -v_{\text{rel}} \qquad v'_y = \left[c^2 - v_{\text{rel}}^2\right]^{1/2} \qquad v'_z = 0 \,.$$

The velocity vector of the light beam has a negative x-component for observer S' (Fig. 3.8a). The magnitude of the velocity of the beam is still c

$$\left[\sum_i v_i'^2\right]^{1/2} = c \,.$$

(iii) For, as shown in Fig. 3.8b,

$$v = (-c/2, 0, 0) \quad \text{and} \quad v_{\text{rel}} = (c/2, 0, 0) \,,$$

the classical addition theorem yields for the x-component of the velocity from the point of view of S' the result $v'_x = -c$. The answer of the relativistic theorem is

$$v'_x = \frac{-c}{1 + 1/4} = -\frac{4}{5}c \qquad v'_y = v'_z = 0 \,.$$

The value for the x-component observed in S' is 20% lower than the classical value.

It is possible to demonstrate via the velocity transformation, that it is not possible to reach velocities that are larger than the velocity of light. The relativistic addition theorem satisfies the Michelson-Morley condition.

- The inverse of the velocity transformation is

$$v_x = (v'_x + v_{rel})\left(1 + \frac{v_{rel}v'_x}{c^2}\right)^{-1}$$

$$v_y = v'_y\left[1 - \frac{v_{rel}^2}{c^2}\right]^{1/2}\left(1 + \frac{v_{rel}v'_x}{c^2}\right)^{-1}$$

$$v_z = v'_z\left[1 - \frac{v_{rel}^2}{c^2}\right]^{1/2}\left(1 + \frac{v_{rel}v'_x}{c^2}\right)^{-1}. \tag{3.11}$$

The inverse transformation can be obtained by interchanging all quantities involved and by replacing v_{rel} by $-v_{rel}$.
- The addition theorem can be tested experimentally. One uses for this purpose a light source, which is at rest with respect to the system S'. The velocity of light as observed in S' (e.g. in the x-direction, (Fig. 3.9)) is

$$v' = (c, 0, 0) \qquad c = 2.9979\ldots \cdot 10^{10}\, \text{cm/s}.$$

For the velocity of light as observed in S one expects $v_{x,\,class} = c + v_{rel}$, in the case of the classical addition theorem, but $v_{x,\,relat} = c$ in the case of the relativistic variant. The difference can only be clear-cut, if the relative velocity is sufficiently large. The experiment can not be done with a moving torch. As an alternative, one should use the decays of elementary particles with a final product in the form of electromagnetic radiation. A suitable example is the decay of a

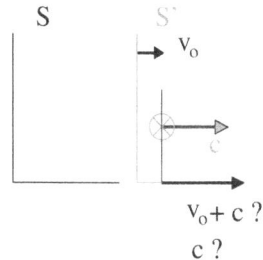

Fig. 3.9 Experiment for a test of the addition theorem

3.2 Implications of the Lorentz Transformation

neutral π-meson (with a half life of about 10^{-16} s) in two γ quanta (a form of electromagnetic radiation)

$$\pi^0 \longrightarrow \gamma + \gamma'.$$

Experiments were performed with π-mesons, which had a velocity of

$$v_{\text{rel}} = 0.99975\, c$$

with respect to the laboratory system. The velocity of the γ-quanta in the laboratory system was measured to be $v_{\text{exper}} = (2.9977 \pm 0.0004) \cdot 10^{10}$ cm/s, in contradiction to the classical expectation of $v_{\text{class}} \approx 6 \cdot 10^{10}$ cm/s.

Differentiation of the velocity transformation (with respect to the relevant time coordinate) yields the transformation between the components of the acceleration

$$a_x = \frac{dv_x}{dt} \quad \ldots \quad a_x' = \frac{dv_x'}{dt'} \quad \ldots .$$

Application of the chain rule is called for, e.g. for

$$\frac{dv_x'}{dt'} = \frac{dv_x'}{dt}\frac{dt}{dt'}.$$

The first factor is calculated with the addition theorem

$$\frac{dv_x'}{dt} = \frac{d}{dt}\left(\frac{v_x - v_{\text{rel}}}{1 - (v_{\text{rel}}v_x)/c^2}\right) = \frac{(1-(v_{\text{rel}}/c)^2)}{(1-(v_{\text{rel}}v_x)/c^2)^2} a_x .$$

The second factor is obtained with the Lorentz transformation of the time

$$\frac{dt}{dt'} = \frac{dt}{\gamma(-(\beta dx)/c + dt)} = \frac{[1 - (v_{\text{rel}}/c)^2]^{1/2}}{(1 - (v_{\text{rel}}v_x)/c^2)}.$$

The multiplication of these results yields the transformation of the x-component of the acceleration (a corresponding result can be obtained for the other components)

$$a_x' = \frac{[1 - (v_{\text{rel}}/c)^2]^{3/2}}{(1 - (v_{\text{rel}}v_x)/c^2)^3} a_x . \tag{3.12}$$

The transformation formula for the components of the acceleration is more complicated than the classical limit. The discussion of the relativistic case will, however, be postponed until the complete relativistic equations of motion are introduced in Chap. 3.4.3.

Fig. 3.10 The Lorentz contraction

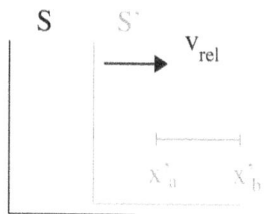

The statements, which can be extracted from the Lorentz transformation and the addition theorem of the velocities, are often termed as astonishing as they seem to contradict the every day experience. They all have, however, been confirmed by experiments. The term 'relativistic' is fully reinforced, if one compares time scales and measuring rods in different inertial systems. The comparison of measuring rods is discussed under the name Lorentz contraction in the next section.

3.2.2 The Lorentz Contraction

A measuring rod, which is at rest with respect to the system S', is oriented along the x'-direction (Fig. 3.10). The coordinates of the end points are x'_a and x'_b. The length of the rod as seen by S' is therefore

$$x'_b - x'_a \equiv l_0(S') \, .$$

The index 0 indicates, that the rod is at rest in this coordinate system. The obvious question is: What is the length of the rod as seen by the observer in S, if the system S' is moving with the velocity v_{rel} with respect to S in the common x-direction? The answer can be given with the Lorentz transformation. The end points of the rod as seen by observer S are

$$x_a = \gamma(x'_a + v_{\text{rel}} t'_a) \qquad x_b = \gamma(x'_b + v_{\text{rel}} t'_b) \, .$$

As coordinates and times are linked together by the Lorentz transformation, one has to consider also the transformation of the times

$$ct_a = \gamma(ct'_a + \beta x'_a) \qquad ct_b = \gamma(ct'_b + \beta x'_b) \, .$$

The formula for the transformation of the length of the rod must be obtained on the basis of the four equations, together with the relevant important observation: In order to measure the length of the rod, observer S has to register the two end points at the same time. The length of the rod from the point of view of S is $l(S)$ (no index of l as the rod is moving) is given by

$$l(S) = x_b - x_a$$

3.2 Implications of the Lorentz Transformation

with the auxiliary condition

$$t_a - t_b = 0 .$$

This condition takes, according to the transformation of the time coordinates the form

$$c(t'_b - t'_a) + \beta(x'_b - x'_a) = 0 .$$

The length of the rod is, according to the transformation of the space coordinates

$$l(S) = x_b - x_a = \gamma(x'_b - x'_a) + \gamma v_{\text{rel}}(t'_b + t'_a) .$$

Insertion of the auxiliary condition gives

$$l(S) = \left[1 - (v_{\text{rel}}/c)^2\right]^{1/2} l_0(S') \qquad (3.13)$$

or in compact form

$$l(S) = \gamma^{-1} l_0(S') .$$

A shorter derivation of this result can be obtained with the inverse transformation

$$x'_b = \gamma(x_b - v_{\text{rel}} t_b) \qquad x'_a = \gamma(x_a - v_{\text{rel}} t_a) .$$

The requirement of simultaneity of the measurement of the end points ($t_b = t_a$) in the system S leads to

$$l_0(S') = x'_b - x'_a = \gamma(x_b - x_a) = \left[1 - (v_{\text{rel}}/c)^2\right]^{-1/2} l(S)$$

or again

$$l(S) = \gamma^{-1} l_0(S') .$$

This result says: A measuring rod, which moves in the longitudinal direction with the velocity v_{rel}, appears to be shortened. This shortening is called the **Lorentz contraction**.

As the statement concerning the transverse coordinates in the case considered is $y' = y$ respectively $z' = z$, there follows, that the contraction can only be observed in the direction of motion. A volume, which has the size V_0 in a rest system, appears

Fig. 3.11 Lorentz contraction in the macro-world?. (**a**) Bicycle at rest. (**b**) Fast moving bicycle

to be diminished, if it moves with a velocity v_{rel}. The contraction of the volume V_0 to a volume V is described by

$$V = V_0 \left[1 - (v_{\text{rel}}/c)^2 \right]^{1/2} . \tag{3.14}$$

The direction of the motion does not play any role. The contraction of the length or the volume is the only consequence of the Lorentz transformation, that has not yet been verified so far. The reason is, that macroscopic objects can not be accelerated to velocities, which are high enough, so that the effect of the contraction can be ascertained. The measurement of the length of microscopic objects is, on the other hand, a difficult matter. The lack of confirmation is the reason for the rather extensive amount of literature on the topic volume contraction. One of the questions, which is discussed in this context, is: What would a real object (for example a bicycle) look like, if it passes an observer with a velocity of $0.99\,c$ (Fig. 3.11a)? The statements above would predict a contraction in the direction of motion. A more detailed discussion in the literature claims, that one has to include an analysis of the process of vision for a complete picture. The eye constructs an image with rays, which are incident at the same time. For an extended object one would have to include the fact, that the points, which are more distant have a longer run-time (Fig. 3.11b). The combination of the effect of the run-time with the Lorentz contraction leads to a contraction and a rotation of the object. Such considerations are nonetheless academic. The eye would not be able to resolve the situation, if an object rushes past at a velocity of $0.99\,c$.

For the inverse problem (Fig. 3.12a), the measuring rod rests in the system S, one finds for the length of the rod[7]

$$l(S') = l_0(S) \left[1 - (v_{\text{rel}}/c)^2 \right]^{1/2} . \tag{3.15}$$

It is always the moving rod, which appears to be shortened.

As space and time are intimately connected in the theory of relativity, there must exist a counterpart concerning the time. This is the time dilatation.

---

[7] The derivation is left to the reader. The crucial point is the fact, that the observer S' has to register the end points of the rods at the same time.

Fig. 3.12 The Time Dilatation. (a) t'_a. (b) t'_b

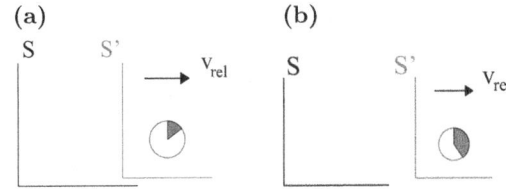

3.2.3 The Time Dilatation

The derivation of the appropriate formula for the comparison of time intervals of two inertial systems with respect to each other will also be discussed in some detail. One discusses a physical process (e.g. the oscillation of a spring), which takes place at a fixed spot in system S'. The process begins at the time t'_a and ends at the time t'_b (Fig. 3.12b).

The duration of the process from the point of view of S' is

$$\tau_0(S') = t'_b - t'_a .$$

The index 0 signifies in this case, that the process takes place in the respective system at a fixed location.

The question is: Which time interval is registered by an observer in the system S for this process? The question is answered with the equations

$$ct_a = \gamma(ct'_a - \beta x'_a) \qquad ct_b = \gamma(ct'_b - \beta x'_a) .$$

This information expresses the fact, that the process takes place at the position x'_a (the view of S'). The two equations state directly

$$c(t_b - t_a) = \gamma c(t'_b - t'_a)$$

or

$$\tau(S) = t_b - t_a = \gamma \tau_0(S') = \frac{\tau_0(S')}{\left[1 - (v_{\text{rel}}/c)^2\right]^{1/2}} . \qquad (3.16)$$

An observer in S measures for the same process a time interval, which is increased by the factor γ. This is the basic statement of **time dilatation**. The remaining equations of the transformation (with $x'_b = x'_a$)

$$x_a = \gamma(x'_a + v_{\text{rel}} t'_a) \qquad x_b = \gamma(x'_a + v_{\text{rel}} t'_b) ,$$

allow the calculation of the distance by which the system has moved in the time interval $\tau(S)$ as viewed by S. As expected one finds

$$x_b - x_a = \gamma\, v_{\rm rel}\, \tau_0(S') = v_{\rm rel}\, \tau(S)\,.$$

In the case of the time dilatation one can also look at the inverse transformation. A process with the duration $\tau_0(S)$ at a fixed position as viewed by S takes the time

$$\tau(S') = \gamma\, \tau_0(S)\,. \tag{3.17}$$

The time dilatation can be verified by experiments. An experiment, which is quoted often, is the prolonged life time of fast moving μ-mesons. The $\mu^{\pm}$-meson is an unstable elementary particle, which decays mainly according to the process

$$\mu^{\pm} \longrightarrow e^{\pm} + \nu_\mu + \tilde{\nu}_e$$

that is into an electron or a positron and two neutrinos (more exactly a neutrino and an antineutrino). The life time of a μ-meson at rest (or a very slow μ-meson) is

$$\tau_0 = 2.2\cdot 10^{-6}\,{\rm s}\,.$$

These mesons are created at the edge of our atmosphere by a chain of reactions, which is initiated by cosmic radiation. In the first step the π-mesons (with a shorter life time at rest of $\tau_0 = 10^{-8}$ s) are created. The π -mesons decay into μ-mesons.

▸ cosm. radiation + matter $\longrightarrow$ $\pi^{\pm}$-mesons $\longrightarrow$ $\mu^{\pm}$-mesons.

The $\mu^{\pm}$-mesons are created about 10 to 20 kilometres above the surface of the earth. The incident cosmic radiation is very energetic, so that the finally created $\mu^{\pm}$-mesons have a high mean velocity

$$\bar{v}_\mu \approx 0.995\, c\,.$$

If the mean distance, which the μ-mesons cover at this velocity is calculated, one obtains

$$x = \bar{v}_\mu \tau_0 \approx c\tau_0 = 3\cdot 10^8 \cdot 2.2\cdot 10^{-6}\,{\rm m} = 660\,{\rm m}\,.$$

A (mean) distance of 15 kilometre corresponds to approximately 23 mean life times. As the radioactive decay proceeds according to

$$n(t) = n_0\, e^{-t/\tau_0}\,,$$

3.2 Implications of the Lorentz Transformation

one can calculate that

$$n(23\tau_0) = n_0\, e^{-23} \approx \frac{n_0}{10^{10}},$$

that is practically zero μ-mesons should be observed at the surface of the earth. A considerable flow of mesons is, however, found experimentally. The apparent contradiction can be resolved by considering the fact, that the life time of the mesons is increased by their high velocity of $v_{\rm rel} = v_\mu \approx 0.995\, c$ to

$$\tau = \frac{\tau_0}{\left[1 - (0.995)^2\right]^{1/2}} \approx 10\,\tau_0\,.$$

The life time of the μ-mesons from the point of view of the earth system is ten times as long, so that they can cover a mean distance of 6.6 km. The number of μ-mesons, which should be observed at the surface of the earth is therefore (in the same unit of time as above)

$$n(2.3\tau_0) = n_0\, e^{-2.3} \approx \frac{n_0}{10},$$

in good accordance with the experimental results.[8]

Corresponding effects were observed for other unstable elementary particles.[9] Time dilatation was also recorded for macroscopic clocks in rockets. This became possible through the high accuracy of Caesium-clocks ($\pm 10^{-9}$ seconds/day,).

Another story concerning time dilatation, which has led to controversial discussions, is the **twin paradox**. This story can be told in the following fashion: As a person is also a kind of clock, one can think of the following scenario: If a person is moving uniformly with a high velocity in a straight line he/she would live longer as a contemporary at rest. There is no advantage for the moving person. Its own life (τ_0) ends at the normal time of life. The situation can be discussed in a controversial manner, if one compares the life time of twins, an astronaut and a terrestrial person. Twin T1 stays on the earth, twin T2 undertakes a voyage into space and returns to earth. The earth twin can say: my brother has moved with the velocity $v_{\rm rel} \approx c$ and claim: my brother is therefore younger. The space travelling twin can make a corresponding remark about his earth bound brother and claim: my brother has moved with a velocity $v_{\rm rel} \approx c$ (in the opposite direction). He is therefore younger.

The argumentation, which leads to the paradox, is not correct. It does not include the fact, that periods of acceleration have been involved for the space twin. These periods are necessary for the start and the landing as well as for the turn-around. The argument on the basis of the special theory of relativity is not conclusive. Twin

---

[8] The flow of μ-mesons at different heights has also been measured.
[9] A summary can e.g. be found in Bailey et al., Nature 268 (1971), p. 301.

T1 is (in good approximation) in an inertial system, twin T2 is not. The symmetry of the motion, that has been assumed in the argumentation does not exist.[10]

Real insight into the structure of the special theory of relativity can only be gained, if the appropriate mathematical language is used. This language is based on the concept of non-Euclidean spaces, in particular the Minkowski space.

3.3 The Minkowski Space

3.3.1 Definition

Space (x, y, z) and time (t) are to be treated on the same level in the theory of relativity. This is the reason, why it is necessary to describe physical processes in a four-dimensional space in which space- and time-coordinates are united. Normally such spaces are defined over the domain of real numbers (with the designation $\mathcal{R}(4)$ or $\mathcal{R}_4$), which is spanned by mutually orthogonal unit vectors

$$e_1, \; e_2, \; e_3, \; e_4 \, .$$

The metric (the usual terminology in this connection) of such spaces is determined by the scalar products

$$g_{ik} = (e_i \cdot e_k) = \delta_{ik} \qquad i, k = 1, \ldots, 4.$$

An arbitrary vector in this space

$$r = \sum_{i=1}^{4} x_i e_i$$

is characterised by the length or the square of the length

$$r^2 = x_1^2 + x_2^2 + x_3^2 + x_4^2 \, .$$

A space with this metric is called **Euclidean**. Euclidean spaces with real coordinates are not suited for the formulation of the theory of relativity. A possibility to reproduce the necessary space-time structure is offered by the use of an alternative

---

[10] Interested students can look at several contributions to the topic of the twin paradox in the journal 'American Journal of Physics', as e.g. the article by R. Perrin, Am. J. Phys.,Vol. **47** (1979) p. 317.

3.3 The Minkowski Space

metric. One identifies the space coordinates, which will finally be called covariant coordinates, by[11]

$$x_0 = ct \quad x_1 = x \quad x_2 = y \quad x_3 = z \tag{3.18}$$

and considers the coordinates

$$(e_\mu \cdot e_\nu) = g_{\mu\nu} \qquad \mu, \nu = 0, 1, 2, 3 \tag{3.19}$$

with the **metric tensor**

$$(g_{\mu\nu}) = \begin{pmatrix} 1 & 0 & 0 & 0 \\ 0 & -1 & 0 & 0 \\ 0 & 0 & -1 & 0 \\ 0 & 0 & 0 & -1 \end{pmatrix}. \tag{3.20}$$

This can be interpreted as: The four basis vectors are still orthogonal but three of the elementary length are negative. A space with such a metric is called **pseudo-Euclidean**. The reason, why one should be interested in a space with such a metric can be answered directly. For a vector with the form

$$r = ct\, e_0 + x\, e_1 + y\, e_2 + z\, e_3$$

one obtains with this metric

$$r^2 = \sum_{\nu,\mu=0}^{3} g_{\mu\nu} x_\mu x_\nu = c^2 t^2 - x^2 - y^2 - z^2.$$

This is exactly the Michelson-Morley condition (3.5), which has been used as the starting point for establishing the Lorentz transformation. A space, which is defined by the metric of Eq. (3.20), is referred to as a four-dimensional **Minkowski space**. Standard classifications are $\mathcal{M}_4$ or $\mathcal{M}(1, 3)$, where the second variant emphasises the sequence and the different nature of the time and space coordinates.[12]

As it is not easy to represent a four dimensional space, the discussion is often restricted to a two-dimensional space-time $\mathcal{M}(1, 1)$. The y- and z-coordinates are ignored and the relative motion takes place in the x-direction. Such a space is

---

[11] For dimensional reasons.
[12] Other variants, as e.g. counting the coordinates with x_1 instead of x_0 and x_2, x_3, x_4 for the space coordinates and a metric $\bar{g}_{ik} = -g_{ik}$ are also used in the literature. It is advisable to check, which definition of the coordinates and which metric is used in a particular case.

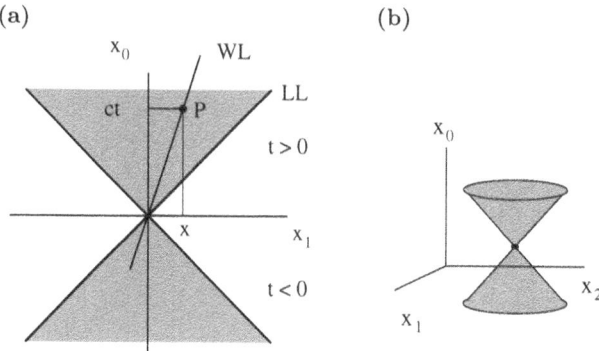

Fig. 3.13 Representation of the Minkowski world. (**a**) The space $\mathcal{M}(1, 1)$. (**b**) The space $\mathcal{M}(1, 2)$

spanned by the coordinates

$$x_0 = ct \qquad x_1 = x \qquad \text{with} \quad (g_{\mu\nu}) = \begin{pmatrix} 1 & 0 \\ 0 & -1 \end{pmatrix}.$$

The two coordinates are then represented by a normal two-dimensional, orthogonal coordinate system, so that the consequences of the metric have to be included by other means (see Chap. 3.3.3). The corresponding graphical representation of this Minkowski space is called a **Minkowski diagram**. Each point of the x_1-x_0 plane is, in the parlance of the theory of relativity, called an **event**, which corresponds to the statement: An object (mass point) is at the time t at the position x (Fig. 3.13a). The time development of a mass point is depicted by a curve in the (two-dimensional) Minkowski space. These curves are called **world lines**. The motion of a particle, which is (as observed in system S) at the time $t = 0$ at the position $x = 0$ and which moves with the uniform velocity $+v$ (Fig. 3.13a) is represented by a straight line through the origin with the inclination

$$\tan \alpha = \frac{ct}{x} = \frac{c}{v} > 1 \ .$$

The world lines are in general curvilinear trajectories.

A beam of light, which passes the origin at the time $t = 0$ and is moving in the $\pm x$-direction, is characterised by a straight line with the inclination ± 1. These lines are the **light lines**. As the velocity of light is a limiting velocity for all objects and signals, one can make the following statement: If the origin represents the present, then all objects and signals, which pass the origin *now*, have to be associated with world lines, which lie in the past ($t_0 < 0$), that is in the lower section of the diagram, which is shaded in grey in Fig. 3.13a. In the future they can only be found in the upper shaded section. Events in the areas on the sides can not be connected with the

3.3 The Minkowski Space

origin. The light lines divide events, which are causal with respect to the present and those which are not.

The light lines of an event $P_0 = (ct_0, x_0, y_0)$ in a $\mathcal{M}(1,2)$ diagram, which is supposed to represent the present at time t, have to be replaced by a **light cone** through P_0 with an opening angle of 45° (Fig. 3.13b). The inner section of the cone contains points, which can be connected causally with P_0, where the lower half represents the past and the upper half the future. The outer part of the cone contains events, which can not be connected causally with P_0. In the four-dimensional space $\mathcal{M}(1,3)$ one would have, referred to a point marking the present $P_0 = (ct_0, x_0, y_0, z_0)$, to divide the Minkowski space by a three-dimensional structure, the **hyper light cone** into events, which can be connected causally or not with P_0.

For a direct characterisation of the fact that a causal connection of two events is possible or not, one uses the square of the distance between two points P and P_0 of the $\mathcal{M}(1,3)$

$$s^2 = c^2(t-t_0)^2 - (x-x_0)^2 - (y-y_0)^2 - (z-z_0)^2 \,. \tag{3.21}$$

The corresponding nomenclature is referred to the point marking the present at P_0:

- Points P, with $s^2 = 0$, lie on the hyper-light cone through P_0.
- For $s^2 > 0$, the points P are called **time-like** (with respect to P_0). They can be causally connected with P_0.
- For $s^2 < 0$, the points P are called **space-like** (with respect to P_0). They can not be causally connected with P_0.

One should keep in mind that these definitions depend on the metric, which has been adopted. For a metric with $g'_{ik} = -g_{ik}$ points with $s^2 > 0$ would be called space-like.

3.3.2 The Lorentz Transformation in the $\mathcal{M}(1,1)$-world

In order to depict the Lorentz transformation, one starts with the equations of the transformation, which have the form

$$x'_0 = \gamma(-\beta x_1 + x_0) \qquad x'_1 = \gamma(x_1 - \beta x_0)$$

in the $\mathcal{M}(1,1)$-world. The coordinate axes of the inertial systems S' can be characterised from the point of view of the system S in the following fashion: The $x'_0 (= ct')$-axis is specified by $x'_1 = 0$, the $x'_1 (= x')$-axis correspondingly by $x'_0 = 0$.

Fig. 3.14 Representation of a Lorentz transformation in the $\mathcal{M}(1, 1)$-world with pseudo-Euclidean coordinates

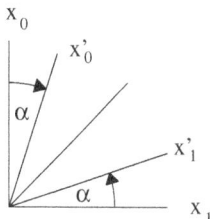

According to the transformation, the x'_0-axis as viewed by S, is represented by the straight line

$$x_0 = \frac{c}{v_{\text{rel}}} x_1 \, .$$

The x'_1-axis is represented by the straight line

$$x_0 = \beta x_1 = \frac{v_{\text{rel}}}{c} x_1$$

with the reciprocal inclination. The Lorentz transformation, which describes the transition from the system S to the system S', is represented by a Minkowski diagram, in which the two axes $x_0 = ct$ and $x_1 = x$ are rotated by an angle α with $\tan \alpha = v_{\text{rel}}/c$ in the direction of the light line (Fig. 3.14). The rotated axes correspond to the coordinate axes $x'_0 = ct'$ and $x'_1 = x'$ of the system S' as viewed by S.

This graphic representation of a (reduced) Lorentz transformation in a Minkowski diagram can be justified as follows: One defines the functions

$$\cosh \alpha = \gamma = \frac{1}{\left[1 - (v_{\text{rel}}/c)^2\right]^{1/2}} \qquad \sinh \alpha = \beta \gamma = \frac{v_{\text{rel}}}{c \left[1 - (v_{\text{rel}}/c)^2\right]^{1/2}}$$

of the rotation angle α, which satisfy the relation

$$\cosh^2 \alpha - \sinh^2 \alpha = 1$$

for the hyperbolic functions and obtains the reformulated transformation equations

$$x'_0 = (\cosh \alpha) x_0 - (\sinh \alpha) x_1 \qquad x'_1 = -(\sinh \alpha) x_0 + (\cosh \alpha) x_1 \, ,$$

which are similar to the standard equations for rotations. The trigonometric functions are replaced by hyperbolic functions. The pseudo-Euclidean metric causes the rotation in the direction of the light line instead of the normal rotation of the two axes in the same direction.

3.3 The Minkowski Space

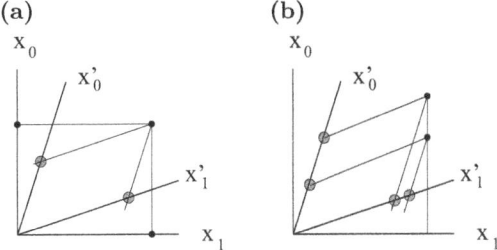

Fig. 3.15 Comparison of events in the $\mathcal{M}(1, 1)$-world. (**a**) Coordinates. (**b**) Same position in S

This graphic representation of the Lorentz transformation can be used to discuss a number of the aspects of the theory of relativity. In order to read off the coordinates of an event for both inertial systems in a Minkowski diagram, one uses appropriate parallel lines to the pairs of coordinate axes (Fig. 3.15a) in the two systems. The intersections of the parallels with the complementary coordinate axes yield the coordinates in the two inertial systems.

The following facts can be stated: Two events, which happen for an observer in S at the same location, are found on a straight line parallel to the x_0-axis (Fig. 3.15b). For an observer in S' they are found (naturally) in different places, as the system S' is moving relative to S. Two events, which happen at the same time for an observer in S', are found on parallel lines to the x_1'-axis (Fig. 3.16a). It is possible to confirm by just looking at the diagram, that they do not happen at the same time for the observer S.

Imagine that one observes a measuring rod with the length $l_0(S')$, which is oriented along to the x_1'-axis and is at rest in the system S'. The world lines of the end points are then parallels the x_0'-axis (Fig. 3.16b). The end points of a rod have to be observed at the same time, if an observer in a system S measures the length of the rod. This condition is realised, if the world lines of the end points of the rod intersect the x_1-axis. The Lorentz contraction can be 'seen' directly in the diagram. The procedure is, however, not complete. The axes have to be supplied with a scale. One can furnish a scale on the x_0-axis and the x_1-axis, but have to make sure, that this information is transferred correctly to the x_0'- and the x_1'-axis. The calibration of these axes can be achieved with the invariant

$$s^2 = {x_0'}^2 - {x_1'}^2 = x_0^2 - x_1^2 \ .$$

Fig. 3.16 Comparison of events in the $\mathcal{M}(1, 1)$-world. (**a**) Same time in S'. (**b**) Measuring rods

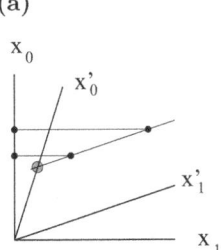 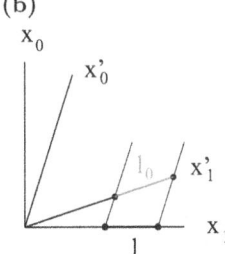

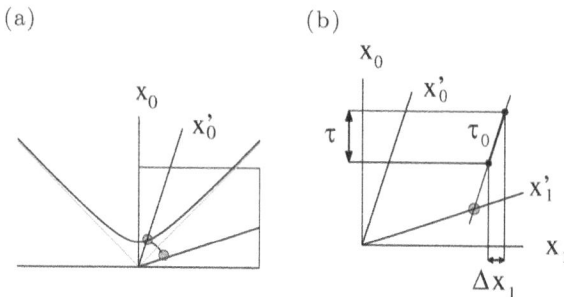

Fig. 3.17 Comparison of events in the $\mathcal{M}(1, 1)$-world. (a) Transfer of scales. (b) Clock resting in S'

If $s = 1$ is chosen, the equation $x_0^2 - x_1^2 = 1$ represents a hyperbola, which passes through the point $(x_0, x_1) = (1, 0)$ and approaches the light lines asymptotically. The intersection of this gauge hyperbola with the x_0'-axis represents, because of $(x_0')^2 - 0 = 1$, a line segment of length 1 on the x_0'-axis. The same scale has to be transferred to the x_1'-axis (Fig. 3.17a). After this transfer of the scales, one can (in principle) extract the Lorentz contraction from the Minkowski diagram in the $\mathcal{M}(1, 1)$-world in a quantitative manner.

A corresponding treatment can be used for the time dilatation e.g. for the situation, that the clock is at rest in S' and that a time interval $\tau_0(S')$ is considered (Fig. 3.17b). It is possible to extract from the Minkowski diagram (again with scale transfer) the length of the time interval $\tau(S)$ in the system S, as well as the line segment Δx_1, by which the clock has moved in the system S in the time $\tau(S)$.

3.3.3 Formal Aspects 1: Co- and Contravariant Coordinates

It is useful, as in the discussion of vector spaces in the Euclidean world, to represent the Minkowski coordinates

$$x_0 = ct \qquad x_1 = x \qquad x_2 = y \qquad x_3 = z$$

as the components of a vector with four components[13]

$$\mathbf{R} = \{x_0, x_1, x_2, x_3\} = \{ct, x, y, z\} . \tag{3.22}$$

This basic **four-vector** is called the **event vector**. In order to represent the Michelson-Morley condition one needs the scalar product of this vector. As a consequence of the pseudo-Euclidean (skew) metric (see Appendix C.8 for a general

---

[13] The components of four-vectors are represented by curly brackets in order to distinguish them from three-vectors.

3.3 The Minkowski Space

overview of this topic)

$$(g_{\mu\nu}) = \begin{pmatrix} 1 & 0 & 0 & 0 \\ 0 & -1 & 0 & 0 \\ 0 & 0 & -1 & 0 \\ 0 & 0 & 0 & -1 \end{pmatrix},$$

one has to distinguish between the **covariant** decomposition (3.22) of this vector $\{x_\mu\}$ and its **contravariant** decomposition $\{x^\mu\}$. The definition of these decompositions leads to a relation with the metric tensor

$$x_\mu = \sum_{\nu=0}^{3} g_{\mu\nu} x^\nu \qquad x^\mu = \sum_{\nu=0}^{3} g^{\mu\nu} x_\nu . \tag{3.23}$$

The metric tensor with upper indices is (in the special theory of relativity) identical with the tensor with lower indices

$$g^{\mu\nu} \equiv g_{\mu\nu} \tag{3.24}$$

and the relation

$$\sum_{\lambda'} g^{\rho\lambda'} g_{\lambda'\rho'} = \delta_{\rho\rho'} \tag{3.25}$$

is valid. In this situation the contravariant coordinates are

$$x^0 = ct \qquad x^1 = -x \qquad x^2 = -y \qquad x^3 = -z .$$

The scalar product can be represented by a **contraction**[14] of the co- and contravariant components of the four-vectors

$$\mathbf{R} \cdot \mathbf{R} = \sum_{\mu,\nu=0}^{3} g^{\mu\nu} x_\mu x_\nu = \sum_{\mu=0}^{3} x_\mu x^\mu = c^2 t^2 - x^2 - y^2 - z^2 , \tag{3.26}$$

so that the Michelson-Morley condition takes the form

$$\sum_{\mu=0}^{3} x'_\mu x'^\mu = \sum_{\mu=0}^{3} x_\mu x^\mu .$$

---

[14] This concept will be explained in more detail on p. 229.

A general, homogeneous and linear transformation (L) in Minkowski space, which connects the coordinates of a system S with the coordinates of a system S', is

$$x'_\mu = \sum_{\lambda=0}^{3} L_{\mu\lambda} x^\lambda \qquad (\mu = 0, 1, 2, 3) \,. \tag{3.27}$$

The contravariant components in the system S (unprimed coordinates) are related with the covariant components in the system S' (primed coordinates). The homogeneous transformation has to be replaced by an inhomogeneous transformation, if the origins of the two systems do not coincide at the time $t_0 = t'_0 = 0$

$$x'_\mu = \sum_{\lambda=0}^{3} L_{\mu\lambda} x^\lambda + b_\mu \,.$$

This transformation is called a **Poincaré transformation**. The equations for the transformation between the covariant components in the system S and the contravariant components in S' can be obtained with the metric tensor

$$x'^\mu = \sum_{\lambda\nu\rho} g^{\mu\nu} L_{\nu\rho} g^{\rho\lambda} x_\lambda + \sum_\nu g^{\mu\nu} b_\nu = \sum_\lambda L^{\mu\lambda} x_\lambda + b^\mu \,. \tag{3.28}$$

A four-vector **B**, which describes the position of the origin of the four-dimensional coordinate system S as viewed from the system S', can be specified by covariant components **B** $= \{b_0, b_1, b_2, b_3\}$ or by contravariant components **B** $= \{b^0, b^1, b^2, b^3\}$. The matrix of the transformation with upper indices is obtained from the original one by multiplication with the metric tensor

$$\sum_{\nu\rho} g^{\mu\nu} L_{\nu\rho} g^{\rho\lambda} = L^{\mu\lambda} \,. \tag{3.29}$$

If Eqs. (3.27) and (3.28) for the homogeneous transformation ($b_\mu = b^\mu = 0$ for all μ) are inserted into the Michelson-Morley condition (3.26), one obtains, with a correct handling of the co-and contravariant components

$$\sum_{\mu\lambda\rho} L_{\mu\lambda} L^{\mu\rho} x^\lambda x_\rho = \sum_\lambda x^\lambda x_\lambda \,.$$

Comparison of the coefficients leads to the statement

$$\sum_\mu L_{\mu\lambda} L^{\mu\rho} = \delta_{\lambda\rho} \,. \tag{3.30}$$

3.3 The Minkowski Space

This result can also be written in a form, which contains the original transformation matrix

$$\sum_{\mu\nu} L_{\mu\lambda} g^{\mu\nu} L_{\nu\sigma} = g_{\lambda\sigma} .$$

For this purpose the relation (3.30) is multiplied by $g_{\rho\sigma}$ and the sum over ρ is executed. Finally one uses the representation (3.29) for $L^{\mu\rho}$ and sums up the resulting expression with the orthogonality relation (3.25).

The condition (3.30) defines, as an extension of the arguments in Chap. 3.1.2, the general Lorentz transformation. It demonstrates, that the Lorentz transformation, an orthogonal transformation in Minkowski space, represents four-dimensional rotations and reflections. The rotations involve pure rotations of the three-dimensional space, but also rotations, in which space coordinates and the time coordinate are transformed together. Reflections in Minkowski space comprise time reversal and reflections at the space part of the coordinate origin.[15] The simple Lorentz transformation with the transformation matrix (3.8)

$$(L_{\lambda\mu}) = \begin{pmatrix} \gamma & \beta\gamma & 0 & 0 \\ -\beta\gamma & \gamma & 0 & 0 \\ 0 & 0 & 1 & 0 \\ 0 & 0 & 0 & 1 \end{pmatrix}$$

describes time reversal and space reflections via the matrices

$$L_{\mu\lambda} = -g_{\mu\lambda} \quad \text{resp.} \quad L_{\mu\lambda} = g_{\mu\lambda} .$$

The Lorentz transformation for arbitrarily oriented inertial systems and an arbitrary direction of the relative velocity can be constructed in the following manner:

- Rotate the original coordinate system, so that the x_1-axis is parallel to the y_1-axis

$$y_\mu = \sum_\lambda D_{\mu\lambda}(\Omega) x^\lambda .$$

---

[15] The fact, that the Lorentz transformation can be interpreted to represent geometric operations in Minkowski space, is the reason for the discussion of Lorentz- and Poincaré transformations in group theory.

The rotation matrix for a space rotation in the Minkowski space has the form

$$(D_{\lambda\mu}) = \begin{pmatrix} 1 & 0 & 0 & 0 \\ 0 & \ddots & & \\ 0 & & -d^{(3)} & \\ 0 & & & \ddots \end{pmatrix}.$$

The rotation matrix $d^{(3)}$ in R_3 is expressed in terms of the Euler angles.[16]
- Execute the simple Lorentz transformation of the rotated system to the primed system

$$y'_\sigma = \sum_\mu L_{\sigma\mu}(\beta) y^\mu.$$

- The system with the coordinates y'_σ is not yet oriented correctly. It has to be rotated, in such a way, that the y'_1-axis coincides with the x'_1-axis

$$x'_\rho = \sum D_{\rho\sigma}(\Omega') y'^\sigma.$$

- If the individual transformations are combined, one finds in matrix form

$$\mathbf{R}' = (D(\Omega'))(L(\beta))(D(\Omega))\mathbf{R}. \tag{3.31}$$

A general Lorentz transformation is a combination of two rotations of space and one simple space-time rotation (in the order shown in Fig. 3.18; see also Detail 3.7.3).

The situation is somewhat more transparent, if the two inertial systems have the same orientation, but different relative velocities. In this case the second rotation is the inverse of the first and one finds for the Cartesian coordinates and time coordinate the transformation

$$\mathbf{r}' = \mathbf{r} + (\gamma - 1) \frac{(\mathbf{r} \cdot \mathbf{v}_{\text{rel}})}{v_{\text{rel}}^2} \mathbf{v}_{\text{rel}} - \gamma \mathbf{v}_{\text{rel}} t$$

Fig. 3.18 Illustration of the operations for setting up the general Lorentz transformation

---

[16] See e.g. Dreizler and Lüdde, 2010, Theoretical Mechanics (Springer Berlin Heidelberg) Chap. 6.3.5.

3.3 The Minkowski Space

$$t' = \gamma \left(t - \frac{(\mathbf{r} \cdot \mathbf{v}_{\text{rel}})}{c^2} \right).$$

It is appropriate to state at this point, that the composition of Lorentz transformations is not a commutable operation, except in the case that the relative velocities are parallel.

One element of the discussion of electrodynamics on the basis of the theory of relativity is the special role of the derivatives of the Minkowski coordinates in the form of four-gradients. Of interest is the question, whether the four-gradient, which is obtained by derivatives with respect to the covariant coordinates

$$\bar{\nabla} = \left(\frac{\partial}{\partial x_0}, \frac{\partial}{\partial x_1}, \frac{\partial}{\partial x_2}, \frac{\partial}{\partial x_3} \right) = \left(\frac{1}{c} \frac{\partial}{\partial t}, \nabla \right),$$

is really a four-vector. If this is the case, one has to be able to show, that the components transform in the same way as the components of the event vector.

In order to discuss the properties of the transformation, one considers the action of this operator on a scalar field f in Minkowski space. Such a field is characterised by the property, that it has the some value for points of the Minkowski space, which are connected by a Lorentz transformation

$$f(x_0, x_1, x_2, x_3) = f(x'_0, x'_1, x'_2, x'_3).$$

The derivatives $f(x'_0, x'_1, x'_2, x'_3)$ with respect to the covariant coordinates are obtained with the chain rule

$$\frac{\partial f}{\partial x_{\mu'}} = \sum_\lambda \frac{\partial f}{\partial x^\lambda} \frac{\partial x^\lambda}{\partial x'_\mu}.$$

One needs the derivative of a contravariant coordinate in S with respect to a covariant coordinate in S'. This can be obtained by the inverse of the transformation (3.27) using the orthogonality relation (3.30)

$$x^\lambda = \sum_\nu L_{\nu\lambda} x'_\nu$$

and partial differentiation. The resulting transformation

$$\frac{\partial f}{\partial x_{\mu'}} = \sum_\lambda L^{\mu\lambda} \frac{\partial f}{\partial x_\lambda} \qquad (3.32)$$

shows, that the derivative with respect to a covariant Minkowski coordinate transforms in the same way as a contravariant coordinate (and vice versa). A much used notation is therefore

$$\frac{\partial}{\partial x_\mu} = \partial^\mu \quad \text{and} \quad \frac{\partial}{\partial x^\mu} = \partial_\mu$$

with the transformation

$$\partial_{\mu'} = \sum_{\lambda=0}^{3} L_{\mu\lambda} \partial^\lambda . \tag{3.33}$$

The four-gradient is indeed a four-vector

$$\tilde{\nabla} = \{\partial^0, \partial^1, \partial^2, \partial^3\} = \left\{ \frac{\partial}{\partial x_0}, \frac{\partial}{\partial x_1}, \frac{\partial}{\partial x_2}, \frac{\partial}{\partial x_3} \right\} . \tag{3.34}$$

The formal version of the Minkowski world is needed for a completely consistent handling of the different quantities, which differ in their transformation properties. It is possible to introduce tensors of second, and higher rank. Tensors of second rank as e.g. $[A] = (A_{\mu\nu})$, are characterised by the fact, that they transform as

$$A'_{\mu\nu} = \sum_{\lambda\sigma} L_{\mu\lambda} L_{\nu\sigma} A^{\lambda\sigma} \tag{3.35}$$

in the transition between two inertial systems. A product with the elements $x^\lambda y^\sigma$ satisfies the transformation (3.35)

$$x'_\mu y'_\nu = \sum_{\lambda\sigma} L_{\mu\lambda} L_{\nu\sigma} x^\lambda y^\sigma .$$

Tensors, which are constructed by multiplication of the components of two vectors (tensors of first rank), are a common, but not the only form of tensors of second rank. Contravariant or partially contravariant tensor elements can be constructed with the metric tensor or corresponding combinations of four-vectors, as e.g.

$$A^{\mu\nu} = \sum_{\lambda\sigma} g^{\mu\lambda} g^{\nu\sigma} A_{\lambda\sigma} \qquad B^{\mu\nu} = x^\mu y^\nu$$

or

$$A_\mu^{\ \nu} = \sum_\lambda g^{\nu\lambda} A_{\mu\lambda} \qquad B_\mu^{\ \nu} = x_\mu y^\nu .$$

3.3 The Minkowski Space

The mathematical operation of contraction allows the reduction of the rank of tensors. It can be viewed as a generalisation of the concept of the scalar product (compare (3.26)). A vector (tensor of first rank) contracted with a second vector gives a scalar (tensor of zero rank). The justification is based on the orthogonality relations for the Lorentz transformation and for the metric tensor as one has

$$\sum_{\mu} x'_\mu y'^\mu = \sum_{\mu\lambda\sigma} L_{\mu\lambda} L^{\mu\sigma} x^\lambda y_\sigma = \sum_\lambda x^\lambda y_\lambda$$

$$= \sum_{\mu\lambda\sigma} g^{\lambda\mu} g_{\lambda\sigma} x_\mu y^\sigma = \sum_\mu x_\mu y^\mu .$$

The contraction transforms like a number. In the co-/contravariant notation one easily recognises the relation

$$\sum_\mu A_{\nu\mu} x^\mu = y_\nu$$

as the contraction of a tensors of second rank with a vector to yield a vector, etc. The application of the **sum convention of Einstein** facilitates the work with equations in the theory of relativity. The rules of the convention are: Write the sums over Minkowski indices without the sign for a sum, but imply summation over all doubly occurring indices, as e.g.

$$\sum_\mu a_\mu b^{\mu\nu} \equiv a_\mu b^{\mu\nu} .$$

For the discussion of mechanics and electrodynamics on the basis of the theory of relativity a different realisation of the Minkowski world is actually sufficient. It will be introduced in the next section.

3.3.4 Formal Variant 2: Imaginary Time Coordinate

The Minkowski space can also be discussed in terms of an Euclidean metric and complex coordinates, e.g. in the $\mathcal{M}(1, 1)$-world, with the coordinates $x_0 = ict$, x_1 and the Lorentz transformation

$$x'_0 = \gamma(x_0 - i\beta x_1) \qquad x'_1 = \gamma(i\beta x_0 + x_1)$$

as well as the corresponding matrix representation

$$(L_{\lambda\mu}) = \begin{pmatrix} \gamma & -i\beta\gamma \\ i\beta\gamma & \gamma \end{pmatrix} .$$

Fig. 3.19 Alternative representation of a Lorentz transformation in the $\mathcal{M}(1,1)$-world: Imaginary time coordinate

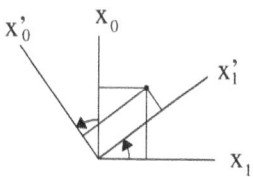

This matrix can be interpreted as a rotation matrix with a complex angle in an Euclidean space. It corresponds to the introduction of the angle φ with the definition

$$\sin\varphi = i\beta\gamma \qquad \cos\varphi = \gamma$$

and the property

$$\sin^2\varphi + \cos^2\varphi = -\frac{\beta^2}{1-\beta^2} + \frac{1}{1-\beta^2} = 1 \ .$$

The matrix of the Lorentz transformation looks like a standard rotation matrix in a two-dimensional Euclidean space

$$(L_{\lambda\mu}) = \begin{pmatrix} \cos\varphi & \sin\varphi \\ -\sin\varphi & \cos\varphi \end{pmatrix} .$$

The graphical representation in a Minkowski diagram of the $\mathcal{M}(1,1)$-world is, as indicated in Fig. 3.19, a rotation of the two coordinate axes. Scaling is not necessary for this variant. The consequences of the Lorentz transformation can be discussed directly:

In the formal discussion one uses the coordinates[17]

$$X_0 = ict \qquad X_1 = x \qquad X_2 = y \qquad X_3 = z \qquad (3.36)$$

in order to distinguish this form from the real ones. The metric tensor has the matrix elements $g_{\mu\nu} = \delta_{\mu\nu}$. The scalar product of an event vector

$$\mathbf{R} = \{X_0, X_1, X_2, X_3\} = \{ict, x, y, z\} \qquad (3.37)$$

with itself is then

$$\sum_{\mu=0}^{3} X_{\mu'} X_{\mu'} = -c^2 t'^2 + r'^2 = \sum_{\lambda=0}^{3} X_\lambda X_\lambda = -c^2 t^2 + r^2 \ , \qquad (3.38)$$

---

[17] Here also the warning: There exist alternative variants in the literature.

3.3 The Minkowski Space

as it is not necessary to distinguish between co- and contravariant vectors. Note however, that events with $S^2 = \mathbf{R} \cdot \mathbf{R} < 0$ have a time-like separation.

The general linear transformation in Minkowski space, which relates the coordinate system S (unprimed coordinates) with S' (primed coordinates) is

$$X_{\mu'} = \sum_{\lambda=0}^{3} L_{\mu\lambda} X_\lambda \qquad (\mu = 0, 1, 2, 3) . \qquad (3.39)$$

Insertion of this transformation into the Michelson-Morley condition (3.38)

$$\sum_\mu X_\mu X_\mu = \sum_\mu X_{\mu'} X_{\mu'}$$

yields

$$\sum_{\mu\lambda\lambda'} L_{\mu\lambda} L_{\mu\lambda'} X_\lambda X_{\lambda'} = \sum_\lambda X_\lambda X_\lambda$$

and the comparison of the coefficients (with the same notation as in the previous section, but different matrix elements)

$$\sum_\mu L_{\mu\lambda} L_{\mu\lambda'} = \delta_{\lambda\lambda'} . \qquad (3.40)$$

The last equation indicates explicitly that use of the complex time coordinate leads to an orthogonal transformation in Minkowski space. The transformation can be divided into proper and improper Lorentz transformations according to

$$\det(L) = 1 \quad \text{and} \quad \det(L) = -1 .$$

The improper transformations are products of proper transformations and a reflection. Proper transformations are rotations in the four-dimensional space. Reflections are time reversal and reflections at the origin in the space section.

The proper Lorentz transformation with

$$X'_0 = \gamma(X_0 - i\beta X_1) \qquad X'_1 = \gamma(i\beta X_0 + X_1)$$

$$X'_2 = X_2 \qquad X'_3 = X_3$$

is described by the matrix L

$$(L_{\lambda\mu}) = \begin{pmatrix} \gamma & -i\beta\gamma & 0 & 0 \\ i\beta\gamma & \gamma & 0 & 0 \\ 0 & 0 & 1 & 0 \\ 0 & 0 & 0 & 1 \end{pmatrix}.$$

The orthogonality relation (3.40) can be confirmed by direct calculation.
    The proof, that the four-gradient

$$\bar{\nabla} = \left(\frac{\partial}{\partial X_0}, \frac{\partial}{\partial X_1}, \frac{\partial}{\partial X_2}, \frac{\partial}{\partial X_3} \right) = \left(\frac{1}{ic} \frac{\partial}{\partial t}, \frac{\partial}{\partial x}, \frac{\partial}{\partial y}, \frac{\partial}{\partial z} \right)$$

$$= \left(\frac{1}{ic} \frac{\partial}{\partial t}, \nabla \right) \tag{3.41}$$

is a four-vector can be demonstrated as before. The derivative of a scalar field f with respect to the primed coordinate is obtained with the chain rule as

$$\frac{\partial f}{\partial X_{\mu'}} = \sum_{\lambda} \frac{\partial f}{\partial X_{\lambda}} \frac{\partial X_{\lambda}}{\partial X_{\mu'}}.$$

The derivative of an unprimed coordinate with respect to the primed coordinates can be found, if one determines the inverse transformation

$$X_{\lambda} = \sum_{\mu} L_{\mu\lambda} X_{\mu'} \tag{3.42}$$

using the transformation (3.39)

$$X_{\mu'} = \sum_{\lambda} L_{\mu\lambda} X_{\lambda}$$

and the orthogonality relation (3.40)

$$\sum_{\mu} L_{\mu\lambda} L_{\mu\lambda'} = \delta_{\lambda\lambda'}.$$

With this transformation one can calculate the derivative of f in the form

$$\frac{\partial f}{\partial X_{\mu'}} = \sum_{\lambda} L_{\mu\lambda} \frac{\partial f}{\partial X_{\lambda}},$$

as well as the formal equation of the transformation

$$\frac{\partial}{\partial X_{\mu'}} = \sum_\lambda L_{\mu\lambda} \frac{\partial}{\partial X_\lambda} \ . \tag{3.43}$$

The four-gradient is indeed a four-vector.

If the inverse transformation (3.42) is inserted into the invariance condition (3.38)—watch out: summation over the first index -one finds a second set of orthogonality relations

$$\sum_\lambda L_{\mu\lambda} L_{\mu'\lambda} = \delta_{\mu\mu'} \ . \tag{3.44}$$

With the presentation of the peculiarities of the Minkowski world and its formal mathematical background all necessary tools for the return to the applications in physics are assembled. Before a discussion of the basic equations of electrodynamics in terms of the theory of relativity, it is useful and necessary to look at relativistic mechanics. This allows an understanding of the basic concepts of physics in a relativistic context. One learns e.g. that the energy is not a scalar quantity but the zero component of a four-vector.

3.4 Relativistic Mechanics

The basic quantities of classical mechanics are classified as

▶   scalars, vectors, tensors (2. rank), ...

according to their behaviour under orthogonal transformations in $R(3)$. Each quantity, which has the same transformation properties as the position vector, can be called a vector. This applies also to the velocity vector

$$v(t) = \frac{d r(t)}{dt} \ .$$

The numerator is a difference of position vectors, it transforms like a vector. The denominator is an invariant under space transformations and under Galilei transformations It does not change under these transformations. One can use the same argumentation to classify the basic quantities of mechanics against the background of the theory of relativity.

3.4.1 The Four-Velocity

The basic vector is the event vector (3.37)

$$\mathbf{R} = \{X_0, X_1, X_2, X_3\} = \{ict, x, y, z\}.$$

For the definition of the velocity in the theory of relativity, the **four-velocity**, one uses in analogy to the classical case the argument

- infinitesimal displacements in Minkowski space

$$d\mathbf{R} = \{ic\,dt, dx, dy, dz\}.$$

- These displacements have to be divided by a time interval, which is invariant with respect to Lorentz transformations. For the construction of an appropriate interval one can look at the invariant

$$d\mathbf{R} \cdot d\mathbf{R} = dx^2 + dy^2 + dz^2 - c^2 dt^2.$$

This scalar product of time-like, infinitesimally connected events is negative $((d\mathbf{R})^2 < 0)$. After multiplication by the invariant quantity $-1/c^2$, and taking the square root, one obtains a positive quantity with the unit of time

$$d\tau = \left[-\frac{1}{c^2}(d\mathbf{R})^2\right]^{1/2} = \left[dt^2 - (dx^2 + dy^2 + dz^2)\frac{1}{c^2}\right]^{1/2}$$

$$= dt\left[1 - \frac{v^2}{c^2}\right]^{1/2} \quad \text{with} \quad (v^2 = v_x^2 + v_y^2 + v_z^2). \tag{3.45}$$

The scalar product (in $R(3)$) $\boldsymbol{v} \equiv \boldsymbol{v}_{\text{class}}$ is the normal classical velocity vector as viewed from an inertial system. The time difference $d\tau$ is per construction invariant against Lorentz transformations

$$d\tau = d\tau' = dt'\left[1 - \frac{v'^2}{c^2}\right]^{1/2}.$$

The time $d\tau$ is called the **eigenzeit** (other names, that are also in use are eigentime and proper time).
- A possible definition of the four-velocity is therefore

$$\mathbf{V} = \frac{d\mathbf{R}}{d\tau} = \{V_0, V_1, V_2, V_3\}$$

3.4 Relativistic Mechanics

$$= \left\{ \frac{ic}{\sqrt{1-v^2/c^2}}, \frac{v_x}{\sqrt{1-v^2/c^2}}, \frac{v_y}{\sqrt{1-v^2/c^2}}, \frac{v_z}{\sqrt{1-v^2/c^2}} \right\}$$

$$= \frac{1}{\sqrt{1-v^2/c^2}} \{ic, \boldsymbol{v}\}, \tag{3.46}$$

where $\boldsymbol{v}$ is the velocity of classical mechanics.

If the four-velocity in the inertial system S is known, one can obtain the components in the system S' with the transformation

$$V_{\mu'} = \sum_\lambda a_{\mu\lambda} V_\lambda,$$

as *all* four-vectors are transformed with the matrix $a_{\mu\lambda}$, e.g. for the simple Lorentz transformation

$$\begin{pmatrix} V'_0 \\ V'_1 \\ V'_2 \\ V'_3 \end{pmatrix} = \begin{pmatrix} \gamma & -i\beta\gamma & 0 & 0 \\ i\beta\gamma & \gamma & 0 & 0 \\ 0 & 0 & 1 & 0 \\ 0 & 0 & 0 & 1 \end{pmatrix} \begin{pmatrix} V_0 \\ V_1 \\ V_2 \\ V_3 \end{pmatrix}.$$

For the component V'_0 one finds in particular

$$\frac{ic}{\sqrt{1-v'^2/c^2}} = \frac{\gamma}{\sqrt{1-v^2/c^2}} \left(ic - i\frac{v_x v_{\text{rel}}}{c} \right)$$

$$\implies \frac{1}{\sqrt{1-v'^2/c^2}} = \frac{\gamma}{\sqrt{1-v^2/c^2}} \left(1 - \frac{v_x v_{\text{rel}}}{c^2} \right),$$

which is a useful relation for conversions. The component V'_1, which relates to the direction of the relative motion, is transformed according to

$$\frac{v'_x}{\sqrt{1-v'^2/c^2}} = \frac{\gamma}{\sqrt{1-v^2/c^2}} (v_x - v_{\text{rel}})$$

$$\implies v'_x = \frac{(v_x - v_{\text{rel}})}{(1 - (v_x v_{\text{rel}})/c^2)}.$$

The perpendicular component V_2' (and a similar result for V_3') is

$$\frac{v_y'}{\sqrt{1-v'^2/c^2}} = \frac{1}{\sqrt{1-v^2/c^2}} v_y$$

$$\Longrightarrow v_y' = \frac{v_y \left[1 - v_{\text{rel}}^2/c^2\right]^{1/2}}{1 - (v_x v_{\text{rel}})/c^2}.$$

The result can be summarised by: The Lorentz transformation of the four-velocity is identical with the addition theorem (obtained also by other means) for the four-velocity of the theory of relativity.

The square of the absolute value of the four-velocity is

$$\mathbf{V} \cdot \mathbf{V} = \frac{v^2 - c^2}{1 - v^2/c^2} = -c^2 = \mathbf{V}' \cdot \mathbf{V}'. \tag{3.47}$$

This quantity is an invariant, which is (naturally) related to c.

In the limit of small velocities ($v \ll c$ in the system S) the result is

$$\mathbf{V}' \longrightarrow (ic, \mathbf{v} - \mathbf{v}_{\text{rel}}).$$

The three space components satisfy the classical relation. The time component does not provide any particular information and can be ignored. If the results are taken together, one can say, that the definition of the four-velocity looks very reasonable.

3.4.2 The Four-Momentum and the Relativistic Energy

The transition from the four-velocity to the four-momentum requires the multiplication of the velocity with a suitable scalar factor, which has the dimension of a mass. In view of the discussion, which has to follow, it will be denominated by m_0. The definition of the four-momentum is thus

$$\mathbf{P} = \{P_0, P_1, P_2, P_3\} = m_0 \mathbf{V} \tag{3.48}$$

$$= \left\{ \frac{m_0 ic}{\sqrt{1-v^2/c^2}}, \frac{m_0 v_x}{\sqrt{1-v^2/c^2}}, \frac{m_0 v_y}{\sqrt{1-v^2/c^2}}, \frac{m_0 v_z}{\sqrt{1-v^2/c^2}} \right\}.$$

The usefulness of this definition and the question of the interpretation of the individual terms has to be investigated now. The first point is the limit of small velocities $v \ll c$, with v being again the velocity of a mass point in a specified inertial system

$$\mathbf{P} \xrightarrow{v/c \ll 1} \{im_0 c, m_0 v_x, m_0 v_y, m_0 v_z\}.$$

3.4 Relativistic Mechanics

Fig. 3.20 Dependence of the mass $m(v)$ on the velocity

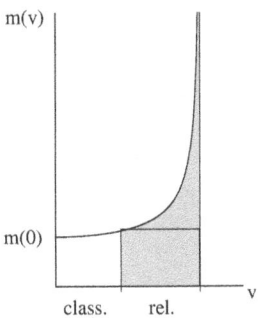

This limit suggests the following interpretation: The factor m_0 corresponds to the mass (of the mass point) in classical physics. It is called the **rest mass**. It has, provided the mass point rests in the inertial system, the same value in every inertial system. The mass, which appears in the expression for the four-momentum above,

$$m_{\text{relat}} \equiv m = m(v) = \frac{m_0}{\sqrt{1 - v^2/c^2}}, \tag{3.49}$$

is the mass of a moving mass point. This **relativistic mass** m_{relat} will be identified here and further on with m. The formula (3.49) for the relativistic mass, which says, that the mass of an object increases according to the square root in the denominator (Fig. 3.20), has been confirmed fully by experiment. Problems would have arisen without a provision of the relativistic increase of the mass in many fields of physics, e.g. the construction of high energy accelerators for elementary particles.

The space part of the four-momentum

$$\boldsymbol{p} = (p_1, p_2, p_3) = \left(m(v)v_x, m(v)v_y, m(v)v_z\right) = m(v)\boldsymbol{v}$$

can be interpreted, with regard to the non-relativistic limit, as the relativistic extension of the classical momentum.

The question of the interpretation of the component p_0

$$p_0 = im(v)c \quad \text{with} \quad p_0 \xrightarrow{v/c \ll 1} im_0 c$$

can be answered by the following argument: Start with a (not the only possible) definition of the relativistic kinetic energy

$$dT_{\text{relat}} \equiv dT = \frac{d\boldsymbol{p}}{dt} \cdot d\boldsymbol{r} = d\boldsymbol{p} \cdot \boldsymbol{v},$$

insert the definition of the classical momentum and obtain the classical form of the kinetic energy

$$dT_{\text{class.}} = m_0 v \cdot dv \qquad T_{\text{class.}} = \int_0^v dT_{\text{class.}} = \frac{m_0}{2} v^2 \;.$$

Using the extension of the relativistic extension of the three momentum yields

$$dT = (dm\, v + m\, dv) \cdot v \;.$$

The total differential of the relativistic mass is according to the chain rule

$$dm = \sum_{i=1}^{3} \frac{\partial m}{\partial v_i} dv_i = \frac{m_0}{c^2} \frac{1}{\left[1 - (v^2/c^2)\right]^{3/2}} v \cdot dv \;,$$

so that

$$dT = m_0 v \cdot dv \left(\frac{v^2/c^2}{\left[1 - (v^2/c^2)\right]^{3/2}} + \frac{1}{\left[1 - (v^2/c^2)\right]^{1/2}} \right)$$

$$= \frac{m_0 v \cdot dv}{\left[1 - (v^2/c^2)\right]^{3/2}} = c^2 dm$$

follows. This equation can be integrated directly

$$\int_0^v dT = c^2 \int_0^v dm$$

or

$$T(v) - T(0) = c^2 (m(v) - m(0)) \;.$$

With definition $T(0) = 0$ and the identification $m(0) = m_0$ one obtains one of the basic equations of the relativistic mechanics

$$T(v) = (m(v) - m_0) c^2 \;. \tag{3.50}$$

With the binomial series one finds in the limit of small velocities for the relativistic mass

$$m(v) = \frac{m_0}{\left[1 - (v^2/c^2)\right]^{1/2}} = m_0 \left(1 + \frac{1}{2} \frac{v^2}{c^2} + \frac{3}{8} \left(\frac{v}{c}\right)^4 + \ldots \right)$$

and thus for the kinetic energy

$$T(v) \xrightarrow{v/c \ll 1} \frac{m_0}{2}v^2 + \frac{3}{8}m_0\frac{v^4}{c^2} + \cdots .$$

In this limit the expression for the relativistic kinetic energy goes over into the expression for the kinetic energy of classical mechanics. The first relativistic correction is of the order $(v/c)^2$.

The interpretation of the result (3.50) is due to Einstein. According to this interpretation: $E_0 = m_0c^2$ is the energy at rest (inner energy) of a mass(point), that is the quantity of energy that could be obtained if the total mass would be transferred into a different form of energy. The energy

$$E = mc^2 \qquad (3.51)$$

is the total energy of a moving mass point, that is the energy at rest and the relativistic kinetic energy (without any potential energy). The kinetic energy can therefore be defined as the difference of the total and the rest energy

$$T = E - E_0 = (m - m_0)c^2 .$$

The statement: An object gains energy is equivalent to the statement an object gains mass. The relation $E = mc^2$ is called the theorem of the equivalence of energy and mass. The correctness of this interpretation of the relativistic energy situation is confirmed by elementary processes. Examples are:

- Pair destruction (Fig. 3.21a)

$$e^+ + e^- \longrightarrow 2\gamma .$$

The mass (rest mass and kinetic energy) of two reaction partners (electron and positron) is transferred into electromagnetic energy. The inverse of this process

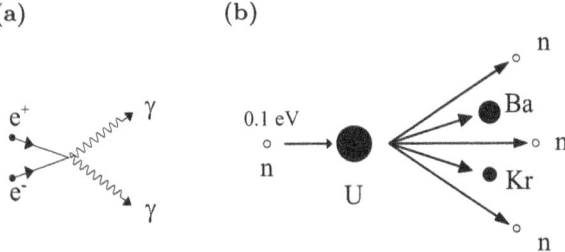

Fig. 3.21 Conversion of mass into energy. (**a**) Destruction of pairs. (**b**) Fission of uranium

can be observed as well

$$\gamma \longrightarrow e^+ + e^- \,.$$

- A much discussed example is the nuclear reaction[18]

$$^{235}_{92}\text{U} + ^{1}_{0}\text{n} + (0.1\,\text{eV}) \longrightarrow \,^{143}_{56}\text{Ba} + ^{90}_{36}\text{Kr} + 3\,^{1}_{0}\text{n} + (2 \cdot 10^8\,\text{eV})\,,$$

the most likely fission process of uranium (Fig. 3.21b). Two massive particles and a small amount of energy are changed (in the average) into 5 massive particles with a lower total mass. The loss of mass manifests itself in the high kinetic energy of the final products.

The four-momentum defined in (3.48) is thus

$$\mathbf{P} = \left\{ i\frac{E}{c},\, \mathbf{p} \right\}\,. \tag{3.52}$$

Energy is not a scalar quantity in relativistic physics. Energy in units of the velocity of light and the three-momentum are the components of a four-vector. At a transition from one inertial system into another

$$P_{\mu'} = \sum_{\lambda=0}^{3} L_{\mu\lambda} P_\lambda \qquad (\mu = 0, 1, 2, 3)$$

energy and momentum are transformed in the same fashion as time and space.

The transformation properties of **P** lead to another much used formula of relativistic mechanics. The scalar product of **P** with itself is an invariant

$$\mathbf{P}' \cdot \mathbf{P}' = \mathbf{P} \cdot \mathbf{P}\,.$$

One chooses the system S' as the system, in which the mass rests

$$\mathbf{P}' = \mathbf{P}_0 = (i\,m_0 c,\, \mathbf{0})\,.$$

From the point of view of S' the relative motion with respect to S can be interpreted as the velocity of the mass. With

$$\mathbf{P} = \left\{ i\frac{E}{c},\, \mathbf{p} \right\}$$

---

[18] 1 eV $\approx 1.60 \cdot 10^{-19}$ Joule

3.4 Relativistic Mechanics

one can conclude from the invariance of the scalar product

$$p^2 - \frac{E^2}{c^2} = -m_0^2 c^2$$

or after rearrangement

$$E = \left[p^2 c^2 + m_0^2 c^4 \right]^{1/2} . \tag{3.53}$$

This relation between the relativistic three-momentum of a particle (massive) and the relativistic total energy is often used.

In the non-relativistic limit, which can also be characterised by $pc/(m_0 c^2) < 1$, one can extract the statement

$$E \xrightarrow{\text{nonrel.}} m_0 c^2 + \frac{p^2}{2m_0} + \dots .$$

The total relativistic energy equals the rest energy plus the non-relativistic kinetic energy plus relativistic corrections.

The energy-momentum relation (3.53) is also used for the discussion of elementary particles with a vanishing rest mass. If the rest mass equals zero, energy and momentum are proportional to each other

$$E = pc \qquad (m_0 = 0) .$$

Such particles can only move with the velocity of light in a vacuum. They can not be decelerated, but loose their energy of motion by their destruction into another form of energy (as the creation of massive particles). The photon (light quantum or γ-quantum) and the neutrinos are such particles.[19]

3.4.3 The Equations of Motion in Relativistic Mechanics

The last topic of this short outline of relativistic mechanics is the equation of motion for mass points. The continuation of the previous discussion is the extension of the classical equation of motion $\dot{p} = F$. A possible ansatz is the four-vector equation

$$\frac{d}{d\tau} \mathbf{P} = \mathbf{K} = \{ K_0, K_1, K_2, K_3 \} . \tag{3.54}$$

---

[19] The question concerning the rest mass of the three neutrinos (ν_e, ν_μ, ν_τ) is still not settled. After years of work, one has so far found, that a small rest mass is, within the accuracy of present measurements, possible.

The derivative of the four-momentum with respect to the eigenzeit is determined by a force with four components. This force has the name **Minkowski force**

$$\frac{d}{d\tau}\mathbf{P} = m_0 \frac{d}{d\tau}\mathbf{V} = m_0\mathbf{B} = m_0\{B_0, B_1, B_2, B_3\}. \tag{3.55}$$

The derivative of the four-momentum can be written as the rest mass multiplied by a four-acceleration. One property of this acceleration, which follows from (3.47), is

$$\mathbf{V} \cdot \mathbf{B} = 0. \tag{3.56}$$

The four-velocity and the four-acceleration are in every inertial system orthogonal with respect to each other.

The explicit calculation of the derivative of each component of the four-velocity with respect to the eigenzeit results in a first step

$$\frac{d}{d\tau}\boldsymbol{p} = \frac{d\boldsymbol{p}}{dt}\frac{dt}{d\tau} = \frac{1}{\sqrt{1-v^2/c^2}}\frac{d\boldsymbol{p}}{dt},$$

as the derivative of the time with respect to the eigenzeit is

$$\frac{dt}{d\tau} = \frac{1}{\sqrt{1-v^2/c^2}}.$$

In addition one has because of $\boldsymbol{p} = m(v)\boldsymbol{v}$

$$\frac{d}{dt}\boldsymbol{p} = m_0 \left(\frac{\boldsymbol{b}}{[1-v^2/c^2]^{1/2}} + \frac{1}{c^2}\frac{(\boldsymbol{v}\cdot\boldsymbol{b})\,\boldsymbol{v}}{[1-v^2/c^2]^{3/2}} \right). \tag{3.57}$$

Here the vector $\boldsymbol{b} = d\boldsymbol{v}/dt \equiv \boldsymbol{b}_{\text{class}}$ is the three-vector of the classical acceleration in the respective inertial system.

The space components of the four-acceleration are therefore

$$(B_1, B_2, B_3) = \frac{\boldsymbol{b}}{(1-v^2/c^2)} + \frac{1}{c^2}\frac{(\boldsymbol{v}\cdot\boldsymbol{b})\,\boldsymbol{v}}{\left(1-v^2/c^2\right)^2}. \tag{3.58}$$

Four small velocities they are

$$(B_1, B_2, B_3) \xrightarrow{v/c \ll 1} \boldsymbol{b} + \left\{\frac{v^2}{c^2}\boldsymbol{b} + \frac{(\boldsymbol{v}\cdot\boldsymbol{b})}{c^2}\boldsymbol{v}\right\} + \ldots.$$

The space components of **B** are in lowest order identical with the classical acceleration $\boldsymbol{b}$. In the order v^2/c^2 there appears, however, a term, which has the

3.4 Relativistic Mechanics

same direction as v. This is the reason, why the discussion of the motion of a mass point is more complicated in the theory of relativity as in classical physics.

The discussion of the component B_0 respectively the corresponding component of the force $K_0 = m_0 B_0$ can proceed in the following fashion. The condition $\mathbf{V} \cdot \mathbf{B} = 0$ implies $\mathbf{V} \cdot \mathbf{K} = 0$ or for the explicit form of the scalar product with the four-velocity

$$\sum_{\mu=1}^{3} V_\mu K_\mu - \frac{ic}{\sqrt{1 - v^2/c^2}} K_0 = 0 .$$

Resolution with respect to K_0 and use of the expression for V_μ, K_μ, ($\mu = 1, 2, 3$) from (3.46) and (3.54)

$$V_1 = \frac{v_x}{\sqrt{1 - v^2/c^2}}, \quad \ldots$$

$$K_1 = \frac{dp_1}{d\tau} = \frac{1}{\sqrt{1 - v^2/c^2}} \frac{dp_1}{dt}, \quad \ldots$$

yields

$$K_0 = \frac{i}{c} \frac{1}{\sqrt{1 - v^2/c^2}} \left(v \cdot \frac{dp}{dt} \right) .$$

The transition from kinematics to dynamics can be implemented, if the generalisation of Newton's equation of motion

$$\frac{dp}{dt} = F \tag{3.59}$$

is used. The *specified* force $F = (F_1, F_2, F_3)$ determines the variation in time of the relativistic three-momentum with the provision for the variation of the relativistic mass with the velocity. This relativistic equation of motion, which is referred to a specific inertial system, turns into the classical equation of motion for small velocities. In the general case one realises that the derivative of the relativistic three-acceleration is not proportional to the acceleration and the force F does not possess any specific transformation properties under Lorentz transformations.

The zero component of the Minkowski force K_0 is therefore

$$K_0 = \frac{i}{c} \frac{v \cdot F}{\sqrt{1 - v^2/c^2}} . \tag{3.60}$$

The role of this component will become clear, if one examines the corresponding component of the equation of motion more closely

$$\frac{i}{c}\frac{1}{\sqrt{1-v^2/c^2}}\frac{dE}{dt} = \frac{i}{c}\frac{\mathbf{v}\cdot\mathbf{F}}{\sqrt{1-v^2/c^2}}$$

or

$$\frac{dE}{dt} = \mathbf{v}\cdot\mathbf{F}\ . \qquad (3.61)$$

The variation of the relativistic total energy equals the power[20] of the external force. The alternative relation

$$dE = \mathbf{F}\cdot d\mathbf{r}$$

shows again, that the ansatz for the relativistic kinetic energy was consistent.

One can recognise a certain duplicated approach for the solution of relativistic problems of motion:

(1) The solution of a specific problem with specified forces in a specified inertial system can best be achieved with the equations

$$\frac{d}{dt}(m(v(t))\mathbf{v}(t)) = \mathbf{F} \qquad \frac{dE}{dt} = \mathbf{F}\cdot\mathbf{v}\ . \qquad (3.62)$$

(2) If one is faced with a transition of a problem of motion from one inertial system to another, one has to consider the transformation of the four-vectors **B** and **K**

$$B_{\mu'} = \sum_\lambda L_{\lambda\mu} B_\mu \qquad K_{\mu'} = \sum_\lambda L_{\lambda\mu} K_\mu\ .$$

The equations of motion in the primed system are

$$m_0\, \mathbf{B}' = \mathbf{K}'\ .$$

The explicit equations of motion in the form (3.62) for the primed system could then be extracted. It turns out, however that it is easier to transform the solutions in the system S.

The discussion of relativistic equations of motion is continued in Chap. 3.5.3 with the discussion of equations of motion for a point charge in electromagnetic fields. The first point to take up is, however, to answer the question about the covariance

---

[20] Recall: Power equals work per time.

of electrodynamics with respect to Lorentz transformations, which has been posed at the start.

3.5 Electrodynamics and the Theory of Relativity

The proof of the covariance of electrodynamics with respect to Lorentz transformations requires the demonstration, that all equations of electrodynamics retain their form under these transformation. This task will accomplished in three steps. For the potentials, for the fields and for the forces of electrodynamics. The discussion will be restricted to the vacuum theory. The representation used is free of the choice of a particular system of units.

3.5.1 The Potential Equations

One of the basic equations of electrodynamics, which expresses charge conservation, is the equation of continuity

$$\nabla \cdot \boldsymbol{j}_{tr} + \frac{\partial \rho_{tr}}{\partial t} = 0 \,.$$

It is convenient to write this equation for the present purpose as a scalar product of two four-vectors.

$$\bar{\nabla} \cdot \mathbf{J} = 0 \,. \tag{3.63}$$

The **four-current density**, which has been introduced in this fashion, has the components

$$\mathbf{J} = \{ i\, c\, \rho_{tr},\ j_{tr,x},\ j_{tr,y},\ j_{tr,z} \} \,. \tag{3.64}$$

If this argument is correct, one can state, that the invariance properties of the scalar product guarantee charge conservation in all inertial systems

$$\bar{\nabla} \cdot \mathbf{J} = \bar{\nabla}' \cdot \mathbf{J}' = 0 \,.$$

The relative strength of the electromagnetic interaction would produce considerable measurable effects, if this were not the case. No effects of this kind have been observed, which hint at a violation of charge conservation. The fact, that the vector **J** transforms like the event vector **J**, follows in the end from the consistency of the transformation properties of all equations of electrodynamics.

The scalar product of two four-gradients has the form

$$\bar{\nabla} \cdot \bar{\nabla} = \text{div}_4 \, \text{grad}_4 = -\frac{1}{c^2}\frac{\partial^2}{\partial t^2} + \Delta \, .$$

This scalar product is the d'Alembert operator, the operator of the wave equation of electrodynamics. This operator is an obvious relativistic invariant.

$$\Box = \Box' \, .$$

A look at the potential equations of electrodynamics

$$V = -4\pi k_e \rho_{tr} \qquad \boldsymbol{A} = -4\pi k_m \boldsymbol{j}_{tr}$$

with the gauge condition (the Lorentz gauge)

$$\nabla \cdot \boldsymbol{A} + \frac{k_m}{k_e}\frac{\partial V}{\partial t} = 0 \, ,$$

invites the following remarks: For reasons of consistency, the potentials V and $\boldsymbol{A}$ (up to factors) have to form a four-vector, if ρ_{tr} and $\boldsymbol{j}_{tr}$ are (up to factors) the components of a four-vector. Define now the four-potential

$$\boldsymbol{A} = \left\{\frac{ick_m V}{k_e}, \boldsymbol{A}\right\} = \left\{\frac{iV}{ck_f}, \boldsymbol{A}\right\} , \tag{3.65}$$

where the last step follows from (1.42). The four-potential equations can then be written in the form

$$\Box \boldsymbol{A}(R) = -4\pi k_m \boldsymbol{J}(R) \, , \tag{3.66}$$

e.g. for the zero component

$$\Box \left(\frac{iV}{ck_f}\right) = -4\pi k_m (ic\,\rho_{tr}) \quad \longrightarrow \quad V = -4\pi k_e \rho_{tr} \, .$$

The gauge condition can be written as the scalar product of two four-vectors

$$\bar{\nabla} \cdot \boldsymbol{A} = \frac{1}{ic}\frac{\partial}{\partial t}\left(\frac{iV}{ck_f}\right) + \nabla \cdot \boldsymbol{A} = 0 \, . \tag{3.67}$$

The Lorentz gauge is a relativistic invariant. It is valid in any inertial system, if it can be used in a particular inertial system.

The potential equation has the property: The four-potential and the four-current density in any inertial system $\boldsymbol{A}'$, $\boldsymbol{J}'$, which are calculated from $\boldsymbol{A}$, $\boldsymbol{J}$ in a specific

3.5 Electrodynamics and the Theory of Relativity

inertial system with a Lorentz transformation, satisfy the equation

$$\mathbf{A}'(\mathbf{R}') = -4\pi k_m \mathbf{J}'(\mathbf{R}') .$$

The potential equations are covariant.

3.5.2 The Field Equations

The potentials are only auxiliary quantities. This is the reason, why there is still the need to answer the question: Is it possible to make the transition from the transformation properties of the potentials to the transformation properties of the fields? And what is the form of the field equations, the Maxwell equations, in a covariant formulation?

In order to answer the first part of these questions one starts with the relations (1.67) and (1.68)

$$\mathbf{B} = \nabla \times \mathbf{A} = \left(\frac{\partial}{\partial y} A_z - \frac{\partial}{\partial z} A_y, \ldots \right)$$

$$\mathbf{E} = -\nabla V - k_f \frac{\partial \mathbf{A}}{\partial t} = -\left(\frac{\partial V}{\partial x} + k_f \frac{\partial A_x}{\partial t}, \ldots \right) .$$

One can expect, that the fields can be constructed from the components of the four-potential and the components of the four-gradient. There are 16 products

$$G_{\mu,\nu} = \frac{\partial A_\nu}{\partial X_\mu} ,$$

which one can construct with these ingredients. Under a Lorentz transformation for a transition between two inertial systems they behave like a product of four-vectors

$$G'_{\mu',\nu'} = \sum_{\mu,\nu} L_{\mu',\mu} L_{\nu',\nu} G_{\mu,\nu} .$$

The equations of this transformation indicate, that the quantity with 16 components $[G] = (G_{\mu,\nu})$ is a tensor of rank 2 in the four-dimensional Minkowski space. On the other hand there exist only 6 field components, which have to be selected from these 16 quantities. There is indeed only one possibility: the antisymmetric tensor $[F]$ with the components

$$F_{\mu,\nu} = G_{\mu,\nu} - G_{\nu,\mu} = \frac{\partial A_\nu}{\partial X_\mu} - \frac{\partial A_\mu}{\partial X_\nu} . \tag{3.68}$$

The components of [F] have the properties

$$F_{\mu,\mu} = 0 \quad \text{(4 tensor elements)}$$

$$F_{\mu,\nu} = -F_{\nu,\mu} \quad \mu \neq \nu \quad \text{(6 conditions)}$$

The tensor [F] is also a tensor of rank 2. In view of the 16 restrictive conditions there are exactly 6 independent elements. In order to prove that the 6 independent tensor elements represent the components of the electric and the magnetic field, one can just list them explicitly

$$F_{01} = +\frac{1}{ic}\frac{\partial A_x}{\partial t} - \frac{\partial}{\partial x}\left(\frac{iV}{ck_f}\right) = -\frac{i}{ck_f}\left(k_f\frac{\partial A_x}{\partial t} + \frac{\partial V}{\partial x}\right) = \frac{i}{ck_f}E_x \quad \checkmark.$$

The relation $k_e/(k_f k_m) = c^2$ has been used for this result (see (1.38) p. 25). In a similar manner one finds

$$F_{02} = -\frac{i}{ck_f}\left(k_f\frac{\partial A_y}{\partial t} + \frac{\partial V}{\partial y}\right) = \frac{i}{ck_f}E_y \quad \checkmark$$

$$F_{03} = -\frac{i}{ck_f}\left(\frac{1}{c}\frac{\partial A_z}{\partial t} + \frac{\partial V}{\partial z}\right) = \frac{i}{ck_f}E_z \quad \checkmark$$

$$F_{12} = \frac{\partial A_y}{\partial x} - \frac{\partial A_x}{\partial y} = (\nabla \times A)_z = B_z \quad \checkmark$$

$$F_{13} = \frac{\partial A_z}{\partial x} - \frac{\partial A_x}{\partial z} = -B_y \quad \checkmark$$

$$F_{23} = \frac{\partial A_z}{\partial y} - \frac{\partial A_y}{\partial x} = B_x. \quad \checkmark$$

The field tensor [F] can thus be summarised in the matrix

$$[F] = \begin{bmatrix} 0 & (iE_x)/(ck_f) & (iE_y)/(ck_f) & (iE_z)/(ck_f) \\ -(iE_x)/(ck_f) & 0 & B_z & -B_y \\ -(iE_y)/(ck_f) & -B_z & 0 & B_x \\ -(iE_z)/(ck_f) & B_y & -B_x & 0 \end{bmatrix}.$$

3.5 Electrodynamics and the Theory of Relativity

The transformation of the tensor elements in two inertial systems S and S'

$$F'_{\mu'\nu'}(\mathbf{R}') = \sum_{\mu\nu} L_{\mu'\mu} L_{\nu'\nu} F_{\mu\nu}(\mathbf{R})$$

yield (see Detail 3.7.4.1) with the simple variant of the Lorentz transformation the tensor

$$[F'] = \begin{bmatrix} 0 & F'_{01} & F'_{02} & F'_{03} \\ F'_{10} & 0 & F'_{12} & F'_{13} \\ F'_{20} & F'_{21} & 0 & F'_{23} \\ F'_{30} & F'_{31} & F'_{32} & 0 \end{bmatrix},$$

in the system S' with the elements

$$F'_{01} = \frac{\mathrm{i} E'_y}{ck_f} = \mathrm{i} \frac{E_x}{ck_f} \qquad F'_{02} = \frac{\mathrm{i} E'_y}{ck_f} = \mathrm{i}\gamma \left(\frac{E_y}{ck_f} - \beta B_z \right)$$

$$F'_{03} = \frac{\mathrm{i} E'_z}{ck_f} = \mathrm{i}\gamma \left(\frac{E_z}{ck_f} - \beta B_y \right)$$

$$F'_{10} = \frac{-\mathrm{i} E'_x}{ck_f} = -\mathrm{i} \frac{E_x}{ck_f} \qquad F'_{12} = B'_z = \gamma \left(B_z - \frac{\beta E_y}{ck_f} \right)$$

$$F'_{13} = -B'_y = -\gamma \left(B_y - \frac{\beta E_z}{ck_f} \right)$$

$$F'_{20} = \frac{-\mathrm{i} E'_y}{ck_f} = -\mathrm{i}\gamma \left(\frac{E_y}{ck_f} - \beta B_z \right) \qquad F'_{21} = -B'_z = -\gamma \left(B_z - \frac{\beta E_y}{ck_f} \right)$$

$$F'_{23} = B'_x = B_x$$

$$F'_{30} = \frac{-\mathrm{i} E'_z}{ck_f} = -\mathrm{i}\gamma \left(\frac{E_z}{ck_f} - \beta B_y \right) \qquad F'_{31} = B'_y = -\gamma \left(B_y - \frac{\beta E_z}{ck_f} \right)$$

$$F'_{32} = -B'_x = -B_x \, .$$

The transformation properties of the electromagnetic fields are more complicated than the properties of the potentials in the covariant formulation, as the electromagnetic fields are elements of a tensor. The magnetic and the electric components of the fields are transformed together in the relativistic theory. An electric field as observed by S ($\mathbf{E} \neq \mathbf{0}$, $\mathbf{B} = \mathbf{0}$) is observed as an electric **and** a magnetic field by

S'. This is in accord with the fact, that the observer S' registers a moving charge with an electromagnetic field, while the charge is resting in the system S.

For the discussion of the Maxwell equations (in vacuum) this set is divided into two groups:

- Equations without source terms (4 homogeneous equations), in which the description in terms of potentials is rather obvious

$$\nabla \cdot \boldsymbol{B} = 0 \qquad \nabla \times \boldsymbol{E} + k_f \frac{\partial \boldsymbol{B}}{\partial t} = \boldsymbol{0} \ .$$

- Equations with source terms (4 inhomogeneous equations)

$$\nabla \times \boldsymbol{B} - \frac{k_m}{k_e} \frac{\partial \boldsymbol{E}}{\partial t} = 4\pi k_m \boldsymbol{j}_{tr}$$

$$\nabla \cdot \boldsymbol{E} = 4\pi k_e \rho_{tr} \ .$$

The Maxwell equations contain derivatives of the fields. This requires the consideration of a tensor of rank 3, which arises through the differentiation of the elements of the tensor [F]. This tensor

$$\tilde{H}_{\lambda\mu\nu} = \frac{\partial}{\partial X_\lambda} F_{\mu\nu}$$

has $4^3 = 64$ elements. One of these tensors of rank 3 is of particular interest. This is the tensor, which arises from the element $\tilde{H}_{\lambda\mu\nu}$ plus all the terms, which are obtained by a cyclic interchange of the indices

$$H_{\lambda\mu\nu} = \frac{\partial}{\partial X_\lambda} F_{\mu\nu} + \frac{\partial}{\partial X_\mu} F_{\nu\lambda} + \frac{\partial}{\partial X_\nu} F_{\lambda\mu} \ . \qquad (3.69)$$

The elements of this tensor have the following properties:

$$H_{\lambda\mu\nu} = H_{\mu\nu\lambda} = H_{\nu\lambda\mu}$$

$$H_{\lambda\mu\nu} = -H_{\lambda\nu\mu} = -H_{\mu\lambda\nu} = -H_{\nu\mu\lambda} \ . \qquad (3.70)$$

This tensor is antisymmetric with respect to the interchange of any pair of the three indices. All elements of [H] with two equal indices have the value zero. The elements, which do not vanish, are those with three different indices. There exist exactly four basic combinations

$$012 \qquad 013 \qquad 023 \qquad 123$$

3.5 Electrodynamics and the Theory of Relativity

and 20 additional permutations. If one inserts e.g. in

$$H_{123} = \frac{\partial}{\partial X_1} F_{23} + \frac{\partial}{\partial X_2} F_{31} + \frac{\partial}{\partial X_3} F_{12}$$

the elements of the tensor [F], one finds for the divergence of the **B**-field

$$H_{123} = \frac{\partial B_1}{\partial x} + \frac{\partial B_y}{\partial y} + \frac{\partial B_z}{\partial z} = \nabla \cdot \boldsymbol{B} \ .$$

For H_{023} one recognises the x-component of the law of induction

$$H_{023} = \frac{\partial}{\partial X_0} F_{23} + \frac{\partial}{\partial X_2} F_{30} + \frac{\partial}{\partial X_3} F_{02}$$

$$= \frac{1}{ic} \frac{\partial B_x}{\partial t} - \frac{\partial}{\partial y} \left(\frac{i E_z}{ck_f} \right) + \frac{\partial}{\partial z} \left(\frac{i E_y}{ck_f} \right)$$

$$= \frac{1}{i c k_f} \left(k_f \frac{\partial B_x}{\partial t} + (\nabla \times \boldsymbol{E})_x \right) \ .$$

The remaining components of the law of induction correspond to the tensor elements H_{012} and H_{013}.

The homogeneous Maxwell equations can therefore written in the form

$$H_{\lambda\mu\nu}(\boldsymbol{R}) = \frac{\partial}{\partial X_\lambda} F_{\mu\nu}(\boldsymbol{R}) + \frac{\partial}{\partial X_\mu} F_{\nu\lambda}(\boldsymbol{R}) + \frac{\partial}{\partial X_\nu} F_{\lambda\mu}(\boldsymbol{R}) = 0$$

$$\text{for all } \lambda, \mu, \nu \ . \quad (3.71)$$

As a consequence of the symmetry of the tensor, one realises, that 4 of the 64 equations are relevant and 40 are identically equal to zero.

The elements of the tensors [H] transform according to

$$H'_{\lambda'\mu'\nu'}(\boldsymbol{R}') = \sum_{\mu\nu\lambda} L_{\lambda'\lambda} \, L_{\mu'\mu} \, L_{\nu'\nu} H_{\lambda\mu\nu}(\boldsymbol{R}) \ .$$

It follows, however, that every term on the right hand side vanishes

$$H'_{\lambda'\mu'\nu'}(\boldsymbol{R}') = 0 \ .$$

The homogeneous Maxwell equations have the same form in each inertial system.

For the inhomogeneous Maxwell equations one can argue as follows: On the left hand side one also finds derivatives of the electromagnetic fields (the tensor [F]).

As the right hand side contains the component of a four-vector, it is necessary to construct four-vectors with a contraction as e.g.

$$\tilde{H}_{\lambda\mu\nu} = \frac{\partial}{\partial X_\lambda} F_{\mu\nu}$$

to obtain

$$\sum_\mu \frac{\partial}{\partial X_\mu} F_{\mu\nu} \ .$$

One can check explicitly, that this contraction transforms like a four-vector

$$\left(\sum_{\mu'} \frac{\partial F'_{\mu'\nu'}}{\partial X'_{\mu'}} \right) = \sum_{\mu'\nu_1\nu_2\nu_3} L_{\mu'\nu_1} L_{\mu'\nu_2} L_{\mu'\nu_3} \frac{\partial F_{\nu_2\nu_3}}{\partial X_{\nu_1}}$$

$$= \sum_{\nu_3} L_{\mu'\nu_3} \left(\sum_{\nu_1} \frac{\partial F_{\nu_1\nu_3}}{\partial X_{\nu_1}} \right) ,$$

as the sum over μ' corresponds to the orthogonality relation

$$\sum_{\mu'} L_{\mu'\nu_1} L_{\mu'\nu_2} = \delta_{\nu_1\nu_2} \ .$$

One can, at this stage, practically guess, that the inhomogeneous Maxwell equations can be summarised as

$$\sum_\mu \frac{\partial F_{\mu\nu}(\mathbf{R})}{\partial X_\mu} = 4\pi k_m J_\nu(\mathbf{R}) \qquad \nu = 0, 1, 2, 3 \ . \tag{3.72}$$

The equation with $\nu = 0$ is indeed Coulomb's law, the equations with $\nu = 1, 2, 3$ are identical with the extended law of Ampère (see Detail 3.7.4.2).

These equations are covariant. If (3.72) is multiplied with $L_{\nu'\nu}$, one obtains by summation over ν with the steps

$$\sum_{\mu\nu} L_{\nu'\nu} \frac{\partial F_{\mu\nu}(\mathbf{R})}{\partial X_\mu} = \sum_{\nu\mu\mu'} L_{\nu'\nu} \delta_{\mu\mu'} \frac{\partial F_{\mu\nu}(\mathbf{R})}{\partial X_{\mu'}}$$

$$= \sum_{\nu\mu\mu'\sigma} L_{\nu'\nu} L_{\sigma\mu} L_{\sigma\mu'} \frac{\partial F_{\mu\nu}(\mathbf{R})}{\partial X_{\mu'}} = \sum_\sigma \frac{\partial F_{\sigma\nu'}(\mathbf{R'})}{\partial X'_\sigma}$$

3.5 Electrodynamics and the Theory of Relativity

the statement

$$\sum_\sigma \frac{\partial F_{\sigma v'}(\mathbf{R}')}{\partial X'_\sigma} = 4\pi k_m J'_{v'}(\mathbf{R}') .$$

The inhomogeneous Maxwell equations have the same form in every inertial system.

Finally, one should ask the question, in how far one can combine the electromagnetic fields in order to win *Lorentz invariant* quantities. The answer is: There exist exactly two combinations, which have the same value in each inertial system. One of these quantities is the complete contraction of the tensor [F]

$$\frac{k_f}{k_m} \sum_{\mu,v=0}^{3} F_{\mu v} F_{\mu v} = 2 \left(\frac{k_f}{k_m} B^2 - \frac{E^2}{k_e} \right) . \qquad (3.73)$$

The second quantity is also a contraction, namely of the product of two tensor elements with an arbitrary combination of the indices. In order to formulate this contraction

$$\frac{k_f}{k_m} \sum_{\mu,v,\lambda,\rho=0}^{3} \epsilon_{\mu v \lambda \rho} F_{\mu v} F_{\lambda \rho} = \frac{8\,\mathrm{i}}{c k_m} (\mathbf{E} \cdot \mathbf{B}) \qquad (3.74)$$

one needs the four-dimensional extension of the Levi-Civita symbol $\epsilon_{\mu v \lambda \rho}$ with the properties

$$\epsilon_{\mu v \lambda \rho} = \begin{cases} 0 & \text{if two indices are equal} \\ +1 & \text{for even permutations of } (1234) \\ -1 & \text{for odd permutations of } (1234) \end{cases} .$$

These invariants are essential quantities, which are used in the foundation of quantum field theories as quantum electrodynamics or the extension quantum chromodynamics.

3.5.3 The Relativistic Forces on Charges

The last equation of electrodynamics, which will be discussed, is the equation of motion of a charge in electric and magnetic fields. The motion of charges in such fields is controlled by the complete Lorentz force. One obtains for the three components of the force

$$f_\mu(\mathbf{R}) = k_f \sum_v F_{\mu v}(\mathbf{R}) J_v(\mathbf{R}) \quad (\mu = 1, 2, 3) \qquad (3.75)$$

if covariant quantities in the form force per volume (see Chap. 1.4.2)

$$f = \rho_{tr} E + k_f (j_{tr} \times B)$$

are introduced into the equation of the force (see Detail 3.7.5.1). The right hand side of (3.75) contains the contraction of a second rank tensor with a four-vector. The three components of the force are the space part of a four-vector, which causes the change of the mechanical momentum density (momentum per volume) with time. If Eq. (3.75) is used for the definition of a time component f_0, one finds

$$f_0 = \frac{i}{c} E \cdot j_{tr} .$$

The time component is up to a factor identical with the Joule heat term on p. 36. This term represents the change of the mechanical energy of a charge with time by the work (per volume) of the fields.

The contraction of the tensors [F] with the four-vector **J** leads to the statement, that the force vector $\mathbf{K} = \{f_0, f_1, f_2, f_3\}$ is a four-vector. The consequence is: The total electromagnetic force is Lorentz covariant

$$f'_\mu(\mathbf{R}') = k_f \sum_\nu F'_{\mu\nu}(\mathbf{R}') J'_\nu(\mathbf{R}') .$$

On the other side it has been found (compare p. 197), that neither the magnetic nor the electric force by themselves are Lorentz or Galilei covariant. This supports the fact, that the ansatz (Dreizler and Lüdde, 2024, Electrostatics and Magnetostatics (Springer Berlin Heidelberg), Eq. (5.46))

$$F_{\text{mag}}(r) = q k_f \left[v(r) \times B(r) \right]$$

for the action of electromagnetic forces on charges is acceptable.

The arguments in Chap. 1.4.2 suggest, that the relation (3.75) could represent a covariant extension of the concept of Maxwell's stress tensor. In order to verify this fact, one replaces the components of the four-current in (3.75) by the inhomogeneous Maxwell equations (3.72). The result

$$f_\mu = \frac{k_f}{4\pi k_m} \sum_{\nu\rho} F_{\mu\nu} \frac{\partial F_{\nu\rho}}{\partial X_\rho}$$

can be written as the four-divergence of a tensor of second rank using the relation

$$\sum_{\nu\rho} \frac{\partial}{\partial X_\rho} (F_{\mu\nu} F_{\nu\rho}) = \sum_{\nu\rho} \left(F_{\mu\nu} \frac{\partial F_{\nu\rho}}{\partial X_\rho} + F_{\nu\rho} \frac{\partial F_{\mu\nu}}{\partial X_\rho} \right) .$$

3.5 Electrodynamics and the Theory of Relativity

The second term on the right hand side is changed with the aid of the homogeneous Maxwell equations and the antisymmetry of the tensor [F] (see Detail 3.7.5.2)

$$\sum_{\nu\rho} F_{\nu\rho} \frac{\partial F_{\mu\nu}}{\partial X_\rho} = -\frac{1}{4} \sum_{\nu\lambda\rho} \delta_{\mu,\rho} \frac{\partial}{\partial X_\rho} (F_{\nu\lambda} F_{\nu\lambda}) \ .$$

With the definition of the symmetric **energy-momentum tensor**

$$T_{\mu\nu} = \frac{k_f}{4\pi k_m} \sum_\lambda \left(F_{\mu\lambda} F_{\lambda\nu} + \frac{1}{4}\delta_{\mu,\nu} \sum_\rho F_{\lambda\rho} F_{\lambda\rho} \right) \qquad (3.76)$$

one can change the equation of the force into the form

$$f_\mu = \sum_\nu \frac{\partial T_{\mu\nu}}{\partial X_\nu} \ . \qquad (3.77)$$

The trace of the tensor [T] vanishes because the antisymmetry of the tensor [F]

$$\sum_\mu T_{\mu\mu} = \frac{k_f}{4\pi k_m} \sum_{\mu\lambda} \left(F_{\mu\lambda} F_{\lambda\nu} + \frac{1}{4} \sum_\rho F_{\lambda\rho} F_{\lambda\rho} \right)$$

$$= \frac{k_f}{4\pi k_m} \sum_{\mu\lambda} \left(F_{\mu\lambda} F_{\lambda\nu} + F_{\lambda\mu} F_{\lambda\mu} \right) = 0 \ . \qquad (3.78)$$

The elements of the tensor [T] can be identified with known quantities: The T_{00}-element is the energy density of the electromagnetic field

$$T_{00} = \frac{k_f}{4\pi k_m} \sum_\lambda \left(F_{0\lambda} F_{\lambda 0} + \frac{1}{4} \sum_\rho F_{\lambda\rho} F_{\lambda\rho} \right) = \frac{1}{8\pi} \left(\frac{E^2}{k_e} + \frac{k_f B^2}{k_m} \right) \ .$$

The elements $T_{0k} = T_{k0}$ with $k = 1, 2, 3$ are up to a factor the components of the Poynting vector, so e.g. for

$$T_{01} = \frac{k_f}{4\pi k_m} \sum_\lambda F_{0\lambda} F_{\lambda 1} = -\frac{i}{4\pi c k_m} (E \times B)_1 \ .$$

The space elements T_{ik} with $i, k = 1, 2, 3$ form the Maxwell stress tensor, which has been introduced in Chap. 1.4.2

$$T_{ik} = \frac{1}{4\pi k_e} \left(E_i E_k - \frac{1}{2} E^2 \delta_{i,k} \right) + \frac{k_f}{4\pi k_m} \left(B_i B_k - \frac{1}{2} B^2 \delta_{i,k} \right) \ .$$

The conservation laws for momentum and energy can, as seen in Chap. 1.4, be won by volume integration of the covariant force equations (see Detail 3.7.5.3).

The concept of the Lagrangian and the Hamiltonian is very useful for the discussion of problems in relativistic mechanics (and the foundation of relativistic field theories). The application of these concepts for the equation of motion of electrodynamics is presented in the next section.

3.5.4 The Lagrange Equations

The motion of a mass point (or a point charge) is governed by Hamilton's principle (Dreizler and Lüdde, 2010, Theoretical Mechanics (Springer Berlin Heidelberg) Chap. 5.4). The principle says, that the motion in the interval $[t_1, t_2]$ proceeds in such a fashion, so that the action integral with the Lagrangian

$$I = \int_{t_1}^{t_2} L(\mathbf{r}, \dot{\mathbf{r}}, t)\, dt$$

is extremal. The extension of this principle into the relativistic regime is based on the principle of relativity, which demands, that the action integral (a pure number for a given Lagrangian) has to be a Lorentz invariant. The extension of the action integral can be written in standard form

$$I = \int_{t_1}^{t_2} L_{\text{rel}}(\mathbf{r}, \dot{\mathbf{r}}, t)\, dt \tag{3.79}$$

or in covariant form. The covariant form can be obtained by transition to Minkowski coordinates and the introduction of the eigenzeit

$$I = \int_{\tau_1}^{\tau_2} \gamma\, L_{\text{rel}}(\mathbf{X}, \mathbf{V}, \tau)\, d\tau \ .$$

As the eigenzeit is Lorentz invariant, the remaining factor, the product of the relativistic Lagrangian and the factor $\gamma = [1 - (v/c)^2]^{-1/2}$ has to be an invariant

$$L_{\text{inv}}(\mathbf{X}, \mathbf{V}, \tau) = \gamma L_{\text{rel}}(\mathbf{X}, \mathbf{V}, \tau)\ , \tag{3.80}$$

so that the invariance of the action integral can be guaranteed.

In order to find the relativistic Lagrangian it is in most cases sufficient to ensure the invariance of L_{inv} by demanding, that it goes over into the non-relativistic limit for small velocities

$$L_{\text{rel}} \xrightarrow{v^2 \ll c^2} L_{\text{nrel}}\ . \tag{3.81}$$

3.5 Electrodynamics and the Theory of Relativity

For a free particle one can use as an ansatz the simplest variant

$$L_{\text{inv,f}} = -m_0 c^2 , \tag{3.82}$$

and find for the corresponding relativistic Lagrangian

$$L_{\text{rel,f}} = -\frac{m_0 c^2}{\gamma} = -m_0 c^2 \left[1 - \frac{v^2}{c^2}\right]^{1/2}$$

$$\xrightarrow{v^2 \ll c^2} -m_0 c^2 \left\{1 - \frac{1}{2}\frac{v^2}{c^2} + \ldots\right\} = \frac{1}{2}m_0 v^2 - m_0 c^2 + \ldots .$$

The ansatz (3.82) yields the result, that is required. The constant, which has appeared, is the negative rest energy, by which the energy scale is fixed.

The non-relativistic part of the Lagrangian for the **interaction** of a point charge q in a time dependent electromagnetic field, which connects with the Lorentz force, is

$$L_{\text{nrel, WW}} = -q\left(V(\mathbf{r},t) - k_f \mathbf{A}(\mathbf{r},t)\cdot \mathbf{v}\right) . \tag{3.83}$$

The Lagrangian

$$L_{\text{nrel, em}} = L_{\text{nrel, f}} + L_{\text{n.rel, WW}}$$

can be changed with the prescription

$$\frac{d}{dt}\left(\frac{\partial L_{\text{nrel, em}}}{\partial \dot{x}_i}\right) - \frac{\partial L_{\text{nrel, em}}}{\partial x_i} = 0$$

into the non-relativistic Lagrangian (see Detail 3.7.6.1)

$$\frac{d\mathbf{p}}{dt} = q\left(\mathbf{E} + k_f(\mathbf{v}\times\mathbf{B})\right) \qquad \mathbf{p} = m_0 \mathbf{v} . \tag{3.84}$$

If the classical potentials are replaced by the components of the four-potential and the velocity is replaced by the components of the four-momentum (3.48), one arrives at the Lagrangian

$$L_{\text{nrel, WW}} = -q\left(V - k_f \mathbf{v}\cdot \mathbf{A}\right)$$

$$= q\frac{k_f}{m_0}\left[(\mathrm{i}\,m_0 c)\left(\frac{k_m}{k_e}\mathrm{i} c V\right) + (m_0 \mathbf{v})\cdot(\mathbf{A})\right]$$

$$= \frac{q}{m_0} k_f (\mathbf{P}\cdot\mathbf{A})\frac{1}{\gamma} .$$

The scalar product $(\mathbf{P} \cdot \mathbf{A})$ is Lorentz invariant, so that it is possible to identify an invariant relativistic Lagrangian as $\gamma L_{\text{nrel, ww}} = L_{\text{inv, ww}}$. This implies the fact, that there is no difference between the relativistic and the standard interaction Lagrangian of a point charge with electromagnetic fields.

The total Lagrangian for a relativistic particle in an electromagnetic field is thus

$$L_{\text{rel, em}} = \frac{1}{\gamma}\left[-m_0 c^2 + \frac{q}{m_0} k_f (\mathbf{P} \cdot \mathbf{A})\right], \tag{3.85}$$

or in an explicit form

$$L_{\text{rel, em}} = -m_0 c^2 \left[1 - \frac{v^2}{c^2}\right]^{1/2} + q\left(k_f \mathbf{A} \cdot \mathbf{v} - V\right).$$

With the Lagrangian it is possible to apply the usual definitions in order to calculate the generalised three-momentum[21]

$$\Pi_i = \frac{\partial L}{\partial v_i} = \gamma m_0 v_i + k_f q A_i = p_i + k_f q A_i \qquad i = 1, 2, 3. \tag{3.86}$$

The generalised momentum differs from the relativistic three-momentum $\mathbf{p} = (p_1, p_2, p_3)$ by an additional electromagnetic contribution. If the Lagrange equations are calculated with

$$\frac{d}{dt}\Pi_i - \frac{\partial L}{\partial x_i} = 0 \qquad i = 1, 2, 3,$$

one finds for the equations of motion (Detail 3.7.6.1)

$$\frac{d}{dt}\mathbf{p} = q\left[\mathbf{E} + k_f(\mathbf{v} \times \mathbf{B})\right] \qquad \mathbf{p} = m\mathbf{v} = \gamma m_0 \mathbf{v}. \tag{3.87}$$

The difference with respect to the non-relativistic equations of motion (3.84) is the presence of a relativistic momentum instead of the relativistic momentum.

In order to derive the relativistic Hamiltonian

$$H(\mathbf{x}, \mathbf{\Pi}, t) = \sum_{i=1}^{3} \Pi_i v_i - L_{\text{rel}}(\mathbf{x}, \mathbf{v}, t)$$

according to the standard definition, one has to eliminate the relativistic three-velocity in favour of the generalised three-momentum (Detail 3.7.6.2). In order to

---

[21] Compare with the remarks concerning the non-covariant form of the relativistic equations of motion in Chap. 3.4.3.

3.5 Electrodynamics and the Theory of Relativity

do this, one has to resolve (3.86) with respect to the velocity components in order to find

$$v = \frac{c\left(\mathbf{\Pi} - k_f q \mathbf{A}\right)}{\left[\left(\mathbf{\Pi} - k_f q \mathbf{A}\right)^2 + m_0^2 c^2\right]^{1/2}} .$$

The transcription of the free Lagrangian and the magnetic interaction is then

$$m_0 c^2 \left[1 - v^2/c^2\right]^{1/2} = \frac{m_0^2 c^3}{\left[\left(\mathbf{\Pi} - k_f q \mathbf{A}\right)^2 + m_0^2 c^2\right]^{1/2}}$$

$$q k_f v \cdot \mathbf{A} = \frac{q k_f c \left(\mathbf{\Pi} \cdot \mathbf{A} - k_f q \mathbf{A}^2\right)}{\left[\left(\mathbf{\Pi} - k_f q \mathbf{A}\right)^2 + m_0^2 c^2\right]^{1/2}} .$$

Collection of all the terms

$$H = \mathbf{\Pi} \cdot v + m_0 c^2 \left[1 - v^2/c^2\right]^{1/2} - q k_f v \cdot \mathbf{A} + q V ,$$

results in the compact relation

$$H = \left[c^2 \left(\mathbf{\Pi} - k_f q \mathbf{A}\right)^2 + m_0^2 c^4\right]^{1/2} + q V . \tag{3.88}$$

If the relativistic Hamiltonian, is to represent the energy, the zero component has to be the zero component of a four-vector. In order to check this statement, one has a look at the electromagnetic four-momentum

$$\mathbf{P}_{em} = \left\{\frac{E_{em}}{ic}, \mathbf{p}_{em}\right\} ,$$

where the energy is replaced by the Hamiltonian minus the electrical energy and the three-momentum by the generalised three-momentum

$$\mathbf{P}_{em} = \left\{\frac{(H - qV)}{ic}, \mathbf{\Pi} - k_f q \mathbf{A}\right\} . \tag{3.89}$$

Calculation of the scalar product

$$\mathbf{P}_{em} \cdot \mathbf{P}_{em} = -\frac{1}{c^2}(H - qV)^2 + \left(\mathbf{\Pi} - k_f q \mathbf{A}\right)^2 = -(m_0^2 c^2)$$

and comparison of this result with the expression

$$(H - qV)^2 = c^2 \left(\mathbf{\Pi} - k_f\, q\, \mathbf{A}\right)^2 + m_0^2 c^4\ ,$$

which can be won directly from (3.88), shows, that $(H - qV)/(\mathrm{i}c)$ is the zero component of a four-momentum.

A covariant formulation is also possible. One can, in order to illustrate this statement, consider the variation

$$\delta \int_{\tau_1}^{\tau_2} L_{\mathrm{inv}}(\mathbf{X},\, \mathbf{V},\, \tau)\, \mathrm{d}\tau = 0\ ,$$

where the variation of the Minkowski coordinates and the components of the four-velocity vanishes at the limits of the integration interval. A complication arises from the constraint (see (3.47))

$$\mathbf{V} \cdot \mathbf{V} = -c^2\ .$$

The variation of the four components of the relativistic velocity are not independent of each other. This auxiliary condition has to be treated with the method of Lagrange multipliers or with an equivalent device. This covariant formulation will not be discussed here.

3.6 A Short History of the Aether and Other Historical Remarks

The question, whether light is a wave phenomenon or whether it involves the motion of particles (corpuscles), has occupied the scientists of the seventeenth century. The best known proponents of these contrary opinions of the nature were C. Huygens on the side of wave theory and I. Newton on the side of particle theory. A decision in favour of waves emerged starting with the year 1801. The results of optical experiments, as e.g. the appearance of Newtonian rings were interpreted consistently by T. Young as interference patterns. Only a few years later the fact, that light can be polarised, was recognised by the double refraction of light (E. Malus, 1808—actually Malus was an adherent of the particle interpretation). This insight and additional optical phenomena served A. Fresnel (1815, 1821) as a basis for a rather complete wave theory of light.

In the second half of the nineteenth century it could be proved on the basis of Maxwell's theory, that light was a special form of electromagnetic waves. R. Kohlrausch and W. Weber determined the velocity of light with the help of the equations of Maxwell's theory from optical data. H. Hertz was able to prove the existence of electromagnetic waves in the range of frequencies around $10^9\,\mathrm{s}^{-1}$ in (1887/1888) (resp. a wave length in the range of decimetre) and showed that the properties of these waves were the same as the properties of light waves.

3.6 A Short History of the Aether and Other Historical Remarks

It is really a twist of nature, that the particle aspect of the electromagnetic radiation was restored again in the twentieth century e.g. with the examination of cavity radiation (M. Planck, 1900) and of the photo-electric effect by A. Einstein (1905). Quantum mechanics was finally able to unite the two contrary aspects under one roof.

The mechanistic view of the world in the nineteenth century initiated the idea, that there must be a carrier of the light waves in analogy to the air as a carrier of sound waves. The hypothesis of the existence of an *aether*, which permeates all matter in complete space, was a central point of the 1821 theory of Fresnel. One had to assume that

- this aether is transparent, as one could not find any suitable other medium between the earth and the sun.
- It has a low viscosity, as the earth as well as the other celestial bodies move without any friction through the aether.
- It is very high rigid (high stiffness, so that one can explain the high value for the velocity of light ($c_L = 3 \cdot 10^8$ m/s versus the velocity of sound $c_S = 3 \cdot 10^2$ m/s). The speed of propagation of waves in a medium is normally $c \propto$ [modul of elasticity]$^{1/2}$.

The Maxwell equations (1865) contain the statement, that electromagnetic waves move in vacuum with a velocity, which is independent of the motion of the source. According to the classic Galilei transformation this is only possible in a special system of reference, which was identified with the aether system. The Maxwell equations should be valid in their original form in this system. Any motion relative to the aether system should therefore be detectable. The negative result of the Michelson-Morley experiments (1881, 1887) showed however, in the words of Michelson, that 'the hypothesis of a stationary aether could not be correct'. This statement was confirmed by all interferometer experiments that followed. A way for the salvation of the aether hypothesis could be found, if the aether is moving with the earth. This possibility is counter acted by the aberration of the light from distant stars.[22] The aberration was discovered before the Michelson-Morley experiment by J. Bradley (1729) and investigated again in later years by G. Airy (1871) and H. Fizeau (1859).

The theoretical efforts to deal with the negative result of the Michelson-Morley experiment, were undertaken mainly by G. Fitzgerald (1893) and by H.A. Lorentz (1892, 1904). Both researchers suggested, independent of each other on the basis of an analysis of the Maxwell equations, that an object, which moves through the aether, contracts. In the same fashion time intervals are dilated. These effects compensate each other, so that, independent of the motion through the aether, always the velocity of light c is found. Lorentz failed, because he tried to save the concept of an aether. He was not able to find an explanation for the contraction with the aid

---

[22] The aberration is e.g. discussed in A. Stewart, Scientific American, Vol. **210** (1964), p. 100.

of models of matter. In addition, he tried to understand the time dilatation as a local phenomenon, which could be different from the concept of the absolute time in the global aether system. The equations for the Lorentz transformation carry his name rightfully, as they appear for the first time in his publication of the year 1904.

The starting point of Einstein was also the theory of Maxwell in his publication with the title *Electrodynamics of moving bodies*. The title refers to the question of the systems of reference, which are used to analyse the theory. The conditions for the analysis were two clear cut hypotheses

- The principle of relativity with the equivalence of all inertial systems.
- The (as stated by Einstein) apparently irreconcilable independence of the velocity of propagation of the electromagnetic radiation) on the state of motion of the source.

With this basis, Einstein was able to derive the (special) relativistic equations for the transformation between inertial systems and all the additional properties, which followed. The introduction of an aether was superfluous.

Two other scientists have contributed to the development of the special theory of relativity. H. Poincaré formulated, before Einstein, in 1904 the principle of general relativity and realised, that it was necessary to develop a new form of mechanics. He did, however, not free himself completely from the analysis of Lorentz. He was also the first to recognise the fact, that the Lorentz transformations form a group and has investigated the relevant aspects with success (1904). The introduction of the representation of the theory of relativity in a four-dimensional Euclidean space with the coordinates $i\,c\,t$, x, y, z was also initiated by Poincaré. The completely covariant formulation of the theory of relativity on the basis of pseudo-Euclidean geometry is the merit of H. Minkowski (1908/1910).

The general theory of relativity, which was developed starting from 1912 by A. Einstein (and others) is based on the idea that the space-time world is not flat but curved in the vicinity of distributions of mass. Otherwise, the basis of this theory is the same as that of the special theory, namely the principle of relativity and the independence of the velocity of light from the inertial system. The curvature of the four-dimensional space is expressed by the replacement of the elements of the metric tensor in the expression for the square of the infinitesimal length of arc

$$ds^2 = \sum_{\mu\nu} g_{\mu\nu}\,dX_\mu dX_\nu$$

by functions of the coordinates in four-space

$$g_{\mu\nu} \longrightarrow g_{\mu\nu}(X_0, X_1, X_2, X_3)\,.$$

Such a metric is called a Riemannian metric. This ansatz is responsible for the appearance of a *force term* in the equations of motion, which is determined by the space-time geometry (the derivatives of the functions $g_{\mu\nu}(X_0, X_1, X_2, X_3)$). The geometry itself is determined by the distribution of the masses. The additional term is (as shown by Einstein in 1916) in lowest order the law of gravitation of Newton. It can be understood as a generalisation of the gravitation, which is produced by the distribution of matter and the resulting change of the space-time structure.

An experimental confirmation of these ideas is provided by the deflection of electromagnetic waves at the edge of the sun, which was first observed in 1919, and by the observed perihel rotation of the orbit of mercury. This rotation is the form of the orbit, which results from a deviation of law of gravitation in the Newtonian form.

There still remains the formal answer to the questions raised at the beginning of this chapter: One can confirm, that the assertions of the special theory of relativity are confirmed by experiment. This theory is based on two simple and transparent postulations: The equality of all inertial system and the equality of the velocity of light in all inertial systems. The theory of relativity has been accepted only with hesitation in spite of this transparency. This could be due to the fact, that it caused a scientific revolution, which has changed the view of the world in a radical manner. Electrodynamics, which was the starting point for the formulation of the theory of relativity, is in all points consistent with the special theory of relativity: The equations of electrodynamics have the same form in all inertial systems. They are form invariant. It is fortunate, that the relativistic modification are only important at high velocities, so that we can live with the simpler classical form of mechanics for most purposes.

3.7 Details

3.7.1 The Aberration

All bodies in the sky seem to be projected on a heavenly sphere (thought to be infinite) if viewed from the earth. If the place of observation on the earth is changed, due to the motion of the earth, then the projected point on the sphere is changed as well. This shift is called the parallax or the parallactic displacement (Fig. 3.22).

Samuel Molyneux, a citizen of London, and the English astronomer James Bradley tried, in the year 1725, to demonstrate the motion of the earth around the

Fig. 3.22 Parallactic shift

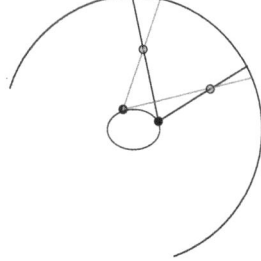

sun by measuring the parallactic displacement of a star in order to prove the theory of Kopernicus. They decided to observe the trajectory of the star γ-Draconis in the constellation of the Dragon, which can always be found above the horizon in the latitude of London. The accuracy of the observation depends on the stability of the mounting of the telescope and on the accuracy with which the alignment of the telescope can be determined. The telescope was mounted on a chimney and special care was taken for the measurement of the angle, which determined its orientation.

The observed displacement did in no way correspond to the values expected by the observers. The displacement should have had the largest possible value in March and in September. It occurred, however, in December and in June. In addition a Southern shift was measured instead of a predicted displacement in the Northern direction.

Bradley and Molyneux concluded that they had observed a different effect. In order to investigate the unexpected result more closely, they investigated additional stars and improved the accuracy of their experimental set up. With new instruments they could collect new data very quickly and confirm their previous result.

The final interpretation of their observations was, according a story told at the time, due to Bradley. On a sailing excursion on the Themse he observed, that flags on a mast depended not only on the direction and velocity of the wind, but also on the velocity of the boat. He concluded, that a moving observer registers a light source at a different spot than an observer at rest (Fig. 3.23). The reason is the large (but finite) velocity of the light. If the velocity of the observer is v and the velocity of the light is c, then the angle α, by which the source appears to be shifted is

$$\tan \alpha = \frac{v}{c}.$$

The value of this angle is about $\alpha = 20.5''$, as the mean speed of the earth around the sun is approx. 29.8 km/s and the speed of light c is approx. $2.998 \cdot 10^8$ m/s

$$\tan \alpha \approx 10^{-4} \approx \alpha \approx 10^{-4} \text{ rad} \approx 20.5''.$$

The value of α changes during the year as the direction of the motion of the earth with respect to the star changes.

Fig. 3.23 Aberration

3.7 Details

The early observations of Molyneux and Bradley are a large step in the development of physics as they

- supported the Kopernican view of the world,
- improved the standards of the astronomical observations,
- started the discussion of the propagation of light.

The history of the measurement of the aberration is represented in the following list:

- The first measurement began in December 1725.
- The data for γ-Draconis had been collected in the Summer of 1727.
- The observations were continued with the improved apparatus in August 1727.
- The interpretation of the results in terms of the aberration is vouched by a letter of Bradley to Halley from January 1729.
- Molyneux died in 1728. He did not live to experience the decoding of the results of his experiments.
- The correct value of the parallax was first obtained in the year 1838 by Wilhelm Bessel. The first measurements were not accurate enough, but allowed the correct interpretation.

3.7.2 The Simple Lorentz Transformation

The derivation of the simplest form of the Lorentz transformation via the Michelson-Morley condition involves only simple algebraic manipulations. The prerequisites are:
One considers two inertial systems S, S', which

- coincide at the time $t = 0$ (and are paraxial) and
- move with the constant relative velocity v_{rel} in the common x-direction (Fig. 3.24).

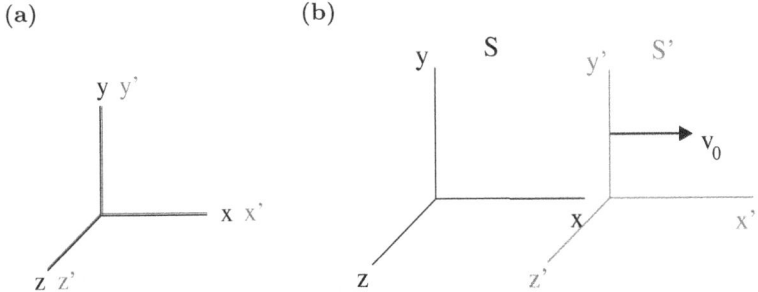

Fig. 3.24 Derivation of the simple form of the Lorentz transformation. (a) $t = 0$. (b) $t > 0$

The linear equation for the transformation is supposed to have the form

$$x' = Ax + Bt \qquad y' = y \qquad z' = z \qquad t' = Cx + Dt.$$

The origin of the system S' ($x' = 0$) with respect to the system S has the coordinate $x = v_{rel}\, t$, so that

$$0 = A v_{rel}\, t + Bt \qquad \text{or} \qquad B = -A v_{rel}.$$

If the Michelson-Morley condition

$$x^2 - c^2 t^2 = x'^2 - c^2 t'^2$$

is now inserted into the equations for the transformation, direct sorting yields

$$x^2 - c^2 t^2 = (Ax - A v_{rel} t)^2 - c^2 (Cx + Dt)^2$$

$$\Longleftrightarrow$$

$$\left(A^2 - c^2 C^2\right) x^2 - 2\left(A^2 v_{rel} + c^2 C D\right) xt$$

$$-c^2 \left(D^2 - \frac{v_{rel}^2}{c^2} A^2\right) t^2 = x^2 - c^2 t^2.$$

The coordinate x and the time t can be chosen, so that comparison of coefficients leads to a set of equations for the unknown coefficients A, C, D:

$$A^2 - c^2 C^2 = 1 \tag{3.90}$$

$$A^2 v_{rel} + c^2 C D = 0 \tag{3.91}$$

$$D^2 - \frac{v_{rel}^2}{c^2} A^2 = 1. \tag{3.92}$$

The solution of this non-linear system of equations can be found in the following fashion: Resolve Eq. (3.91) with respect to C

$$C = -\frac{v_{rel} A^2}{c^2 D}$$

and insert the result into (3.90)

$$A^2 - \frac{v_{rel}^2 A^4}{c^2 D^2} = 1. \tag{3.93}$$

3.7 Details

Equation (3.92) is a second relation between D^2 and A^2, which yields after insertion into (3.93) a quadratic equation for A

$$A^2 - \frac{v_{\text{rel}}^2}{c^2} \frac{A^4}{1 + v_{\text{rel}}^2 A^2 / c^2} = 1$$

respectively

$$A^2 + \frac{v_{\text{rel}}^2}{c^2} A^4 - \frac{v_{\text{rel}}^2}{c^2} A^4 = 1 + \frac{v_{\text{rel}}^2}{c^2} A^2 \, .$$

A sign of the double solution

$$A = \pm \left[1 - \frac{v_{\text{rel}}^2}{c^2}\right]^{-1/2}$$

is selected by the condition, that the limit $v_{\text{rel}}/c \to 0$ should give the Galilei transformation. The quantity A is then positive, as one has

$$x' = Ax + Bt \quad \Longrightarrow \quad x' \xrightarrow{v_{\text{rel}}/c \to 0} x - v_{\text{rel}} t \, .$$

The quantity D can be gained from (3.92)

$$D = \pm \left[1 + \frac{A^2 v_{\text{rel}}^2}{c^2}\right]^{1/2} = \pm \left[1 + \frac{v_{\text{rel}}^2}{c^2 - v_{\text{rel}}^2}\right]^{1/2} = \pm \left[1 - \frac{v_{\text{rel}}^2}{c^2}\right]^{-1/2} \, .$$

Again the positive sign is chosen by the condition, that the limit $v_{\text{rel}}/c \to 0$ should lead to the Galilei transformation,

The quantity C can be determined directly from (3.91)

$$C = -\frac{A^2 v_{\text{rel}}}{c^2 D} = -\left(\frac{v_{\text{rel}}}{c^2}\right)\left[1 - \frac{v_{\text{rel}}^2}{c^2}\right]^{-1/2} \, .$$

3.7.3 A More General Form of the Lorentz Transformation

The simplest form of the Lorentz transformation applies to a situation, in which two inertial systems S and S' are orientated in the same way, which coincide for the time $t = 0$ and move uniformly with respect to each other in one of the coordinate directions (e.g. the x-direction). The equations for the transformation are in this case

$$(ct') = \frac{1}{\left[1 - \left(\frac{v_{rel}}{c}\right)^2\right]^{1/2}} \left((ct) - \frac{v_{rel}}{c} x\right) \tag{3.94}$$

$$x' = \frac{1}{\left[1 - \left(\frac{v_{rel}}{c}\right)^2\right]^{1/2}} \left(-\frac{v_{rel}}{c} (ct) + x\right) \qquad y' = y \qquad z' = z$$

respectively in compact form for the coordinates of interest

$$(ct') = \gamma ((ct) - \beta x) \qquad x' = \gamma (-\beta(ct) + x) \ . \tag{3.95}$$

It is none the less possible to discuss, irrespective of the restricted form, all essential consequences of the transformation as length contraction, time dilatation etc.

More general forms of the Lorentz transformation have to be considered, if

- the coordinate systems are oriented in the same fashion and coincide for $t = 0$, but the velocity of the relative motion is arbitrary $\boldsymbol{v}_{rel}$ (Fig. 3.25).

or

- if the relative orientation and the situation at $t = 0$ as well as the relative velocity are arbitrary.

The equations for the generalisation of the Lorentz transformation in the first case are

$$(ct') = \gamma \left((ct) - \frac{(\boldsymbol{r} \cdot \boldsymbol{v}_{rel})}{c}\right) \tag{3.96}$$

$$\boldsymbol{r}' = -\gamma \frac{\boldsymbol{v}_{rel}}{c} (ct) + [\boldsymbol{r} + (\gamma - 1) \frac{(\boldsymbol{r} \cdot \boldsymbol{v}_{rel})}{v_{rel}^2} \boldsymbol{v}_{rel}] \ .$$

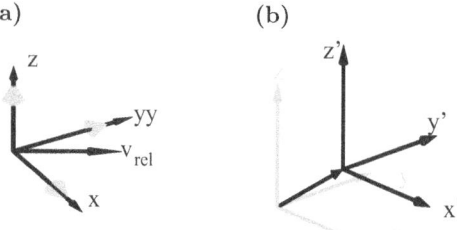

Fig. 3.25 The two inertial systems S and S'. (**a**) $t = 0$. (**b**) $t > 0$

3.7 Details

Fig. 3.26 Transformation of the two inertial systems S and S'. (a) $t = 0$. (b) First rotation

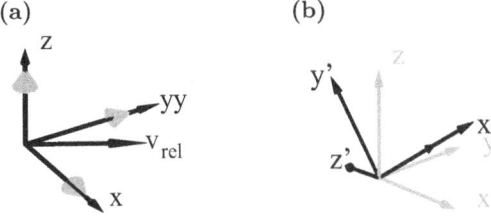

The derivation of these relations will be discussed for the $\mathcal{M}(1, 2)$- and for the $\mathcal{M}(1, 3)$-world.

The following argument within the co-/contravariant formulation should be useful as an introduction of the actual derivation: The two (equally oriented) coordinate systems are rotated in such a manner, that the x-direction coincides with the direction of the vector $\boldsymbol{v}_{\text{rel}}$ (Fig. 3.26). The rotation of the two systems S and S' is represented in the co-/contravariant notation in the form

$$x'_{\text{rot},\mu} = \sum_{\nu=0}^{3} D_{\mu\nu}(x')^{\nu} \qquad x_{\text{rot},\mu} = \sum_{\nu=0}^{3} D_{\mu\nu}x^{\nu},$$

if the matrix for rotations of the space part of the Minkowski space is $(D_{\mu\nu})$ and the event vector in covariant form is $\boldsymbol{R} = \{ct, \boldsymbol{r}\} = \{x_0, x_1, x_2, x_3\}$. The event vectors in the two rotated systems are related by a simple Lorentz transformation (Fig. 3.27a), which is represented by the matrix $(L_{s,\mu\lambda})$,

$$x'_{\text{rot},\mu} = \sum_{\lambda} L_{s,\mu\lambda} x^{\lambda}_{\text{rot}} .$$

The two coordinate systems are brought back to their original position by application of the matrix for the inverse rotation $((D^{-1})_{\mu\lambda})$ (Fig. 3.27b).

The following equation

$$\sum_{\lambda} (D^{-1})_{\mu\lambda} D^{\lambda\rho} = \delta_{\mu\rho}$$

Fig. 3.27 Transformation of the two inertial systems S and S'. (a) Lorentz transformation. (b) Second rotation

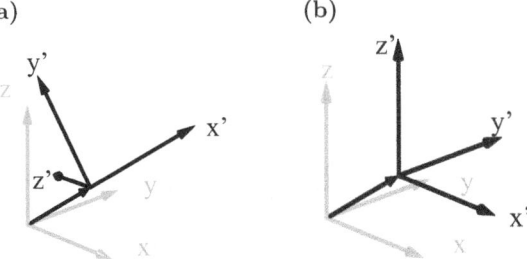

leads to

$$x'_\mu = \sum_\lambda (D^{-1})_{\mu\lambda} (x')^\lambda_{\text{rot}} = \sum_{\lambda\rho} (D^{-1})_{\mu\lambda} L_s^{\lambda\rho} x_{\text{rot},\rho}$$

$$= \sum_{\lambda\rho\nu} (D^{-1})_{\mu\lambda} L_s^{\lambda\rho} D_{\rho\nu} x^\nu .$$

Here one can extract the fact, that the matrix $(L_{\mu\nu})$ with the elements

$$L_{\mu\nu} = \sum_{\lambda\rho} (D^{-1})_{\mu\lambda} L_s^{\lambda\rho} D_{\rho\nu} \qquad (3.97)$$

represents a more general Lorentz transformation.

The matrices, which feature here, are related to the matrices in the simple notation, which has been used in (3.95), in the following way:

- The transformation matrix $(\tilde{M}_{\mu\nu})$ represents a Lorentz transformation (rotation of the true space, simple Lorentz transformation, general Lorentz transformation) in the simple notation, in which only covariant coordinates

$$y_\mu = \sum_\nu \tilde{M}_{\mu\nu} x_\nu \qquad (3.98)$$

are used. The matrix elements of the formally correct and the simple equation of the transformations are connected by

$$\sum_\nu \tilde{M}_{\mu\nu} g_{\nu\rho} = M_{\mu\rho} .$$

If one uses the relation

$$x_\nu = \sum_\rho g_{\nu\rho} x^\rho ,$$

one finds indeed the statement

$$y_\mu = \sum_{\nu\rho} \tilde{M}_{\mu\nu} g_{\nu\rho} x^\rho = \sum_\rho M_{\mu\rho} x^\rho$$

instead of (3.98). Equation (3.97) leads to

$$L_{\mu\nu} = \sum_{\nu'} \tilde{L}_{\mu\nu'} g_{\nu'\nu} = \sum_{\lambda\lambda'\rho\rho'} (D^{-1})_{\mu\lambda} g^{\lambda\lambda'} L_{s,\lambda'\rho'} g^{\rho'\rho} D_{\rho\nu}$$

3.7 Details

$$= \sum_{\lambda\lambda'\lambda''\rho\rho'\rho''\nu'} (\tilde{D}^{-1})_{\mu\lambda''} g_{\lambda''\lambda} g^{\lambda\lambda'} \tilde{L}_{s,\lambda'\rho''} g_{\rho''\rho'} g^{\rho'\rho} \tilde{D}_{\rho\nu'} g_{\nu'\nu}$$

$$= \sum_{\lambda'\rho\nu'} (\tilde{D}^{-1})_{\mu\lambda'} \tilde{L}_{s,\lambda'\rho} \tilde{D}_{\rho\nu'} g_{\nu'\nu} .$$

One can learn here, that the general Lorentz transformation can be combined in the same manner for the formal and in the simple notation

$$\tilde{L}_{\mu\nu} = \sum_{\lambda\rho} (\tilde{D}^{-1})_{\mu\lambda} \tilde{L}_{s,\lambda\rho} \tilde{D}_{\rho\nu} . \tag{3.99}$$

- In order to see this statement in action, one can look at the relation

$$\sum_{\nu} \tilde{L}_{\mu\nu} g_{\nu\rho} = L_{\mu\rho} ,$$

which combines the matrix elements in the simple and the co-/contravariant notations, in explicit matrix form

$$\begin{pmatrix} L_{00} & L_{10} & L_{20} & L_{30} \\ L_{01} & L_{11} & L_{21} & L_{31} \\ L_{02} & L_{12} & L_{22} & L_{32} \\ L_{03} & L_{13} & L_{23} & L_{33} \end{pmatrix} = \begin{pmatrix} \tilde{L}_{00} & \tilde{L}_{10} & \tilde{L}_{20} & \tilde{L}_{30} \\ \tilde{L}_{01} & \tilde{L}_{11} & \tilde{L}_{21} & \tilde{L}_{31} \\ \tilde{L}_{02} & \tilde{L}_{12} & \tilde{L}_{22} & \tilde{L}_{32} \\ \tilde{L}_{03} & \tilde{L}_{13} & \tilde{L}_{23} & \tilde{L}_{33} \end{pmatrix} \begin{pmatrix} -1 & 0 & 0 & 0 \\ 0 & 1 & 0 & 0 \\ 0 & 0 & 1 & 0 \\ 0 & 0 & 0 & 1 \end{pmatrix}$$

$$= \begin{pmatrix} -\tilde{L}_{00} & \tilde{L}_{10} & \tilde{L}_{20} & \tilde{L}_{30} \\ -\tilde{L}_{01} & \tilde{L}_{11} & \tilde{L}_{21} & \tilde{L}_{31} \\ -\tilde{L}_{02} & \tilde{L}_{12} & \tilde{L}_{22} & \tilde{L}_{32} \\ -\tilde{L}_{03} & \tilde{L}_{13} & \tilde{L}_{23} & \tilde{L}_{33} \end{pmatrix} .$$

- The transformation matrix elements $(L_{\mu\nu})$ and $(\tilde{L}_{\mu\nu})$ are different, but the complete equations of the transformation are equivalent.

$$y_\mu = \sum_\nu L_{\mu\nu} x^\nu = \sum_{\nu\rho} L_{\mu\nu} g^{\nu\rho} x_\rho = \sum_{\nu\rho\lambda} \tilde{L}_{\mu\lambda} g_{\lambda\nu} g^{\nu\rho} x_\rho = \sum_\nu \tilde{L}_{\mu\nu} x_\nu .$$

The simple notation, for which (3.99) holds, will be used from here on.

3.7.3.1 Two Space Coordinates, e.g. x and y

The two (3×3) matrices of the Lorentz transformation are in this case

$$(\tilde{L}_{s,\mu\lambda}) = \begin{pmatrix} \gamma & -\beta\gamma & 0 \\ -\beta\gamma & \gamma & 0 \\ 0 & 0 & 1 \end{pmatrix}$$

for the simple Lorentz transformation and

$$(\tilde{D}_{\mu\lambda}) = \begin{pmatrix} 1 & 0 & 0 \\ 0 & \cos\varphi & \sin\varphi \\ 0 & -\sin\varphi & \cos\varphi \end{pmatrix}$$

for a rotation about the (not participating) z-axis. For the calculation of the matrix product the intermediate result is

$$((\tilde{L}_s \tilde{D})_{\mu\lambda}) = \begin{pmatrix} \gamma & -\beta\gamma\cos\varphi & -\beta\gamma\sin\varphi \\ -\beta\gamma & \gamma\cos\varphi & \gamma\sin\varphi \\ 0 & -\sin\varphi & \cos\varphi \end{pmatrix}$$

and the final result

$$((\tilde{D}^{-1}\tilde{L}_s\tilde{D})_{\mu\lambda}) =$$

$$\begin{pmatrix} \gamma & -\beta\gamma\cos\varphi & -\beta\gamma\sin\varphi \\ -\beta\gamma\cos\varphi & \gamma\cos^2\varphi + \sin^2\varphi & (\gamma-1)\cos\varphi\sin\varphi \\ -\beta\gamma\sin\varphi & (\gamma-1)\cos\varphi\sin\varphi & \gamma\sin^2\varphi + \cos^2\varphi \end{pmatrix}.$$

(3.100)

In order to interpret this result one considers the multiplication $\tilde{L}\mathbf{R}$ and looks at the three elements of the product.

- For the time coordinate (the first line of the matrix (3.100)) in the system S' one obtains

$$(ct') = \gamma((ct) - \beta x\cos\varphi - \beta y\sin\varphi).$$

The last two terms on the right hand side are the projection of the vector v_{rel}/c on the coordinate axes $(\cos(90° - \varphi) = \sin\varphi)$ and therefore the scalar product

3.7 Details

of this vector with the position vector $r = (x, y)$. In summary one has

$$(ct') = \gamma\left((ct) - \frac{(v_{rel} \cdot r)}{c}\right). \tag{3.101}$$

- The x'-component is related to the Minkowski coordinates in the system S by

$$x' = -\beta\gamma(ct)\cos\varphi + (\gamma\cos^2\varphi + \sin^2\varphi)x + (\gamma - 1)y\cos\varphi\sin\varphi.$$

The first term on the right hand side contains the x-component of the relative velocity $v_{rel,x} = v_{rel}\cos\varphi$. The remaining terms can be sorted in the form

$$(\gamma\cos^2\varphi + \sin^2\varphi)x + (\gamma - 1)y\cos\varphi\sin\varphi$$
$$= x + (\gamma - 1)(x\cos\varphi + y\sin\varphi)\cos\varphi.$$

Again one finds, up to the factor v_{rel}, the scalar product of position and relative velocity, as well as the projection of a unit vector on the x-axis. Extension with v_{rel}^2 gives the result

$$x' = -\gamma\frac{v_{rel,x}}{c}(ct) + x + (\gamma - 1)\frac{(v_{rel} \cdot r)}{v_{rel}^2}v_{rel,x}.$$

- The corresponding equation for the y'-component

$$y' = -\beta\gamma(ct)\sin\varphi + (\gamma - 1)x\cos\varphi\sin\varphi + (\gamma\sin^2\varphi + \cos^2\varphi)y$$

leads to a corresponding result (replace $\cos^2\varphi$ by $1 - \sin^2\varphi$)

$$y' = -\gamma\frac{v_{rel,y}}{c}(ct) + y + (\gamma - 1)\frac{(v_{rel} \cdot r)}{v_{rel}^2}v_{rel,y}.$$

The individual equations for the space coordinates in S' can, as done in (3.96), be summarised in vector form

$$r' = -\gamma\frac{v_{rel}}{c}(ct) + r + (\gamma - 1)\frac{(v_{rel} \cdot r)}{v_{rel}^2}v_{rel}. \tag{3.102}$$

This result can be interpreted in the form: The quantity

$$r_{parallel} = \frac{(v_{rel} \cdot r)}{v_{rel}^2}v_{rel}$$

is the component of the position vector parallel to the vector of the relative velocity. The component perpendicular to the relative velocity is

$$r_{\text{perpend}} = r - r_{\text{parallel}} \, .$$

Equation (3.102) corresponds to the vector addition of the parallel component, which is given by (3.95)

$$r'_{\text{parallel}} = \gamma \left(-\frac{v_{\text{rel}}}{c}(ct) + r_{\text{parallel}} \right) ,$$

and the equation

$$r'_{\text{perpend}} = r_{\text{perpend}} = r - r_{\text{parallel}} \, ,$$

which expresses the fact, that the perpendicular component has not been transformed. This expresses the fact, that the equations of the transformation (3.101) and (3.102) will also be valid in the $\mathcal{M}(1,3)$-world, which will now be indicated explicitly.

3.7.3.2 Three Space Coordinates

In order to evaluate the relation (3.99) in the Minkowski space $\mathcal{M}(1,3)$ with the coordinates

$$x_0 = ct, \ x_1 = x, \ x_2 = y, \ x_3 = z$$

one needs the dependence of the rotation matrix, embedded into the four dimensional world, on the Euler angles

$$(\tilde{D}_{\mu\lambda}) = \begin{pmatrix} 1 & 0 & 0 & 0 \\ 0 & \tilde{D}_{11} & \tilde{D}_{12} & \tilde{D}_{13} \\ 0 & \tilde{D}_{21} & \tilde{D}_{22} & \tilde{D}_{23} \\ 0 & \tilde{D}_{31} & \tilde{D}_{32} & \tilde{D}_{33} \end{pmatrix} .$$

The matrix elements $\tilde{D}_{ik}$ can be found e.g. in Dreizler and Lüdde, 2010, Theoretical Mechanics (Springer Berlin Heidelberg) Chap. 6.3.5, but are not be needed here. The relevant information for the present discussion is the orthogonality relation $((\tilde{D}^{-1}\tilde{D})_{\mu\lambda}) = (E_{\mu\lambda})$, respectively explicitly

$$\sum_{l=1}^{3} \tilde{D}_{li} \tilde{D}_{lk} = \delta_{ik} \qquad i, k = 1, 2, 3 \, . \tag{3.103}$$

3.7 Details

The matrix of the simple Lorentz transformation ($\tilde{L}$) is identical with the matrix ($\mathcal{L}$) in Chap. 3.3.4

$$(\tilde{L}_{s,\mu\lambda}) = (\mathcal{L}_{s,\mu\lambda}) = \begin{pmatrix} \gamma & -\beta\gamma & 0 & 0 \\ -\beta\gamma & \gamma & 0 & 0 \\ 0 & 0 & 1 & 0 \\ 0 & 0 & 0 & 1 \end{pmatrix}.$$

The evaluation of the matrix product $((\tilde{L}_s \tilde{D})_{\mu\lambda})$ yields

$$((\tilde{L}_s \tilde{D})_{\mu\lambda}) = \begin{pmatrix} \gamma & -\beta\gamma\tilde{D}_{11} & -\beta\gamma\tilde{D}_{12} & -\beta\gamma\tilde{D}_{13} \\ -\beta\gamma & \gamma\tilde{D}_{11} & \gamma\tilde{D}_{12} & \gamma\tilde{D}_{13} \\ 0 & \tilde{D}_{21} & \tilde{D}_{22} & \tilde{D}_{23} \\ 0 & \tilde{D}_{31} & \tilde{D}_{32} & \tilde{D}_{33} \end{pmatrix}.$$

The transformation $(\tilde{L}_{\mu\lambda}) = ((D^{-1}\tilde{L}_s\tilde{D})_{\mu\lambda})$ can be given in the form

$$(\tilde{L}_{\mu\lambda}) = \begin{pmatrix} \gamma & -\beta\gamma\tilde{D}_{11} & -\beta\gamma\tilde{D}_{12} & -\beta\gamma\tilde{D}_{13} \\ -\beta\gamma\tilde{D}_{11} & \tilde{L}_{11} & \tilde{L}_{12} & \tilde{L}_{13} \\ -\beta\gamma\tilde{D}_{12} & \tilde{L}_{21} & \tilde{L}_{22} & \tilde{L}_{23} \\ -\beta\gamma\tilde{D}_{13} & \tilde{L}_{31} & \tilde{L}_{32} & \tilde{L}_{33} \end{pmatrix}$$

The individual elements $\tilde{L}_{ik}$ are

$$\tilde{L}_{11} = \gamma\tilde{D}_{11}\tilde{D}_{11} + \tilde{D}_{21}\tilde{D}_{21} + \tilde{D}_{31}\tilde{D}_{31}$$
$$\tilde{L}_{12} = \gamma\tilde{D}_{11}\tilde{D}_{12} + \tilde{D}_{21}\tilde{D}_{22} + \tilde{D}_{31}\tilde{D}_{32}$$
$$\tilde{L}_{13} = \gamma\tilde{D}_{11}\tilde{D}_{13} + \tilde{D}_{21}\tilde{D}_{23} + \tilde{D}_{31}\tilde{D}_{33}$$

$$\tilde{L}_{21} = \gamma\tilde{D}_{12}\tilde{D}_{11} + \tilde{D}_{22}\tilde{D}_{21} + \tilde{D}_{32}\tilde{D}_{31}$$
$$\tilde{L}_{22} = \gamma\tilde{D}_{12}\tilde{D}_{12} + \tilde{D}_{22}\tilde{D}_{22} + \tilde{D}_{32}\tilde{D}_{32}$$
$$\tilde{L}_{23} = \gamma\tilde{D}_{12}\tilde{D}_{13} + \tilde{D}_{22}\tilde{D}_{23} + \tilde{D}_{32}\tilde{D}_{33}$$

$$\tilde{L}_{31} = \gamma\tilde{D}_{13}\tilde{D}_{11} + \tilde{D}_{23}\tilde{D}_{21} + \tilde{D}_{33}\tilde{D}_{31}$$
$$\tilde{L}_{32} = \gamma\tilde{D}_{13}\tilde{D}_{12} + \tilde{D}_{23}\tilde{D}_{22} + \tilde{D}_{33}\tilde{D}_{32}$$
$$\tilde{L}_{23} = \gamma\tilde{D}_{13}\tilde{D}_{13} + \tilde{D}_{23}\tilde{D}_{23} + \tilde{D}_{33}\tilde{D}_{33}.$$

This list can be written in the compact form

$$\tilde{L}_{ik} = \sum_{l=1}^{3}(1 + (\gamma - 1)\delta_{l1})\tilde{D}_{li}\tilde{D}_{lk}$$

and be simplified with the orthogonality relation (3.103)

$$\tilde{L}_{ik} = \delta_{ik} + (\gamma - 1)\tilde{D}_{1i}\tilde{D}_{1k} \ .$$

For the interpretation one looks again at the equation for the transformation $\mathbf{R}' = (\tilde{L})\mathbf{R}$. The Minkowski coordinates in the system S' are

$$x_0' = \gamma x_0 - \beta\gamma \sum_{l=1}^{3}\tilde{D}_{1l}x_l$$

$$x_1' = -\beta\gamma x_0 \tilde{D}_{11} + (\gamma - 1)\tilde{D}_{11}\sum_{l=1}^{3}\tilde{D}_{1l}x_l + x_1$$

$$x_2' = -\beta\gamma x_0 \tilde{D}_{12} + (\gamma - 1)\tilde{D}_{12}\sum_{l=1}^{3}\tilde{D}_{1l}x_l + x_2$$

$$x_3' = -\beta\gamma x_0 \tilde{D}_{13} + (\gamma - 1)\tilde{D}_{13}\sum_{l=1}^{3}\tilde{D}_{1l}x_l + x_3 \ .$$

The part $\sum_{l=1}^{3}\tilde{D}_{1l}x_l$ represents the projection of the vector $\mathbf{r} = (x, y, z)$ in the direction of the vector of the relative velocity (the 1-direction after the rotation of the original coordinate system). Multiplication of this expression with v_{rel} yields the scalar product $(\mathbf{v}_{\text{rel}} \cdot \mathbf{r})$. The matrix elements multiplied with the magnitude of the relative velocity $\tilde{D}_{1k}v_{\text{rel}}$ can be interpreted as the k-th component of this velocity. With a similar extension and rewriting as in Chap. 3.7.3.1 one finds then

$$x_0' = \gamma\left(x_0 - \frac{1}{c}(\mathbf{v}_{\text{rel}} \cdot \mathbf{r})\right)$$

$$x_k' = -\gamma \frac{v_{\text{rel},k}}{c} x_0 + x_k + (\gamma - 1)\frac{(\mathbf{v}_{\text{rel}} \cdot \mathbf{r})}{v_{\text{rel}}^2} v_{\text{rel},k} \qquad k = 1, 2, 3 \ .$$

The vectorial summary of the space part corresponds to the previous result.

3.7.4 The Maxwell Equations in Relativistic Notation

3.7.4.1 Lorentz Transformation of the Field Tensors
The evaluation of the transformation (Chap. 3.5.2, p. 249)

$$F'_{\mu'\nu'}(\mathbf{R'}) = \sum_{\mu\nu} \mathcal{L}_{\mu'\mu} \mathcal{L}_{\nu'\nu} F_{\mu\nu}(\mathbf{R}) \qquad \mu', \nu' = 0, 1, 2, 3$$

for the elements of the field tensors [**F**] has to deal first with sums of 16 terms each, as e.g.

$$\begin{aligned}
F'_{00} = &\ \mathcal{L}_{00}\,\mathcal{L}_{00}\,F_{00} + \mathcal{L}_{01}\,\mathcal{L}_{00}\,F_{10} + \mathcal{L}_{02}\,\mathcal{L}_{00}\,F_{20} + \mathcal{L}_{03}\,\mathcal{L}_{00}\,F_{30}+ \\
&\ \mathcal{L}_{00}\,\mathcal{L}_{01}\,F_{01} + \mathcal{L}_{01}\,\mathcal{L}_{01}\,F_{11} + \mathcal{L}_{02}\,\mathcal{L}_{01}\,F_{21} + \mathcal{L}_{03}\,\mathcal{L}_{01}\,F_{31}+ \\
&\ \mathcal{L}_{00}\,\mathcal{L}_{02}\,F_{02} + \mathcal{L}_{01}\,\mathcal{L}_{02}\,F_{12} + \mathcal{L}_{02}\,\mathcal{L}_{02}\,F_{22} + \mathcal{L}_{03}\,\mathcal{L}_{02}\,F_{32}+ \\
&\ \mathcal{L}_{00}\,\mathcal{L}_{03}\,F_{03} + \mathcal{L}_{01}\,\mathcal{L}_{03}\,F_{13} + \mathcal{L}_{02}\,\mathcal{L}_{03}\,F_{23} + \mathcal{L}_{03}\,\mathcal{L}_{03}\,F_{33}
\end{aligned}$$

$$\begin{aligned}
F'_{12} = &\ \mathcal{L}_{10}\,\mathcal{L}_{20}\,F_{00} + \mathcal{L}_{11}\,\mathcal{L}_{20}\,F_{10} + \mathcal{L}_{12}\,\mathcal{L}_{20}\,F_{20} + \mathcal{L}_{13}\,\mathcal{L}_{20}\,F_{30}+ \\
&\ \mathcal{L}_{10}\,\mathcal{L}_{21}\,F_{01} + \mathcal{L}_{11}\,\mathcal{L}_{21}\,F_{11} + \mathcal{L}_{12}\,\mathcal{L}_{21}\,F_{21} + \mathcal{L}_{13}\,\mathcal{L}_{21}\,F_{31}+ \\
&\ \mathcal{L}_{10}\,\mathcal{L}_{22}\,F_{02} + \mathcal{L}_{11}\,\mathcal{L}_{22}\,F_{12} + \mathcal{L}_{12}\,\mathcal{L}_{22}\,F_{22} + \mathcal{L}_{13}\,\mathcal{L}_{22}\,F_{32}+ \\
&\ \mathcal{L}_{10}\,\mathcal{L}_{23}\,F_{03} + \mathcal{L}_{11}\,\mathcal{L}_{23}\,F_{13} + \mathcal{L}_{12}\,\mathcal{L}_{23}\,F_{23} + \mathcal{L}_{13}\,\mathcal{L}_{23}\,F_{33}\,.
\end{aligned}$$

The results for the remaining 14 matrix elements follow this pattern. As the expressions for the diagonal elements of the original tensor have the value zero

$$F_{\mu\mu} = 0 \quad \text{with} \quad \mu = 0, 1, 2, 3\,,$$

the sums are reduced to 12 terms. If only the simple Lorentz transformation is considered, the number of terms in each sum is reduced due to

$$\mathcal{L}_{02} = \mathcal{L}_{03} = \mathcal{L}_{12} = \mathcal{L}_{13} = \mathcal{L}_{20} = \mathcal{L}_{21} = \mathcal{L}_{23} = \mathcal{L}_{30} = \mathcal{L}_{31} = \mathcal{L}_{32} = 0$$

to maximal 2. The 16 matrix elements, which represent the field tensor from the point of view of the (primed) coordinate system S', which is moving with respect to the system S, are listed below. The following steps have to be executed:

- Use the matrix of the simple Lorentz transformation $(1 - \beta^2 = 1/\gamma^2)$

$$(\mathcal{L}_{\lambda\mu}) = \begin{pmatrix} \gamma & -i\beta\gamma & 0 & 0 \\ i\beta\gamma & \gamma & 0 & 0 \\ 0 & 0 & 1 & 0 \\ 0 & 0 & 0 & 1 \end{pmatrix}.$$

- The matrix of the field tensor in the system S is

$$[F] = \begin{bmatrix} 0 & (iE_x)/(ck_f) & (iE_y)/(ck_f) & (iE_z)/(ck_f) \\ -(iE_x)/(ck_f) & 0 & B_z & -B_y \\ -(iE_y)/(ck_f) & -B_z & 0 & B_x \\ -(iE_z)/(ck_f) & B_y & -B_x & 0 \end{bmatrix}.$$

The individual elements of the field tensor of the system S' are:

- $F'_{00} = \mathcal{L}_{01}\,\mathcal{L}_{00}\,F_{10} + \mathcal{L}_{00}\,\mathcal{L}_{01}\,F_{01}$

$$= (-i\beta\gamma)(\gamma)\left(-\frac{iE_x}{ck_f}\right) + (\gamma)(-i\beta\gamma)\left(\frac{iE_x}{ck_f}\right) = 0$$

$F'_{01} = \mathcal{L}_{01}\,\mathcal{L}_{10}\,F_{10} + \mathcal{L}_{00}\,\mathcal{L}_{11}\,F_{01}$

$$= (-i\beta\gamma)(i\beta\gamma)\left(-\frac{iE_x}{ck_f}\right) + (\gamma)(\gamma)\left(\frac{iE_x}{ck_f}\right)$$

$$= \frac{iE_x}{ck_f}\gamma^2\left(1 - \beta^2\right) = i\frac{E_x}{ck_f}$$

$F'_{02} = \mathcal{L}_{00}\,\mathcal{L}_{22}\,F_{02} + \mathcal{L}_{01}\,\mathcal{L}_{22}\,F_{12}$

$$= (\gamma)\left(\frac{iE_y}{ck_f}\right) + (-i\beta\gamma)\,B_z = i\gamma\left(\frac{E_y}{ck_f} - \beta B_z\right)$$

$F'_{03} = \mathcal{L}_{00}\,\mathcal{L}_{33}\,F_{03} + \mathcal{L}_{01}\,\mathcal{L}_{33}\,F_{13}$

$$= (\gamma)\left(\frac{iE_z}{ck_f}\right) + (-i\beta\gamma)(-B_y) = i\gamma\left(\frac{E_z}{ck_f} + \beta B_y\right)$$

3.7 Details

$$F'_{10} = \mathcal{L}_{11} \mathcal{L}_{00} F_{10} + \mathcal{L}_{10} \mathcal{L}_{01} F_{01}$$

$$= (\gamma)(\gamma)\left(-\frac{iE_x}{ck_f}\right) + (i\beta\gamma)(-i\beta\gamma)\left(\frac{iE_x}{ck_f}\right) = -i\frac{E_x}{ck_f}$$

$$F'_{11} = \mathcal{L}_{11} \mathcal{L}_{10} F_{10} + \mathcal{L}_{10} \mathcal{L}_{11} F_{01}$$

$$= (\gamma)(i\beta\gamma)\left(-\frac{iE_x}{ck_f}\right) + (i\beta\gamma)(\gamma)\left(\frac{iE_x}{ck_f}\right) = 0$$

$$\bullet F'_{12} = \mathcal{L}_{10} \mathcal{L}_{22} F_{02} + \mathcal{L}_{11} \mathcal{L}_{22} F_{12}$$

$$= (i\beta\gamma)\left(\frac{iE_y}{ck_f}\right) + (\gamma) B_z = \gamma\left(B_z - \beta\frac{E_y}{ck_f}\right)$$

$$F'_{13} = \mathcal{L}_{10} \mathcal{L}_{33} F_{03} + \mathcal{L}_{11} \mathcal{L}_{33} F_{13}$$

$$= (i\beta\gamma)\left(\frac{iE_z}{ck_f}\right) + (\gamma)(-B_y) = -\gamma\left(B_y + \beta\frac{E_z}{ck_f}\right)$$

$$F'_{20} = \mathcal{L}_{22} \mathcal{L}_{00} F_{20} + \mathcal{L}_{22} \mathcal{L}_{01} F_{21}$$

$$= (\gamma)\left(-\frac{iE_y}{ck_f}\right) + (-i\beta\gamma)(-B_z) = -i\gamma\left(\frac{E_y}{ck_f} - \beta B_z\right)$$

$$F'_{21} = \mathcal{L}_{22} \mathcal{L}_{10} F_{20} + \mathcal{L}_{22} \mathcal{L}_{11} F_{21}$$

$$= (i\beta\gamma)\left(-\frac{iE_y}{ck_f}\right) + (\gamma)(-B_z) = -\gamma\left(B_z - \beta\frac{E_y}{ck_f}\right)$$

$$F'_{22} = 0$$

$$F'_{23} = \mathcal{L}_{22} \mathcal{L}_{33} F_{23} = B_x$$

$$F'_{30} = \mathcal{L}_{33} \mathcal{L}_{00} F_{30} + \mathcal{L}_{33} \mathcal{L}_{01} F_{31}$$

$$= (\gamma)\left(-\frac{iE_z}{ck_f}\right) + (-i\beta\gamma) B_y = -i\gamma\left(\frac{E_z}{ck_f} + \beta B_y\right)$$

$$F'_{31} = \mathcal{L}_{33} \mathcal{L}_{10} F_{30} + \mathcal{L}_{33} \mathcal{L}_{11} F_{31}$$

$$= (i\beta\gamma)\left(-\frac{iE_z}{ck_f}\right) + (\gamma) B_y = \gamma\left(B_y + \beta\frac{E_z}{ck_f}\right)$$

$$F'_{32} = \mathcal{L}_{33}\,\mathcal{L}_{22}\,F_{32} = -B_x$$

$$F'_{33} = 0\,.$$

One can recognise, that a pure electric (magnetic) field as seen by S is seen as a combined electromagnetic field by S' (and vice versa). A charge at rest in S is a moving charge as seen by S'.

3.7.4.2 Check of the Covariant Form of the Maxwell Equations

In order to demonstrate, that the inhomogeneous Maxwell equations (in vacuum) can be summarised in the form

$$\sum_{\nu=0}^{3}\frac{\partial F_{\mu\nu}(\mathbf{R})}{\partial X_\nu} = 4\pi k_m J_\mu(\mathbf{R}) \qquad \mu = 0, 1, 2, 3 \tag{3.104}$$

which implies, that the equation with $\nu = 0$ corresponds to the Coulomb law and the equations with $\nu = 1, 2, 3$ to the extended Ampère law, the covariant form has to be translated into the standard form. In order to do this, one needs the following statements on the Minkowski coordinates and the four-current

$$\mathbf{X} = \{ict,\ x,\ y,\ z\}$$

$$\mathbf{J} = \{ic\rho_{tr},\ j_{tr.x},\ j_{tr.y},\ j_{tr.z}\}\,,$$

as well as (once again) the elements of the field tensor

$$[\mathbf{F}] = \begin{bmatrix} 0 & (iE_x)/(ck_f) & (iE_y)/(ck_f) & (iE_z)/(ck_f) \\ -(iE_x)/(ck_f) & 0 & B_z & -B_y \\ -(iE_y)/(ck_f) & -B_z & 0 & B_x \\ -(iE_z)/(ck_f) & B_y & -B_x & 0 \end{bmatrix}.$$

Equation (3.104) yields for $\mu = 0$

$$\frac{\partial F_{00}}{\partial X_0} + \frac{\partial F_{01}}{\partial X_1} + \frac{\partial F_{02}}{\partial X_2} + \frac{\partial F_{03}}{\partial X_3} = 4\pi k_m J_0$$

$$0 + \frac{\partial}{\partial x}\left(\frac{iE_x}{ck_f}\right) + \frac{\partial}{\partial y}\left(\frac{iE_y}{ck_f}\right) + \frac{\partial}{\partial z}\left(\frac{iE_z}{ck_f}\right) = 4\pi k_m ic\rho_{tr}\,,$$

3.7 Details

with the relation

$$\sqrt{\frac{k_e}{k_m k_f}} = c \qquad (3.105)$$

that is

$$\nabla \cdot \mathbf{E} = 4\pi k_e \rho_{tr} \,. \quad \checkmark$$

For $\mu = 1$ one finds

$$\frac{\partial F_{10}}{\partial X_0} + \frac{\partial F_{11}}{\partial X_1} + \frac{\partial F_{12}}{\partial X_2} + \frac{\partial F_{13}}{\partial X_3} = 4\pi k_m J_1$$

$$\frac{1}{ic}\frac{\partial}{\partial t}\left(-\frac{iE_x}{ck_f}\right) + 0 + \frac{\partial}{\partial y}(B_z) + \frac{\partial}{\partial z}(-B_y) = 4\pi k_m j_{tr,x} \,,$$

for $\mu = 2$

$$\frac{\partial F_{20}}{\partial X_0} + \frac{\partial F_{21}}{\partial X_1} + \frac{\partial F_{22}}{\partial X_2} + \frac{\partial F_{23}}{\partial X_3} = 4\pi k_m J_2$$

$$\frac{1}{ic}\frac{\partial}{\partial t}\left(-\frac{iE_y}{ck_f}\right) + \frac{\partial}{\partial x}(-B_z) + 0 + \frac{\partial}{\partial z}(B_x) = 4\pi k_m j_{tr,y} \,,$$

and finally for $\mu = 3$

$$\frac{\partial F_{30}}{\partial X_0} + \frac{\partial F_{31}}{\partial X_1} + \frac{\partial F_{32}}{\partial X_2} + \frac{\partial F_{33}}{\partial X_3} = 4\pi k_m J_3$$

$$\frac{1}{ic}\frac{\partial}{\partial t}\left(-\frac{iE_z}{ck_f}\right) + \frac{\partial}{\partial x}(B_y) + \frac{\partial}{\partial y}(-B_x) + 0 = 4\pi k_m j_{tr,z} \,.$$

Using (3.105) one confirms the statement

$$\nabla \times \mathbf{B} - \frac{k_m}{k_e}\frac{\partial \mathbf{E}}{\partial t} = 4\pi k_m \mathbf{j}_{tr} \,. \quad \checkmark$$

3.7.5 The Energy-Momentum Tensor

If one writes the equations of motion for a point charge in an electromagnetic field in covariant form, one arrives at the energy-momentum tensor, a four-dimensional extension of Maxwell's stress tensor. Three separate points need to be discussed in this context:

1. The formulation of the equations of motion in a covariant form.
2. The transition to a formulation of the Minkowski force of a problem of motion in terms of the energy-momentum tensor.
3. The discussion of the conservation laws of electrodynamics in terms of the covariant formulation.

3.7.5.1 The Covariant Formulation

It is convenient to show that the force per volume element on a point charge in a combined electric and magnetic field

$$f = \rho_{tr} E + k_f (j_{tr} \times B) \tag{3.106}$$

can be expressed in terms of covariant quantities by the equations

$$f_\mu(\mathbf{R}) = k_f \sum_\nu F_{\mu\nu}(\mathbf{R}) J_\nu(\mathbf{R}) \qquad (\mu = 1, 2, 3) . \tag{3.107}$$

The equation in terms of the force (3.106) is written component by component in order to obtain this result

$$\begin{pmatrix} f_x \\ f_y \\ f_z \end{pmatrix} = \rho_{tr} \begin{pmatrix} E_x \\ E_y \\ E_z \end{pmatrix} + k_f \begin{vmatrix} e_x & e_y & e_z \\ j_{tr,x} & j_{tr,y} & j_{tr,z} \\ B_x & B_y & B_z \end{vmatrix}$$

$$= \rho_{tr} \begin{pmatrix} E_x \\ E_y \\ E_z \end{pmatrix} + k_f \begin{pmatrix} j_{tr,y} B_z - j_{tr,z} B_y \\ j_{tr,z} B_x - j_{tr,x} B_z \\ j_{tr,x} B_y - j_{tr,y} B_x \end{pmatrix} .$$

The equations for the individual components are then compared with the corresponding statements in covariant form (3.107). The four-current density is

$$\mathbf{J} = \{J_0, J_1, J_2, J_3\} = \{ic\rho_{tr}, j_{tr,x}, j_{tr,y}, j_{tr,z}\} ,$$

3.7 Details

the electric and the magnetic field components are combined in the field tensor [F] (Chap. 3.5.2)

$$[F] = \begin{bmatrix} F_{00} & F_{01} & F_{02} & F_{03} \\ F_{10} & F_{11} & F_{12} & F_{13} \\ F_{20} & F_{21} & F_{22} & F_{23} \\ F_{30} & F_{31} & F_{32} & F_{33} \end{bmatrix}$$

$$= \begin{bmatrix} 0 & (iE_x)/(ck_f) & (iE_y)/(ck_f) & (iE_z)/(ck_f) \\ -(iE_x)/(ck_f) & 0 & B_z & -B_y \\ -(iE_y)/(ck_f) & -B_z & 0 & B_x \\ -(iE_z)/(ck_f) & B_y & -B_x & 0 \end{bmatrix}.$$

One finds in detail the following three statements, where each set of equations can be read from top to bottom or from bottom to top

$$f_1 = k_f \left(F_{10} J_0 + F_{11} J_1 + F_{12} J_2 + F_{13} J_3 \right)$$

$$= k_f \left(-\frac{iE_x}{ck_f} (ic\rho_{tr}) + 0 + B_z j_{tr,y} - B_y j_{tr,z} \right)$$

$$= \rho_{tr} E_x + k_f (B_z j_{tr,y} - B_y j_{tr,z})$$

$$f_2 = k_f \left(F_{20} J_0 + F_{21} J_1 + F_{22} J_2 + F_{23} J_3 \right)$$

$$= k_f \left(-\frac{iE_y}{ck_f} (ic\rho_{tr}) - B_z j_{tr,x} + 0 + B_x j_{tr,z} \right)$$

$$= \rho_{tr} E_y + k_f (B_x j_{tr,z} - B_z j_{tr,x})$$

$$f_3 = k_f \left(F_{30} J_0 + F_{31} J_1 + F_{32} J_2 + F_{33} J_3 \right)$$

$$= k_f \left(-\frac{iE_z}{ck_f} (ic\rho_{tr}) + B_y j_{tr,x} - B_x j_{tr,y} + 0 \right)$$

$$= \rho_{tr} E_z + k_f (B_y j_{tr,x} - B_x j_{tr,y}) \,.$$

One arrives (up to a factor) at the Joule heat term, if the component f_0 is calculated in the same fashion

$$f_0 = k_f \left(F_{00} J_0 + F_{01} J_1 + F_{02} J_2 + F_{03} J_3 \right)$$

$$= k_f \left(0 + \frac{\mathrm{i} E_x}{c k_f} j_{tr,x} + \frac{\mathrm{i} E_y}{c k_f} j_{tr,y} + \frac{\mathrm{i} E_z}{c k_f} j_{tr,z} \right)$$

$$= \frac{\mathrm{i}}{c} (\mathbf{E} \cdot \mathbf{j}_{tr}) \,. \tag{3.108}$$

3.7.5.2 The Energy-Momentum Tensor

The relation with the energy-momentum tensor can be established with some steps, if the relations (3.107)

$$f_\mu(\mathbf{R}) = k_f \sum_\nu F_{\mu\nu}(\mathbf{R}) J_\nu(\mathbf{R})$$

(with $\mu = 0, \ldots, 3$) are replaced by the components of the four current density in the corresponding terms of the inhomogeneous Maxwell equations (Chap. 3.5.2, (3.72), p. 252)

$$f_\mu = \frac{k_f}{4\pi k_m} \sum_{\nu\rho} F_{\mu\nu} \frac{\partial F_{\nu\rho}}{\partial X_\rho} \,. \tag{3.109}$$

One starts with the derivative of a product of elements of the field tensor

$$\sum_{\nu\rho} \frac{\partial}{\partial X_\rho} \left(F_{\mu\nu} F_{\nu\rho} \right) = \sum_{\nu\rho} \left\{ F_{\mu\nu} \frac{\partial F_{\nu\rho}}{\partial X_\rho} + F_{\nu\rho} \frac{\partial F_{\mu\nu}}{\partial X_\rho} \right\} \,. \tag{3.110}$$

The second term on the right hand side can be changed due to the antisymmetry of the tensor elements by interchange of the indices. In the next step rename the variables over which one evaluates finally the sums

$$F_{\nu\rho} \frac{\partial F_{\mu\nu}}{\partial X_\rho} = F_{\rho\nu} \frac{\partial F_{\nu\mu}}{\partial X_\rho} = F_{\nu\rho} \frac{\partial F_{\rho\mu}}{\partial X_\nu} \,.$$

The second term is separated with this result into two contributions

$$\sum_{\nu\rho} F_{\nu\rho} \frac{\partial F_{\mu\nu}}{\partial X_\rho} = \frac{1}{2} \sum_{\nu\rho} F_{\nu\rho} \left(\frac{\partial F_{\mu\nu}}{\partial X_\rho} + \frac{\partial F_{\rho\mu}}{\partial X_\nu} \right)$$

3.7 Details

and the sum in the bracket is replaced with the aid of the homogeneous Maxwell equation Chap. 3.5.2, (3.71) by the cyclic compliment

$$= -\frac{1}{2} \sum_{\nu\rho} F_{\nu\rho} \frac{\partial F_{\nu\rho}}{\partial X_\mu} .$$

This result can be written as the derivative of a product of two equal terms

$$= -\frac{1}{4} \sum_{\nu\rho} \frac{\partial}{\partial X_\mu} \left(F_{\nu\rho} F_{\nu\rho} \right) .$$

Finally one renames the indices in two steps

$$= -\frac{1}{4} \sum_{\nu\rho} \frac{\partial}{\partial X_\mu} \left(F_{\nu\sigma} F_{\nu\sigma} \right) = -\frac{1}{4} \sum_{\nu\rho\sigma} \delta_{\mu\rho} \frac{\partial}{\partial X_\rho} \left(F_{\nu\sigma} F_{\nu\sigma} \right) .$$

If the expression on the right hand side of Eq. (3.110) is expressed by

$$\sum_{\nu\rho} F_{\mu\nu} \frac{\partial F_{\nu\rho}}{\partial X_\rho} = \sum_{\nu\rho} \frac{\partial}{\partial X_\rho} \left(F_{\mu\nu} F_{\nu\rho} \right) - \sum_{\nu\rho} F_{\nu\rho} \frac{\partial F_{\mu\nu}}{\partial X_\rho}$$

respectively

$$\sum_{\nu\rho} F_{\mu\nu} \frac{\partial F_{\nu\rho}}{\partial X_\rho} = \sum_{\nu\rho} \frac{\partial}{\partial X_\rho} \left(F_{\mu\nu} F_{\nu\rho} \right) + \frac{1}{4} \sum_{\nu\rho} \delta_{\mu\rho\sigma} \frac{\partial}{\partial X_\rho} \left(F_{\nu\sigma} F_{\nu\sigma} \right) ,$$

one finds with the definition of the energy-momentum tensor (3.76) and another change of the notation

$$T_{\mu\rho} = \frac{k_f}{4\pi k_m} \sum_\nu \left(F_{\mu\nu} F_{\nu\rho} + \frac{1}{4} \delta_{\mu\rho} \sum_\sigma F_{\nu\sigma} F_{\nu\sigma} \right) \quad (3.111)$$

the result

$$f_\mu = \sum_\rho \frac{\partial T_{\mu\rho}}{\partial X_\rho} . \quad (3.112)$$

The explicit representation of the tensor elements by the electric and the magnetic components of the field can be calculated directly with (3.111). One first considers the term

$$T_S = \frac{k_f}{16\pi k_m} \sum_{\lambda\rho} F_{\lambda\rho} F_{\lambda\rho}$$

and finds by calculating the square of all elements of the tensor [F]

$$T_S = -\frac{1}{8\pi k_e} E^2 + \frac{k_f}{8\pi k_m} B^2 ,$$

where the relation $c^2 k_m k_f = k_e$ has been used. The element T_{00} is

$$T_{00} = \frac{k_f}{4\pi k_m} (F_{00}F_{00} + F_{01}F_{10} + F_{02}F_{20} + F_{03}F_{30}) + T_S$$

$$= \frac{1}{4\pi k_e} \left(E_x^2 + E_y^2 + E_z^2 \right) + T_S$$

$$= \frac{1}{8\pi} \left(\frac{1}{k_e} E^2 + \frac{k_f}{k_m} B^2 \right) .$$

The elements $T_{0k} = T_{k0}$ are found by a simple sum

$$T_{0k} = \frac{k_f}{4\pi k_m} (F_{00}F_{0k} + F_{01}F_{1k} + F_{02}F_{2k} + F_{03}F_{3k}) .$$

If one inserts here the necessary tensor elements, one finds e.g. for T_{01}

$$T_{01} = \frac{k_f}{4\pi k_m} \left(\frac{i}{ck_f} E_y(-B_z) + \frac{i}{ck_f} E_z B_y \right)$$

$$= -\frac{i}{4\pi c k_m} (E \times B)_1$$

and a corresponding result for all elements of this kind. The elements with the 'spatial' indices are found with

$$T_{ik} = \frac{k_f}{4\pi k_m} (F_{i0}F_{0k} + F_{i1}F_{1k} + F_{i2}F_{2k} + F_{i3}F_{3k}) + \delta_{i,k} T_S .$$

An example for a diagonal element is

$$T_{11} = \frac{k_f}{4\pi k_m} (F_{10}F_{01} + F_{11}F_{11} + F_{12}F_{21} + F_{13}F_{31}) + T_S$$

$$= \frac{1}{4\pi k_e} E_x^2 - \frac{k_f}{4\pi k_m} \left(B_z^2 + B_y^2 \right) - \frac{1}{8\pi k_e} E^2 + \frac{k_f}{8\pi k_m} B^2$$

$$= \frac{1}{4\pi k_e} \left(E_x^2 - \frac{1}{2} E^2 \right) + \frac{k_f}{4\pi k_m} \left(B_x^2 - \frac{1}{2} B^2 \right) ,$$

3.7 Details

an example for a non-diagonal element

$$T_{12} = \frac{k_f}{4\pi k_m}(F_{10}F_{02} + F_{11}F_{12} + F_{12}F_{22} + F_{13}F_{32})$$

$$= \frac{1}{4\pi k_e} E_x E_y + \frac{k_f}{4\pi k_m} B_x B_y \ .$$

The individual results can be collected in the form

$$T_{ik} = \frac{1}{4\pi k_e}\left(E_i E_k - \frac{1}{2}E^2 \delta_{i,k}\right) + \frac{k_f}{4\pi k_m}\left(B_i B_k - \frac{1}{2}B^2 \delta_{i,k}\right) \ .$$

This is Maxwell's stress tensor, which has been quoted in Chap. 1.4.2, (1.65), p. 41 (with $\varepsilon = \mu = 1$).

3.7.5.3 The Conservation Laws
The energy and the momentum conservation laws of electrodynamics

$$\frac{dP_\mu}{dt} = \iiint dV \, f_\mu = \iiint dV \sum_\nu \frac{\partial T_{\mu\nu}}{\partial X_\nu}$$

can be derived from the equations of motion. The component for $\mu = 0$ gives

$$\iiint dV \left(f_0 - \frac{\partial T_{00}}{\partial X_0} - \frac{\partial T_{01}}{\partial X_1} - \frac{\partial T_{02}}{\partial X_2} - \frac{\partial T_{03}}{\partial X_3}\right) = 0$$

and corresponds thus ($X_0 = ict$) to the law of energy conservation in vacuum (Chap. 1.4.1, (1.58), p. 38)

$$\frac{i}{c}\iiint dV \left\{(E \cdot j_{tr}) + \frac{1}{8\pi}\frac{d}{dt}\left(\frac{1}{k_e}E^2 + \frac{k_f}{k_m}B^2\right)\right.$$

$$\left. + \frac{1}{4\pi k_m}\nabla \cdot (E \times B)\right\} = 0 \ .$$

For the space components one has

$$\frac{dp_k}{dt} = \iiint dV \left(\frac{\partial T_{k0}}{\partial X_0} + \frac{\partial T_{k1}}{\partial X_1} + \frac{\partial T_{k2}}{\partial X_2} + \frac{\partial T_{k3}}{\partial X_3}\right) \qquad k = 1, 2, 3 \quad ,$$

or explicitly

$$\frac{dp_k}{dt} = \iiint dV \left\{ -\frac{1}{4\pi c^2 k_m} \frac{d}{dt}(E \times B)_k + \frac{1}{4\pi k_e} \nabla \cdot (E_k E) \right.$$

$$\left. + \frac{k_f}{4\pi k_m} \nabla \cdot (B_k B) - \frac{1}{8\pi k_e} \frac{\partial}{\partial x_k} E^2 - \frac{k_f}{8\pi k_m} \frac{\partial}{\partial x_k} B^2 \right\} .$$

If one uses here (compare Chap. 1.4.2 (1.63)) the definition of the momentum of the electromagnetic field with the components

$$p_{\text{field},k} = \frac{1}{4\pi c^2 k_m} \iiint (E \times B)_k \, dV ,$$

one recognises the conservation law of the momentum of electrodynamics (in the vacuum)

$$\frac{d}{dt}\left[p_{\text{mech}}(t) + p_{\text{field}}(t)\right]_k = \frac{1}{4\pi} \iiint dV \left\{ \nabla \cdot \left(\frac{1}{k_e}(E_k E) + \frac{k_f}{k_m}(B_k B)\right) \right.$$

$$\left. -\frac{1}{2}\frac{\partial}{\partial x_k}\left(\frac{1}{k_e}E^2 - \frac{k_f}{k_m}B^2\right) \right\} ,$$

which has been presented in Chap. 1.4.2, (1.64) in a more compact form.

3.7.6 The Relativistic Lagrange/Hamilton Formalism

The motion of a charged mass point in an electromagnetic field can either be treated with an equation with the Lorentz force or by a corresponding Lagrange function. The relation between the two approaches will be investigated for the relativistic and the non-relativistic situation. Of special interest in different areas of theoretical physics is the Hamilton function of a relativistic point charge in an electromagnetic field. This function will be addressed in Chap. 3.7.6.2.

3.7.6.1 The Lagrange Equations

The non-relativistic Lagrange function of a charged mass point (q, m_0) in an electromagnetic field is quoted in Chap. 3.5.4

$$L_{\text{nrel, em}} \equiv L = \frac{1}{2}m_0 v^2 - q\left(V(r,t) - k_f A(r,t) \cdot v\right) .$$

3.7 Details

The generalised (Cartesian) momenta are

$$p_i = \frac{\partial L}{\partial \dot{x}_i} = m_0 \dot{x}_i + q\, k_f\, A_i = p_i + q\, k_f\, A_i \qquad i = 1, 2, 3 \ .$$

For the construction of the Lagrange equations one needs also

$$\frac{d}{dt}(p_i) = \dot{p}_i + q\, k_f \left(\frac{\partial A_i}{\partial t} + \sum_{k=1}^{3} \frac{\partial A_i}{\partial x_k} \dot{x}_k \right)$$

and

$$\frac{\partial L}{\partial x_i} = q\, k_f \left(\frac{\partial A_1}{\partial x_i} \dot{x}_1 + \frac{\partial A_2}{\partial x_i} \dot{x}_2 + \frac{\partial A_3}{\partial x_i} \dot{x}_3 - \frac{\partial V}{\partial x_i} \right) \ .$$

The evaluation of

$$\frac{d}{dt}\left(\frac{\partial L}{\partial \dot{x}_i}\right) - \frac{\partial L}{\partial x_i} = 0$$

yields e.g. for $i = 1$

$$\frac{d}{dt}\left(\frac{\partial L}{\partial \dot{x}_1}\right) - \frac{\partial L}{\partial x_1} = \dot{p}_1 + q\left(\frac{\partial V}{\partial x_1} + k_f \left(\frac{\partial A_1}{\partial t} - \left(\frac{\partial A_2}{\partial x_1} - \frac{\partial A_1}{\partial x_2}\right) \dot{x}_2 \right.\right.$$

$$\left.\left. + \left(\frac{\partial A_1}{\partial x_3} - \frac{\partial A_3}{\partial x_1}\right) \dot{x}_3 \right) \right) \ ,$$

with the definitions (1.67) and (1.68) in Chap. 1.5 (p. 42) thus

$$= \dot{p}_1 - q E_1 - q\, k_f (B_3 \dot{x}_2 - B_2 \dot{x}_3) = 0 \ .$$

A corresponding statement holds for the 2 and 3 components.
The vectorial form is therefore

$$\dot{p} = q\, E + q\, k_f (v \times B) \ ,$$

where p is the classical momentum $p = m_0 v$.

The relativistic Lagrange function differs from the non-relativistic function only by the term for the kinetic energy

$$T_{\text{rel}} = L_{\text{rel,f}} = -m_0 c^2 \left[1 - \frac{v^2}{c^2} \right]^{1/2} \ .$$

One obtains for this reason the relativistic part of the generalised momentum

$$\frac{\partial T_{\text{rel}}}{\partial \dot{x}_i} = -m_0 c^2 \frac{-\dot{x}_i}{c^2 \left[1 - v^2/c^2\right]^{1/2}} = m_0 \gamma \dot{x}_i = m \dot{x}_i = p_{i,\text{rel}}$$

instead of the classical momentum.

3.7.6.2 The Hamilton Function

In order to calculate the relativistic Hamilton function (Chap. 3.5.4) according to the definition

$$H = \sum_{i=1}^{3} \pi_i v_i - L ,$$

one has to resolve first the expression for the components of the generalised three-momentum (3.86) (p. 258)

$$\boldsymbol{\pi} = \gamma m_0 \boldsymbol{v} + k_f q \boldsymbol{A} \tag{3.113}$$

$$\Longleftrightarrow$$

$$\pi_i = \gamma m_0 v_i + k_f q A_i = p_i + k_f q A_i \qquad i = 1, 2, 3$$

with respect to the velocity components. The first step is: Resolve the relation

$$\frac{m_0 \boldsymbol{v}}{\left[1 - v^2/c^2\right]^{1/2}} = \boldsymbol{\pi} - k_f q \boldsymbol{A}$$

via

$$m_0^2 v^2 c^2 = \left(\boldsymbol{\pi} - k_f q \boldsymbol{A}\right)^2 \left(c^2 - v^2\right)$$

with respect to

$$v^2 = \frac{\left(\boldsymbol{\pi} - k_f q \boldsymbol{A}\right)^2}{\left(\boldsymbol{\pi} - k_f q \boldsymbol{A}\right)^2 + m_0^2 c^2}$$

and calculates

$$\left[1 - v^2/c^2\right]^{1/2} = \frac{m_0 c}{\left[\left(\boldsymbol{\pi} - k_f q \boldsymbol{A}\right)^2 + m_0^2 c^2\right]^{1/2}} .$$

3.7 Details

Insertion into the initial equation (3.113) yields for the components of the three-velocity

$$v = \frac{c\left(\pi - k_f\, q\, A\right)}{\left[\left(\pi - k_f\, q\, A\right)^2 + m_0^2 c^2\right]^{1/2}} \,.$$

It remains to rephrase the free Lagrange function and the magnetic interaction term (3.83)

$$L_{\text{rel, em}} = -m_0\, c^2 \left[1 - \frac{v^2}{c^2}\right]^{1/2} + q\,(k_f\, A \cdot v - V)\,,$$

in detail

$$m_0\, c^2 \left[1 - v^2/c^2\right]^{1/2} = \frac{m_0^2 c^3}{\left[\left(\pi - k_f\, q\, A\right)^2 + m_0^2 c^2\right]^{1/2}}$$

$$-q\, k_f\, v \cdot A = -\frac{q\, k_f\, c\left(\pi \cdot A - k_f\, q\, A^2\right)}{\left[\left(\pi - k_f\, q\, A\right)^2 + m_0^2 c^2\right]^{1/2}} \,.$$

If all the terms are combined

$$H = \pi \cdot v + m_0\, c^2 \left[1 - v^2/c^2\right]^{1/2} - q\, k_f\, v \cdot A + qV\,,$$

one finds

$$H = \pi \cdot \frac{c\left(\pi - k_f\, q\, A\right)}{\left[\left(\pi - k_f\, q\, A\right)^2 + m_0^2 c^2\right]^{1/2}} + \frac{m_0^2 c^3}{\left[\left(\pi - k_f\, q\, A\right)^2 + m_0^2 c^2\right]^{1/2}}$$

$$-\frac{q\, k_f\, c\left(\pi \cdot A - k_f\, q\, A^2\right)}{\left[\left(\pi - k_f\, q\, A\right)^2 + m_0^2 c^2\right]^{1/2}} + qV\,,$$

which can be written as

$$H = \frac{c\left(\pi^2 - k_f\, q\, \mathbf{A} \cdot \boldsymbol{\pi} + m_0^2 c^2 - q\, k_f\, \boldsymbol{\pi} \cdot \mathbf{A} + k_f^2\, q^2\, \mathbf{A}^2\right)}{\left[(\boldsymbol{\pi} - k_f\, q\, \mathbf{A})^2 + m_0^2 c^2\right]^{1/2}} + qV$$

and finally to the compact result

$$H = \left[c^2\,(\boldsymbol{\pi} - k_f\, q\, \mathbf{A})^2 + m_0^2 c^4\right]^{1/2} + qV \ .$$

Literature

A.1 Books and Literature Quoted in the Text

- Bailey et al.
  Nature 268 (1971)
  p. 301
- A. Einstein
  Annalen der Physik, Vol. 17 (1905)
  p. 891
- H.A. Lorentz
  Proceedings of the Academy of Science, Amsterdam, Vol. 6 (1904)
  p. 809
- A.A. Michelson and E.W. Morley
  Philosophical Magazine, Vol. 24 (1987)
  p. 449
- R. Perrin
  American Journal of Physics Vol. 47 (1979)
  p. 317.
- A. Stewart
  Scientific American, Vol.210 (1964)
  p. 100
- N.W. Ashcroft and N.D. Mermin
  Solid state physics
  Saunders Publications, Philadelphia (1976)
- N.W. Ashcroft and N.D. Mermin
  Solid state physics
  Brooks/Cole (2008)

- J.C. Maxwell
  Treatise on Electricity and Magnetism
  Clarendon Press, Oxford (1873)
- J.C. Maxwell
  The scientific letters and papers of James Clerk Maxwell 1862–1873
  Cambridge University Press (1995)
- P. Moon and D.E. Spencer
  Field theory handbook : including coordinate systems, differential equations and their solutions
  Springer (1961)
- P. Morse and H. Feshbach
  Methods of theoretical physics Chapters 1 to 8
  Mc Graw-Hill (1953)
- M.E. Rose
  Elementary theory of angular momentum
  Dover (1995)

A.2 Introductory Texts

- R.P. Feynman and R.B. Leighton and M.L. Sands
  The Feynman lectures on physics : definitive edition
  Pearson Addison-Wesley (2006)
- D.J. Morin and E.M. Purcell
  Electricity and Magnetism
  Cambridge University Press (2013)
- E.M. Purcell
  Berkeley physics course Electricity and Magnetism
  Mc Graw-Hill (1966)

A.3 Electrodynamics

- V.D. Barger and M.G. Olsson
  Classical Electricity and Magnetism
  Allyn & Bacon (1987)
- D.J. Griffith
  Introduction to Electrodynamics
  Cambridge University Press (2023)
- J.D. Jackson
  Classical Electrodynamics
  Wiley (2021)
- K. Milton and J. Schwinger
  Classical Electrodynamics: Second edition
  CRC Press (2024)

- G.L. Pollack and D.R. Stump
  Electromagnetism
  Addison-Wesley (2002)
- J. Schwinger and K.A. Milton et al.
  Classical Electrodynamics
  Westview Press (1998)

A.4 Special Relativity

- W. Rindler
  Introduction to Special Relativity
  Clarendon Press (1991)
- U.E. Schröder
  Special Relativity
  World Scientific Press (1990)

A.5 Mathematics

- H. Cartan
  The Elementary Theory of Analytic Functions of One or Several Complex Variables
  Dover Pub. (1995)
- F.J. Flanigan
  Complex Variables: Harmonics and Analytic Functions
  Dover Pub. (1983)
- K. Kodaira
  Complex Analysis
  Cambridge University Press (2007)
- A. Kyrala
  Applied Functions of a Complex Variable
  Wiley-Interscience (1972)

A.6 Special Functions and Handbooks

- M. Abramovitz and I. Stegun
  Handbook of Mathematical Functions
  National Bureau of Standards applied mathematics series
  Martino Publishing (2014)
- I.S. Gradstejn and I.M. Ryzik
  Table of Integrals, Series and Products
  Elsevier (2007)

- A. Jeffrey and H. Dai
  Handbook of Mathematical Formulas and Integrals
  Elsevier (2004)
- A.F. Nikiforov and V.B. Uvarov
  Special Functions of Mathematical Physics
  Springer Basel (2013)
- A.D. Polyanin and V. Nazaikinskii
  Handbook of Linear Partial Differential Equations for Engineers and Scientists, 2 nd Edition
  Chapman and Hall/CRC (2015)
- A.D. Polyanin and A.I. Chernoutsan
  A Concise Handbook of Mathematics, Physics, and Engineering Sciences
  CRC Press LLC (2010)
- N.M. Temme
  Special Functions: An Introduction to the Functions of Mathematical Physics
  Wiley (1996)
- M. Tenenbaum and H. Pollard
  Ordinary differential equations : an elementary textbook for students of mathematics, engineering, and the sciences
  Dover Publ. (1985)

Systems of Units in Electrodynamics   B

B.1 The Systems

Three parameters, which can in principle be selected freely, are needed to define the units for all quantities of electrodynamics in the vacuum. These constants are usually chosen to be the units of the electric field E, the magnetic induction B and the law of induction, which contains these two experimentally accessible quantities. The equations, which are relevant for this choice, are:

- The electric field E of a point charge q at a distance r from this charge

$$E(r) = k_e \frac{q}{r^2} \; .$$

- The magnetic induction B of a long, thin, straight conducting wire with a stationary current i at a distance r from the axis of the wire

$$B(r) = 2 k_m \frac{i}{r} \; .$$

  The extracted factor 2 is a useful but not a compelling option.
- The law of induction in the differential form

$$\nabla \times E(r,t) + k_f \frac{\partial}{\partial t} B(r,t) = 0 \; .$$

A constraint is due to the wave equations for the two fields in a vacuum. On the basis of the Maxwell equations one finds

$$\Delta E(r,t) + \frac{k_m k_f}{k_e} \frac{\partial^2}{\partial t^2} E(r,t) = 0$$

and

$$\Delta B(r,t) + \frac{k_m k_f}{k_e} \frac{\partial^2}{\partial t^2} B(r,t) = 0 \,.$$

As the factor in front of the time derivatives corresponds to the inverse of the square of the speed of propagation, which has been determined experimentally for electromagnetic waves in a vacuum as

$$c = 2.997925\ldots \cdot 10^{10} \,\frac{\text{cm}}{\text{s}} = 2.997925\ldots \cdot 10^{8} \,\frac{\text{m}}{\text{s}} \,,$$

the restriction is

$$\left[\frac{k_e}{k_m k_f}\right]^{1/2} = c \,.$$

This implies, that only two of the three constants can be chosen freely.

The explicit specification of two constants by measurement (traditionally k_e and k_m are chosen) calls for a somewhat longer consideration.

The constant k_e is introduced first in the Coulomb law

$$F_{12} = k_e \frac{q_1 q_2}{r_{12}^2} \,,$$

which addresses the magnitude of the force between two point charges q_1 and q_2 with a separation of r_{12}. With this choice[1] of the dimensionless constant

$$k_{e,\text{CGS}} = 1$$

one connects the unit for the charge in the CGS-system with mechanical units

$$[q]_{\text{CGS}} = \frac{\text{g}^{1/2}\,\text{cm}^{3/2}}{\text{s}} = \text{statcoul} \,.$$

The unit for the electrical field is therefore, because of $E_{q_1} = F_{12}/q_2$,

$$[E]_{\text{CGS}} = \frac{\text{g}^{1/2}}{\text{cm}^{1/2}\,\text{s}} = \frac{\text{statvolt}}{\text{cm}} \,.$$

---

[1] Only the Gaussian CGS-system and the rationalised MKSA-system will be considered in detail. For two additional systems, the constants are given in Table B.1.

B Systems of Units in Electrodynamics

The choice of the units of the magnetic fields (more precisely magnetic induction) is based on the relation

$$D = m \times B .$$

This relation, which connects the magnetic moment m of a test dipole and the magnetic induction B with a torque D shows, that the units of the magnetic moment are connected with the units for the magnetic induction. In the CGS-system one chooses for the constant k_m the value

$$k_{m,\mathrm{CGS}} = \frac{1}{c} = 0.333564\ldots \cdot 10^{-10} \frac{\mathrm{s}}{\mathrm{cm}} ,$$

so that

$$k_{f,\mathrm{CGS}} = \frac{k_{e,\mathrm{CGS}}}{c^2 k_{m,\mathrm{CGS}}} \implies k_{f,\mathrm{CGS}} = \frac{1}{c}$$

follows. The magnetic induction is measured in the units

$$[B]_{\mathrm{CGS}} = \frac{\mathrm{g}^{1/2}}{\mathrm{cm}^{1/2}\,\mathrm{s}} = [E]_{\mathrm{CGS}} ,$$

the magnetic moment in the units

$$[m]_{\mathrm{CGS}} = \frac{\mathrm{g}^{1/2}\mathrm{cm}^{5/2}}{\mathrm{s}} = [p]_{\mathrm{CGS}} .$$

The electric and the magnetic fields as well as the electric and magnetic dipole moments are associated with the same unit in the CGS-system.

An additional unit, the unit of the strength of the current *Ampère* (short A), is introduced in the rationalised MKSA-system. The definition of the electric current allows an alternative unit for the charge, the *Coulomb* (short C)

$$i = \frac{dq}{dt} \iff 1\,\mathrm{A} = 1\,\frac{\mathrm{C}}{\mathrm{s}} .$$

The technical definition of the unit Ampère includes the measurement of the action of the magnetic forces:

▶ The flow of a current of the same strength in two long, thin, straight parallel conducting wires in the vacuum is produced by a current of the strength of 1 Ampère, if the wires attract each other with a force F of $2 \cdot 10^{-7}$ N per meter of the length of the wires.

The appropriate equation is

$$i^2 = \frac{|F|d}{2k_f k_m l} .$$

If the values $F = 2 \cdot 10^{-7}$ N, $i = 1$ A and $d = l = 1$ m are inserted one finds for the constant k_f

$$k_{f,\text{SI}} k_{m,\text{SI}} = 10^{-7} \frac{\text{kg m}}{\text{C}^2} .$$

The constant $k_{f,\text{SI}}$ is usually chosen, so that this quantity is dimensionless

$$k_{f,\text{SI}} = 1 .$$

The unit *Ampère* corresponds therefore to the constant

$$k_{m,\text{SI}} = 10^{-7} \frac{\text{kg m}}{\text{C}^2} = \frac{\mu_0}{4\pi} ,$$

which is often replaced by the permeability of the vacuum

$$\mu_0 = 4\pi \cdot 10^{-7} \frac{\text{kg m}}{\text{C}^2} = 1.25663\ldots \cdot 10^{-6} \frac{\text{kg m}}{\text{C}^2} .$$

The additional restrictive condition for the constants yields therefore

$$k_{e,\text{SI}} = 10^{-7} c^2 \frac{\text{kg m}}{\text{C}^2} = 8.98755.. \cdot 10^9 \frac{\text{kg m}^3}{\text{C}^2 \text{s}^2} = \frac{1}{4\pi \varepsilon_0} ,$$

so that the dielectric constant of the vacuum is

$$\varepsilon_0 = 8.85418\ldots \cdot 10^{-12} \frac{\text{C}^2 \text{s}^2}{\text{kg m}^3} .$$

For the units of the two fields and the magnetic moment in the SI-system one finds

$$[E]_{\text{SI}} = \frac{\text{kg m}}{\text{C s}^2} = \frac{\text{Volt}}{\text{cm}}$$

$$[B]_{\text{SI}} = \frac{\text{kg}}{\text{C s}}$$

$$[m]_{\text{SI}} = \frac{\text{C m}^2}{\text{s}} .$$

B Systems of Units in Electrodynamics

It is usual to employ

$$m_{\text{loop,SI}} = i \cdot \text{surface}$$

for the magnetic moment of a plane, current carrying, conducting loop in the SI-system. The unit, which is defined by this definition, agrees with the initially stated unit of the magnetic moment. This is not the case for the CGS-system. The definition in the two system of units is compatible with the general definition of the magnetic moment, if one uses

$$m_{\text{loop}} = k_f (i \cdot \text{surface}) .$$

The discussion of the fields in materials demands the inclusion of the auxiliary fields D and H, the dielectric displacement and the magnetic field strength, as well as the quantities, which address the averaged properties of the material, the polarisation P and the magnetisation M. The D-field is produced by the true charges. This implies the relation

$$\nabla \cdot D = 4\pi k_d \rho_{tr}$$

if the question of the units for this field is left open for the moment. The polarisation is generated by the distribution of the polarisation charges, which suggests the (not compelling) definition

$$\nabla \cdot P = -\rho_{\text{pol}} .$$

The fact, that the electric field is produced by the true and the polarisation charges, leads to the relation

$$D = \frac{k_d}{k_e} E - 4\pi k_d P .$$

In the CGS-system the constant k_d, which can still be chosen, is given the value 1

$$k_{d,\text{CGS}} = k_{e,\text{CGS}} = 1 .$$

The resulting equation

$$D = E - 4\pi P$$

states, that all three electric fields are measured in the same units in this system.

This constant is also without a dimension in the SI-system

$$k_{d,\text{SI}} = \frac{1}{4\pi} ,$$

so that the relation between the three fields is

$$D = \varepsilon_0 E - P .$$

The units of the two auxiliary fields are equal, but differ from that of the E-field.[2]

The corresponding statements for magnetic materials are: The H-field is produced by the true currents

$$\nabla \times H = 4\pi k_h j_{tr} .$$

The magnetisation M incorporates the averaging process over the loop-like magnetisation currents in the surface of the material

$$\nabla \times M = k_f j_{tr} .$$

The statement, that the B-field is produced by both current densities, is the reason for the combination

$$H = \frac{k_h}{k_m} B - 4\pi \frac{k_h}{k_f} M .$$

If one wants, that all magnetic fields are measured in the same units—as for the CGS-system—one has to choose

$$k_{h,\text{CGS}} = \frac{1}{c}$$

and finds

$$H = B - 4\pi M .$$

The rationalised SI-system is realised by the choice

$$k_{h,\text{SI}} = \frac{1}{4\pi} .$$

This leads to

$$H = \frac{1}{\mu_0} B - M .$$

---

[2] The name *rationalised* MKSA-system indicates the elimination of the factor 4π in the basic equations.

B Systems of Units in Electrodynamics

The simple material equations for isotropic materials with linear response are used in both systems of units in the form

$$D = \varepsilon E \qquad H = \frac{1}{\mu} B ,$$

with a vacuum limit given by

$$\left.\begin{array}{c}\varepsilon_{CGS}\\ \\ \mu_{CGS}\end{array}\right\} \longrightarrow 1$$

and

$$\varepsilon_{SI} \longrightarrow \varepsilon_0 \qquad \mu_{SI} \longrightarrow \mu_0 .$$

The relative dielectric constant and permeability in the SI-system agree with the dielectric constant and permeability in the CGS-system

$$\left(\frac{\varepsilon}{\varepsilon_0}\right)_{SI} = \varepsilon_{CGS} \qquad \left(\frac{\mu}{\mu_0}\right)_{SI} = \mu_{CGS} .$$

The coefficients

$$k_e, k_m, k_f, k_d, k_h$$

- for the two most used system of units and the electrostatic (electrostatic units - esu) and the electromagnetic system of units (electromagnetic units - emu) are assembled in Table B.1. The associated numerical values are given in Table B.2.
- The Tables B.3 and B.4 contain a list of the physical quantities, which play a role in electrodynamics.
  The formulae, which are used to introduce these quantities, are included the list in order to provide a better orientation besides the units in the SI- and the CGS-system.
- The Tables B.5 and B.6 show the factors, which are needed for the conversion of the physical quantities in the SI-system and the CGS-system, which are listed in Tables B.3 and B.4.

B.2 Tables

Table B.1 Definition of constants

Name	SI	CGS	esu	emu
k_e	$\dfrac{1}{4\pi\varepsilon_0}$	1	1	c^2
k_d	$\dfrac{1}{4\pi}$	1	1	1
k_m	$\dfrac{\mu_0}{4\pi}$	$\dfrac{1}{c}$	$\dfrac{1}{c^2}$	1
k_f	1	$\dfrac{1}{c}$	1	1
k_h	$\dfrac{1}{4\pi}$	$\dfrac{1}{c}$	1	1

Table B.2 Numerical values for Table B.1

Name	Value (SI)
c	$= 2.997925 \cdot 10^8 \, \dfrac{m}{s}$
e_0	$= 1.602192 \cdot 10^{-19} \, C$
ε_0	$= 8.85418 \cdot 10^{-12} \, \dfrac{C^2}{Nm^2}$
μ_0	$= 4\pi \cdot 10^{-7} \, \dfrac{kg\,m}{C^2}$
$\dfrac{1}{4\pi\varepsilon_0}$	$= 8.98755 \cdot 10^9 \, \dfrac{N\,m^2}{C^2}$
$\dfrac{1}{4\pi}$	$= 7.957747 \cdot 10^{-2}$
$\dfrac{\mu_0}{4\pi}$	$= 1 \cdot 10^{-7} \, \dfrac{kg\,m}{C^2}$
$e_{0,CGS}$	$= 4.803250 \cdot 10^{-10} \, esu$

B Systems of Units in Electrodynamics

Table B.3 Systems of units: definition and dimension of physical quantities I

Name		Definition	SI		CGS	
Charge	q	$= r\sqrt{\dfrac{F}{k_e}}$	C (Coulomb)		$\dfrac{g^{1/2} cm^{3/2}}{s}$	$=$ statcoul $=$ esu
E-field	E	$= \dfrac{F}{q}$	$\dfrac{N}{C} = \dfrac{kg\, m}{s^2 C} = \dfrac{V}{m}$		$\dfrac{g^{1/2}}{cm^{1/2} s}$	$= \dfrac{dyn}{statcoul} = \dfrac{statvolt}{cm}$
Electr. flux	Φ_e	$= \oiint E \cdot d\boldsymbol{f}$	$\dfrac{kg\, m^3}{s^2 C} = V\, m$		$\dfrac{g^{1/2} cm^{3/2}}{s}$	$=$ statcoul
Electr. dipole moment	p	$= 2aq$	$C\, m$		$\dfrac{g^{1/2} cm^{5/2}}{s}$	$=$ statcoul cm
Potential	V	$= \int E \cdot d\boldsymbol{s}$	$\dfrac{N\, m}{C} = \dfrac{kg\, m^2}{s^2 C} = V$		$\dfrac{g^{1/2} cm^{1/2}}{s}$	$=$ statvolt $= \dfrac{erg}{statcoul}$
Voltage	U	$= V_2 - V_1$	V (Volt)		statvolt	
Capacity	C	$= \dfrac{q}{U}$	$\dfrac{C^2 s^2}{kg\, m^2} = \dfrac{C}{V}$	$=$ F (Farad)	cm	
Dielectric displacement	$\boldsymbol{D}$	$\oiint \boldsymbol{D} \cdot d\boldsymbol{f} = 4\pi k_d q_{tr}$	$\dfrac{C}{m^2}$	$= \dfrac{g^{1/2}}{m^2 cm^{1/2} s}$		$= \dfrac{statvolt}{cm}$
Polarisation	$\boldsymbol{P}$	$\oiint \boldsymbol{P} \cdot d\boldsymbol{f} = q_{pol}$	$\dfrac{C}{m^2}$		$\dfrac{g^{1/2}}{cm^{1/2} s}$	$= \dfrac{statvolt}{cm}$
Energy density	w_{el}	$= \dfrac{1}{8\pi k_d} \boldsymbol{E} \cdot \boldsymbol{D}$	$\dfrac{N\, m}{m^3} = \dfrac{J}{m^3}$		$\dfrac{dyn\, cm}{cm^3}$	$= \dfrac{erg}{cm^3}$
Current	i	$= \dfrac{dq}{dt}$	$\dfrac{C}{s}$	$=$ A (Ampère)	$\dfrac{g^{1/2} cm^{3/2}}{s^2}$	$=$ statamp
Current density	$\boldsymbol{j}$	$i = \iint \boldsymbol{j} \cdot d\boldsymbol{f}$	$\dfrac{C}{m^2 s} = \dfrac{A}{m^2}$		$\dfrac{g^{1/2}}{cm^{1/2} s^2}$	$= \dfrac{statamp}{cm^2}$

Table B.4 Systems of units: definition and dimension of physical quantities I

Name		Definition	SI	CGS
Magnet. induction	B	$\int B \cdot ds = 4\pi\, i\, k_m$	$\dfrac{\text{kg}}{\text{C s}} = \dfrac{\text{V s}}{\text{m}^2} = \text{T (Tesla)}$	$\dfrac{\text{g}^{1/2}}{\text{cm}^{1/2}\text{s}} = \text{G (Gauss)}$
Magn. moment	m	$= \dfrac{F \cdot r}{B}$	$\dfrac{\text{C m}^2}{\text{s}} = \text{A m}^2$	$\dfrac{\text{g}^{1/2}\text{cm}^{5/2}}{\text{s}} = \text{G cm}^3 = \text{emu}$
Vector potential	A	$B = \nabla \times A$	$\dfrac{\text{kg m}}{\text{C s}} = \text{T m}$	$\dfrac{\text{g}^{1/2}\text{cm}^{1/2}}{\text{s}} = \text{G cm}$
Magn. flux	Φ_m	$= \iint B \cdot df$	$\dfrac{\text{kg m}^2}{\text{C s}} = \text{V s} = \text{Wb (Weber)}$	$\dfrac{\text{g}^{1/2}\text{cm}^{3/2}}{\text{s}} = \text{G cm}^2 = \text{Mx (Maxwell)}$
Magnetisation	M	$j_M = \dfrac{1}{k_f}(\nabla \times M)$	$\dfrac{\text{C}}{\text{m s}} = \dfrac{\text{A}}{\text{m}}$	$\dfrac{\text{g}^{1/2}}{\text{cm}^{1/2}\text{s}} = \text{Oe (Ørstedt)}$
Magn. field strength	H	$\text{rot } H = 4\pi k_h j_w$	$\dfrac{\text{C}}{\text{m s}} = \dfrac{\text{A}}{\text{m}}$	$\dfrac{\text{g}^{1/2}}{\text{cm}^{1/2}\text{s}} = \text{Oe}$
Induction coefficient	L	$U = -L\dfrac{di}{dt}$	$\dfrac{\text{kg m}^2}{\text{C}^2} = \dfrac{\text{V s}}{\text{A}} = \text{H (Henry)}$	$\dfrac{\text{s}^2}{\text{cm}} = \dfrac{\text{statvolt s}}{\text{statamp}}$
Magn. energy density	w_m	$w_m = \dfrac{k_f}{8\pi k_h} B \cdot H$	$\dfrac{\text{kg}}{\text{m s}^2} = \dfrac{\text{J}}{\text{m}^3}$	$\dfrac{\text{g}}{\text{cm s}^2} = \dfrac{\text{erg}}{\text{cm}^3}$
Poynting vektor	S	$S = \dfrac{1}{4\pi k_h}(E \times H)$	$\dfrac{\text{kg}}{\text{s}^3} = \dfrac{\text{J}}{\text{m}^2\text{s}}$	$\dfrac{\text{g}}{\text{s}^3} = \dfrac{\text{erg}}{\text{cm}^2\text{s}}$
Resistance	R	$U = R \cdot i$	$\dfrac{\text{kg m}^2}{\text{s C}^2} = \dfrac{\text{V}}{\text{A}} = \Omega \text{ (Ohm)}$	$\dfrac{\text{s}}{\text{cm}} = \dfrac{\text{statvolt}}{\text{statamp}}$
Power	P	$P = UI$	$\dfrac{\text{m}^2\text{kg}}{\text{s}^3} = \dfrac{\text{J}}{\text{s}} = \text{W (Watt)}$	$\dfrac{\text{cm}^2\text{g}}{\text{s}^3}$

B Systems of Units in Electrodynamics

Table B.5 Conversion factors I

Name		SI	= factor · CGS
Charge	q	1 C	$\dfrac{\tilde{c}}{10}$ · statcoul
E-field	E	$1\,\dfrac{N}{C} = 1\,\dfrac{V}{m}$	$\dfrac{10^6}{\tilde{c}}$ · $\dfrac{\text{statvolt}}{\text{cm}}$
Electr. flux	Φ_e	1 V m	$\dfrac{\tilde{c}}{10}$ · statcoul
Electr. dipole moment	p	1 C m	$10\,\tilde{c}$ · statcoul cm
Potential	V	$1\,\dfrac{Nm}{C} = 1\,V$ (Volt)	$\dfrac{10^8}{\tilde{c}}$ · statvolt
Voltage	U	1 V	$\dfrac{10^8}{\tilde{c}}$ · statvolt
Capacity	C	$1\,\dfrac{C}{V} = 1\,F$ (Farad)	$\dfrac{\tilde{c}^2}{10^9}$ · cm
Dielectric displacement	D	$1\,\dfrac{C}{m^2}$	$\dfrac{4\pi\,\tilde{c}}{10^5}$ · $\dfrac{\text{statvolt}}{\text{cm}}$
Polarisation	P	$1\,\dfrac{C}{m^2}$	$\dfrac{\tilde{c}}{10^5}$ · $\dfrac{\text{statvolt}}{\text{cm}}$
Energy density	w_{el}	$1\,\dfrac{Nm}{m^3} = 1\,\dfrac{J}{m^3}$	10 · $\dfrac{\text{erg}}{\text{cm}^3}$
Current	i	$1\,\dfrac{C}{s} = 1\,A$ (Ampère)	$\dfrac{\tilde{c}}{10}$ · statamp
Current density	j	$1\,\dfrac{C}{m^2 s} = 1\,\dfrac{A}{m^2}$	$\dfrac{\tilde{c}}{10^5}$ · $\dfrac{\text{statamp}}{\text{cm}^2}$

$\tilde{c}$ is the numerical value of the velocity of light in the CGS-system
$\tilde{c} = 2.997925 \cdot 10^{10}$

Table B.6 Conversion factors II

Name		SI	= Factor · CGS
Magnetic induction	B	$1 \frac{\text{kg}}{\text{C s}} = 1\,\text{T}$	$10^4 \cdot \text{G}$
Magn. moment	m	$1\,\text{A m}^2$	$10^3 \cdot \text{G cm}^3 = \text{emu}$
Vector potential	A	$1 \frac{\text{kg m}}{\text{C s}} = 1\,\text{T m}$	$10^6 \cdot \text{G cm}$
Magn. flux	Φ_m	$1\,\text{V s} = 1\,\text{Wb}$	$10^8 \cdot \text{G cm}^2 = \text{Mx}$
Magnetisation	M	$1 \frac{\text{A}}{\text{m}}$	$\frac{\tilde{c}}{10^3} \cdot \text{Oe}$
Magn. field strength	H	$1 \frac{\text{A}}{\text{m}}$	$\frac{4\pi}{10^3} \cdot \text{Oe}$
Induction coefficient	L	$1 \frac{\text{V s}}{\text{A}} = 1\,\text{H}$	$\frac{10^9}{\tilde{c}^2} \cdot \frac{\text{statvolt s}}{\text{statamp}}$
Magn. energy density	w_m	$1 \frac{\text{J}}{\text{m}^3}$	$10 \cdot \frac{\text{erg}}{\text{cm}^3}$
Poynting vector	S	$1 \frac{\text{J}}{\text{m}^2 \text{s}}$	$10^3 \cdot \frac{\text{erg}}{\text{cm}^2 \text{s}}$
Resistance	R	$1 \frac{\text{V}}{\text{A}} = 1\,\Omega$	$\frac{10^9}{\tilde{c}^2} \cdot \frac{\text{statvolt}}{\text{statamp}}$
Power	P	$1 \frac{\text{J}}{\text{s}} = 1\,\text{W}$	$10^7 \cdot \frac{\text{erg}}{\text{s}}$

$\tilde{c}$ is the numerical value of velocity of light in the CGS-system
$\tilde{c} = 2.997925 \cdot 10^{10}$

Additional Mathematical Topics

C.1 Equations of Vector Analysis

In this collection of formulae one finds some multiple products with vectors (Chap. C.2), a set of formulae for the application of the ∇-operator on products of scalar and vector functions (in formal and explicit notation (Chap. C.3)), expressions for the multiple applications of the ∇-operator (Chap. C.4) and the four differential operators ∇, $\nabla\cdot$, $\nabla\times$ and Δ in spherical and cylinder coordinates (Chap. C.5). In this chapter one also finds the derivation of the formulae quoted for the two sets of coordinates.

The notation is

$\qquad a, b, \ldots \qquad$ vectors

$\qquad \varphi(r) \qquad$ scalar function

$\qquad A(r), B(r) \qquad$ vector functions.

C.2 Multiple Products of Vectors

$$a \cdot (b \times c) = b \cdot (c \times a) = c \cdot (a \times b)$$

$$a \times (b \times c) = (a \cdot c)\, b - (a \cdot b)\, c$$

$$(a \times b) \cdot (c \times d) = (a \cdot c)(b \cdot d) - (a \cdot d)(b \cdot c)$$

© The Author(s), under exclusive license to Springer-Verlag GmbH, DE, part of Springer Nature 2025
R. M. Dreizler, C. S. Lüdde, *Electrodynamics and Special Theory of Relativity*, https://doi.org/10.1007/978-3-662-69942-3

C.3 Product Rules for the Application of the ∇-Operator

$$\nabla \cdot (\varphi A) = \varphi(\nabla \cdot A) + \nabla\varphi \cdot A$$

$$\text{div}(\varphi A) = \varphi \,\text{div}\, A + \text{grad}\,\varphi \cdot A$$

$$\nabla \times (\varphi A) = \varphi(\nabla \times A) + \nabla\varphi \times A$$

$$\text{rot}(\varphi A) = \varphi \,\text{rot}\, A + \text{grad}\,\varphi \times A$$

$$\nabla \cdot (A \times B) = B \cdot (\nabla \times A) - A \cdot (\nabla \times B)$$

$$\text{div}(A \times B) = B \cdot \text{rot}\, A - A \cdot \text{rot}\, B$$

$$\nabla \times (A \times B) = A(\nabla \cdot B) - B(\nabla \cdot A) + (B \cdot \nabla)A - (A \cdot \nabla)B$$

$$\text{rot}(A \times B) = A(\text{div}\, B) - B(\text{div}\, A) + (B \cdot \mathbf{grad})A - (A \cdot \mathbf{grad})B$$

C.4 Double Application of ∇

$$\nabla \cdot (\nabla \times A) = \text{div}(\text{rot}\, A) = 0$$

$$\nabla \times (\nabla\varphi) = \text{rot}(\text{grad}\,\varphi) = 0$$

$$\nabla \times (\nabla \times A) = \nabla(\nabla \cdot A) - \Delta A$$

$$\text{rot}(\text{rot}\, A) = \text{grad}(\text{div}\, A) - (\text{div}\,\text{grad})A$$

(The last formula is only valid for a representation of A in Cartesian coordinates!)

C Additional Mathematical Topics

C.5 Differential Operators in Spherical and Cylinder Coordinates

C.5.1 Spherical Coordinates

$$\nabla V = \frac{\partial V}{\partial r} e_r + \frac{1}{r} \frac{\partial V}{\partial \theta} e_\theta + \frac{1}{r \sin \theta} \frac{\partial V}{\partial \varphi} e_\varphi$$

$$\nabla \cdot A = \frac{1}{r^2} \frac{\partial}{\partial r} \left(r^2 A_r \right) + \frac{1}{r \sin \theta} \frac{\partial}{\partial \theta} (\sin \theta A_\theta) + \frac{1}{r \sin \theta} \frac{\partial A_\varphi}{\partial \varphi}$$

$$\nabla \times A = \frac{1}{r \sin \theta} \left[\frac{\partial}{\partial \theta} (\sin \theta A_\varphi) - \frac{\partial A_\theta}{\partial \varphi} \right] e_r$$
$$+ \left[\frac{1}{r \sin \theta} \frac{\partial A_r}{\partial \varphi} - \frac{1}{r} \frac{\partial}{\partial r} (r A_\varphi) \right] e_\theta + \frac{1}{r} \left[\frac{\partial}{\partial r} (r A_\theta) - \frac{\partial A_r}{\partial \theta} \right] e_\varphi$$

$$\Delta V = \frac{1}{r^2} \frac{\partial}{\partial r} \left(r^2 \frac{\partial V}{\partial r} \right) + \frac{1}{r^2 \sin \theta} \frac{\partial}{\partial \theta} \left(\sin \theta \frac{\partial V}{\partial \theta} \right) + \frac{1}{r^2 \sin^2 \theta} \frac{\partial^2 V}{\partial \varphi^2}$$

C.5.2 Cylinder Coordinates

$$\nabla V = \frac{\partial V}{\partial \rho} e_\rho + \frac{1}{\rho} \frac{\partial V}{\partial \varphi} e_\varphi + \frac{\partial V}{\partial z} e_z$$

$$\nabla \cdot A = \frac{1}{\rho} \frac{\partial}{\partial \rho} (\rho A_\rho) + \frac{1}{\rho} \frac{\partial A_\varphi}{\partial \varphi} + \frac{\partial A_z}{\partial z}$$

$$\nabla \times A = \left(\frac{1}{\rho} \frac{\partial A_z}{\partial \varphi} - \frac{\partial A_\varphi}{\partial z} \right) e_\rho + \left(\frac{\partial A_\rho}{\partial z} - \frac{\partial A_z}{\partial \rho} \right) e_\varphi$$
$$+ \frac{1}{\rho} \left(\frac{\partial}{\partial \rho} (\rho A_\varphi) - \frac{\partial A_\rho}{\partial \varphi} \right) e_z$$

$$\Delta V = \frac{1}{\rho} \frac{\partial}{\partial \rho} \left(\rho \frac{\partial V}{\partial \rho} \right) + \frac{1}{\rho^2} \frac{\partial^2 V}{\partial \varphi^2} + \frac{\partial^2 V}{\partial z^2}$$

C.6 Angular Functions

This section of the appendix contains often used formulae with the Legendre polynomials, the Legendre functions and the spherical harmonics.

C.6.1 Legendre Polynomials $P_l(x)$

- Differential equation:

$$(1-x^2)\frac{d^2 S_l(x)}{dx^2} - 2x\frac{dS_l(x)}{dx} + l(l+1)S_l(x) = 0$$

$$x = \cos\theta \qquad l = 0, 1, 2, \ldots$$

- Fundamental system:

$$S_l(x) \longrightarrow P_l(x) = \sum_{n=0}^{l/2} a_{2n,l} x^{2n} \quad l = \text{even}$$

$$P_l(x) = \sum_{n=0}^{(l-1)/2} a_{2n+1,l} x^{2n+1} \quad l = \text{odd}$$

$$Q_l(x) = \sum_{n=0}^{\infty} a_{2n+1,l} x^{2n+1} \quad l = \text{even}$$

$$Q_l(x) = \sum_{n=0}^{\infty} a_{2n,l} x^{2n} \quad l = \text{odd}$$

- Recursion formula for the expansion coefficients:

$$a_{n+2,l} = \frac{n(n+1) - l(l+1)}{(n+1)(n+2)} a_{n,l}$$

- Symmetry:

$$P_l(x) = (-1)^l P_l(-x)$$

- Polynomials in lowest order, (normalisation $P_l(1) = 1$):

$$P_0(x) = 1$$
$$P_1(x) = x$$

C Additional Mathematical Topics

$$P_2(x) = \frac{1}{2}\left(3x^2 - 1\right)$$

$$P_3(x) = \frac{1}{2}\left(5x^3 - 3x\right)$$

$$P_4(x) = \frac{1}{8}\left(35x^4 - 30x^2 + 3\right)$$

$$P_5(x) = \frac{1}{8}\left(63x^5 - 70x^3 + 15x\right)$$

- Generating function:

$$\frac{1}{\left[1 - 2hx + h^2\right]^{1/2}} = \sum_{l=0}^{\infty} h^l P_l(x)$$

- Recursion relations, selection:

$$(l+1)P_{l+1}(x) = (2l+1)xP_l(x) - lP_{l-1}(x)$$

$$x\frac{dP_l(x)}{dx} = \frac{dP_{l-1}(x)}{dx} + lP_l(x)$$

$$\frac{dP_{l+1}(x)}{dx} = \frac{dP_{l-1}(x)}{dx} + (2l+1)P_l(x)$$

- The formula of Rodriguez:

$$P_l(x) = \frac{1}{2^l\, l!}\frac{d^l}{dx^l}\left[(x^2-1)^l\right]$$

- Integrals with Legendre polynomials, selection: A basic formula:

$$\int_{-1}^{1} dx\, f(x)\, P_l(x) = \frac{(-1)^l}{2^l\, l!}\int_{-1}^{1} dx\, \frac{d^l f(x)}{dx^l}\, (x^2-1)^l$$

Some special cases:

$$\int_{-1}^{1} dx\, P_l(x)\, P_m(x) = \delta_{l,m}\frac{2}{(2l+1)}$$

$$\int_{-1}^{1} dx\, x^m\, P_l(x) = \begin{cases} 0 & l > m \\ 2\dfrac{m!}{(m-l)!}\dfrac{(m-l-1)!!}{(m+l+1)!!} & (m-l)\text{ even},\ m \geq l \\ 0 & (m-l)\text{ odd} \end{cases}$$

$$\int_0^1 dx\, P_0(x) = 1 \qquad \int_0^1 dx\, P_1(x) = \frac{1}{2}$$

$$\int_0^1 dx\, P_l(x) = \begin{cases} 0 & l > 0,\ \text{even} \\ (-1)^{(l-1)/2}\dfrac{(l-2)!!}{2^{(l+1)/2}((l+1)/2)!} & l > 1,\ \text{odd} \end{cases}$$

- Legendre series:
  $\longrightarrow$ Representation of a function $f(x)$ in the interval $[-1, 1]$

$$f(x) = \sum_{l=0}^{\infty} A_l P_l(x) \qquad A_l = \frac{(2l+1)}{2}\int_{-1}^{1} f(x) P_l(x)\, dx$$

C.6.2 The Functions $Q_l(x)$

- Functions in low order:

$$Q_0(x) = \frac{1}{2}\ln\left(\frac{1+x}{1-x}\right) = \sum_{n=0}^{\infty} \frac{x^{2n+1}}{(2n+1)}$$

$$Q_1(x) = \frac{1}{2}x\ln\left(\frac{1+x}{1-x}\right) - 1$$

- Recursion formula:
  The functions Q_l satisfy the same recursion formulae as the polynomials P_l, e.g.

$$Q_{l+1}(x) = (2l+1)x\, Q_l(x) - l\, Q_{l-1}(x).$$

C.6.3 Associated Legendre Functions

- Differential equation:

$$(1-x^2)\frac{d^2 P_l^m(x)}{dx^2} - 2x\frac{dP_l^m(x)}{dx} + \left(l(l+1) - \frac{m^2}{(1-x^2)}\right)P_l^m(x) = 0$$

- Fundamental system:

$$P_l^m(x) = (-1)^m(1-x^2)^{m/2}\frac{d^m P_l(x)}{dx^m} \qquad m \geq 0$$

$$Q_l^m(x) = (-1)^m(1-x^2)^{m/2}\frac{d^m Q_l(x)}{dx^m} \qquad m \geq 0$$

- Functions in lowest order:

$$P_1^1(x) = -(1-x^2)^{1/2} = -\sin\theta$$

$$P_2^1(x) = -3x(1-x^2)^{1/2} = -\frac{3}{2}\sin 2\theta$$

$$P_2^2(x) = 3(1-x^2) = \frac{3}{2}(1-\cos 2\theta)$$

$$P_3^1(x) = -\frac{3}{2}(5x^2-1)(1-x^2)^{1/2} = -\frac{3}{8}(\sin\theta + 5\cos 3\theta)$$

$$P_3^2(x) = 15x(1-x^2) = \frac{15}{4}(\cos\theta - \cos 3\theta)$$

$$P_3^3(x) = -15(1-x^2)^{3/2} = -\frac{15}{4}(3\sin\theta - \sin 3\theta)$$

In addition one has

$$P_l^0(x) = P_l(x)$$

- Recursion formulae, selection:

$$(l+1)P_{l+1}^m(x) = (2l+1)[x\, P_l^m(x) - m\sqrt{1-x^2}\, P_l^{m-1}(x)] - l P_{l-1}^m(x)$$

$$x P_l^m(x) = P_{l-1}^m(x) - (l-m+1)\sqrt{1-x^2}\, P_l^{m-1}(x)$$

$$(l-m+1)P_{l+1}^m(x) - (2l+1)x\, P_l^m(x) + (l+m)P_{l-1}^m(x) = 0$$

- The formula of Rodriguez:

$$P_l^m(x) = \frac{(-1)^m}{2^l l!} (1-x^2)^{m/2} \frac{d^{l+m}}{dx^{l+m}} \left((x^2-1)^l \right)$$

This formula is only valid for $-l \le m \le l$. The functions with $+|m|$ and $-|m|$ are linearly dependent. The symmetry relation

$$P_l^{-m}(x) = (-1)^m \frac{(l-m)!}{(l+m)!} P_l^m(x) \qquad m > 0$$

is valid.
- An integral with P_l^m:

$$\int_{-1}^{1} dx\, P_l^m(x)\, P_{l'}^m(x) = \frac{(l+m)!}{(l-m)!} \int_{-1}^{1} dx\, P_l(x)\, P_{l'}(x)$$

$$= \delta_{l,l'} \frac{(l+m)!}{(l-m)!} \frac{2}{(2l+1)}$$

Note:

$$\int_{-1}^{1} dx\, P_l^m(x)\, P_{l'}^{m'}(x) \ne \delta_{l,l'}\, \delta_{m,m'}\, I(l,m)$$

C.6.4 Spherical Harmonics

- Definition:

$$Y_{l,m}(\theta,\varphi) \equiv Y_{l,m}(\Omega) = \left[\frac{(2l+1)}{4\pi} \frac{(l-m)!}{(l+m)!} \right]^{1/2} P_l^m(\cos\theta)\, e^{i m \varphi}$$

in the region

$$0 \le \theta \le \pi \qquad 0 \le \varphi \le 2\pi\,.$$

Note: The real form of these functions

$$P_l^m(\cos\theta)\cos m\varphi \quad \text{and} \quad P_l^m(\cos\theta)\cos m\varphi\,, \quad m \ge 0$$

can also be used

C Additional Mathematical Topics

- Special cases:

$$Y_{l,0}(\theta, \varphi) = \left[\frac{(2l+1)}{4\pi}\right]^{1/2} P_l(\cos\theta)$$

$$Y_{l,m}(0, \varphi) = \left[\frac{(2l+1)}{4\pi}\right]^{1/2} \delta_{m,0}$$

- Symmetry relation:

$$Y_{l,-m}(\theta, \varphi) = (-1)^m Y_{l,m}^*(\theta, \varphi)$$

The spherical harmonics

$$Y_{l,m}(\theta, \varphi) \quad \text{and} \quad Y_{l,-m}(\theta, \varphi)$$

are linearly independent!
- Basic integral:

$$\iint d\Omega\, Y_{l,m}^*(\theta, \varphi) Y_{l',m'}(\theta, \varphi)$$

$$= \int_0^{2\pi} d\varphi \int_0^{\pi} \sin\theta\, d\theta\, Y_{l,m}^*(\theta, \varphi) Y_{l',m'}(\theta, \varphi)$$

$$= \delta_{l,l'} \delta_{m,m'}$$

- Differential equation:

$$\frac{1}{\sin\theta} \frac{\partial}{\partial\theta}\left(\sin\theta \frac{\partial Y_{l,m}(\theta, \varphi)}{\partial\theta}\right) + \frac{1}{\sin^2\theta} \frac{\partial^2 Y_{l,m}(\theta, \varphi)}{\partial\varphi^2}$$

$$+ \quad l(l+1) Y_{l,m}(\theta, \varphi) = 0$$

- Functions in low order:

$$Y_{0,0} = \sqrt{\frac{1}{4\pi}}$$

$$Y_{1,-1} = \sqrt{\frac{3}{8\pi}}\sin\theta\, e^{-i\varphi} \qquad Y_{1,0} = \sqrt{\frac{3}{4\pi}}\cos\theta$$

$$Y_{1,1} = -\sqrt{\frac{3}{8\pi}}\sin\theta\, e^{i\varphi} \qquad Y_{2,-2} = \sqrt{\frac{15}{32\pi}}\sin^2\theta\, e^{-2i\varphi}$$

$$Y_{2,-1} = \sqrt{\frac{15}{8\pi}}\sin\theta\cos\theta\, e^{-i\varphi} \qquad Y_{2,0} = \sqrt{\frac{5}{16\pi}}(3\cos^2\theta - 1)$$

$$Y_{2,1} = -\sqrt{\frac{15}{8\pi}}\sin\theta\cos\theta\, e^{i\varphi} \qquad Y_{2,2} = \sqrt{\frac{15}{32\pi}}\sin^2\theta\, e^{2i\varphi}$$

- Addition theorem:

$$P_l(\cos\alpha) = \frac{4\pi}{(2l+1)} \sum_{m=-l}^{l} Y_{l,m}(\theta,\varphi) Y^*_{l,m}(\theta',\varphi')$$

$$= \frac{4\pi}{(2l+1)} \sum_{m=-l}^{l} Y^*_{l,m}(\theta,\varphi) Y_{l,m}(\theta',\varphi')$$

with the angle α between the two vectors $\boldsymbol{r}$ and $\boldsymbol{r}'$

$$\cos\alpha = \cos(\varphi - \varphi')\sin\theta\sin\theta' + \cos\theta\cos\theta'.$$

C.7 Radial Functions

The following two functions can represent the radial solution of the Poisson equation, after the substitution $x = $ const. r, as many elementary and not so elementary functions are related to them. The properties of these functions are collected here without the related proofs, which can be found e.g. in M. Abramovitz, I. Stegun: 'Handbook of Mathematical Functions' Appendix A.6.

C.7.1 The Hypergeometric Functions $F(a,b,c;x)$

The solutions of the differential equation

$$x(1-x)\frac{d^2 F(x)}{dx^2} + [c - (a+b+1)x]\frac{dF(x)}{dx} - abF(x) = 0 \qquad \text{(C.1)}$$

C Additional Mathematical Topics

are the **hypergeometric functions**. The solution, which is regular at $x = 0$, is given by the Gauss hypergeometric series

$$F(a, b; c; x) = 1 + \frac{ab}{c}x + \frac{a(a+1)b(b+1)}{c(c+1)}\frac{x^2}{2!} + \ldots$$

$$= \frac{\Gamma(c)}{\Gamma(a)\Gamma(b)} \sum_{0}^{\infty} \frac{\Gamma(a+n)\Gamma(b+n)}{\Gamma(c+n)} \frac{x^n}{n!} . \quad (C.2)$$

The series terminates, if a or b are a negative integer, it is not defined if $c = -n$, except if a or b is a negative integer $-m$ with $m < n$. The radius of convergence of the series is $|x| = 1$ with the explicit statements (a, b, c are assumed to be real)

- $(a + b - c) < 0$: absolute convergence on the complete unit circle.
- $0 \le (a+b-c) < 1$: convergence on the unit circle with the exception of $x = 1$.
- $1 \le (a+b-c)$: divergence of the complete unit circle.

The formulae for the derivatives of the hypergeometric series

$$\frac{dF(a, b; c; x)}{dx} = \frac{ab}{c} F(a+1, b+1; c+1; x)$$

$$\frac{d^2 F(a, b; c; x)}{dx^2} = \frac{a(a+1)b(b+1)}{c(c+1)} F(a+2, b+2; c+2; x)$$

etc.

can be derived directly on the basis of the definition, as well as a substantial number of recursion relations as e.g.

$$cF(a, b; c; x) - (c-b)F(a, b; c+1; x) - bF(a, b+1; c+1; x) = 0$$

$$cF(a, b; c; x) - (c-a)F(a, b; c+1; x) - aF(a+1, b; c+1; x) = 0 .$$

The integral representation

$$F(a, b; c; x) = \frac{\Gamma(c)}{\Gamma(b)\Gamma(c-b)} \int_0^1 dt\, t^{b-1}(1-t)^{c-b-1}(1-xt)^{-a}$$

is useful to derive several formulae for transformations of the function as e.g.

$$F(a, b; c; x) = (1-x)^{-a} F\left(a, c-b; c; \frac{x}{x-1}\right)$$

$$F(a, b; c; x) = (1-x)^{c-a-b} F(c-a, c-b; c; x) .$$

The list of functions, which can be represented by the hypergeometric series, contains among others

- the elementary functions

$$\ln(1+x) = xF(1, 1; 2; -x)$$

$$\arcsin x = xF\left(\frac{1}{2}, \frac{1}{2}; \frac{3}{2}; x^2\right),$$

- the complete elliptic integrals

$$K(k) = \int_0^{\pi/2} d\varphi \, (1 - k^2 \sin^2 \varphi)^{-1/2} = \frac{\pi}{2} F\left(\frac{1}{2}, \frac{1}{2}; 1; k^2\right)$$

$$E(k) = \int_0^{\pi/2} d\varphi \, (1 - k^2 \sin^2 \varphi)^{1/2} = \frac{\pi}{2} F\left(-\frac{1}{2}, \frac{1}{2}; 1; k^2\right)$$

and the Legendre polynomials

$$P_l(x) = F(-l, l+1; 1; \frac{(1-x)}{2}).$$

C.7.2 The Confluent Hypergeometric Functions $F(a, c; x)$

The differential equation of the **confluent hypergeometric functions**

$$x\frac{d^2 F(x)}{dx^2} + [c - x]\frac{dF(x)}{dx} - aF(x) = 0 \qquad (C.3)$$

has a regular singularity at $x = 0$ and an irregular singularity at $x = \infty$. The similarity with the differential equation of the hypergeometric functions shows, that both differential equations are special cases of a higher level differential equation, Riemann's differential equation.

The solution, which is regular at the origin $x = 0$

$$F(a, c; x) = 1 + \frac{a}{c}x + \frac{a(a+1)}{c(c+1)}\frac{x^2}{2!} + \ldots \qquad (C.4)$$

$$= \frac{\Gamma(c)}{\Gamma(a)} \sum_0^\infty \frac{\Gamma(a+n)}{\Gamma(c+n)} \frac{x^n}{n!}$$

is called the **Kummer's function** or confluent hypergeometric series. The behaviour of the convergence depends on the properties of the parameters a and c, that is (m and n are positive integers)

- $c \neq -m$, $a \neq -n$: The series converges for all values of x.
- $c \neq -m$, $a = -n$: The solution is a polynomial of degree m.
- $c = -m$, $a \neq -n$ or $c = -m$, $a \neq -n$ with $n \leq m$: There is a simple pole for the parameter value $c = -m$.

For this function exist formulae for the derivatives, recursion relations and integral representations. Examples are

$$\frac{dF(a, c; x)}{dx} = \frac{a}{c} F(a+1, c+1; x)$$

and

$$cF(a, c; x) - (c-a)F(a, c+1; x) - aF(a+1, c+1; x) = 0,$$

as well as the integral representation

$$F(a, c; x) = \frac{\Gamma(c)}{\Gamma(a)\Gamma(c-a)} \int_0^1 dt\, e^{xt} t^{a-1}(1-t)^{c-b-1}.$$

The asymptotic behaviour of the function is

$$\lim_{x \to +\infty} F(a, c; x) = \frac{\Gamma(c)}{\Gamma(a)} e^x x^{a-c}$$

and

$$\lim_{x \to -\infty} F(a, c; x) = \frac{\Gamma(c)}{\Gamma(c-a)} (-x)^{-a}.$$

The confluent hypergeometric series can also represent a good number of functions, as for instance the exponential function

$$e^x = F(a, a; x),$$

the trigonometric functions as

$$\sin x = x\, e^{-ix} F(1, 2; -2ix)$$

and the Bessel functions

$$J_\nu(x) = \frac{x^\nu e^{-ix}}{2^\nu \Gamma(\nu+1)} F\left(\nu + \frac{1}{2}, 2\nu + 1; 2ix\right)$$

$$j_l(x) = \frac{x^{l-1} e^{-ix}}{2^l \Gamma(l+3/2)} F(l+1, 2l+2; 2ix).$$

C.8 Linear Spaces with Non-Euclidean Metric

Euclidean (and unitarian) vector spaces are characterised by the fact, that the scalar product of vectors, especially of basis vectors, with themselves are positive definite. It is, however, possible, that this is not the case. Scalar product of basis vectors with themselves can be negative or zero

$$a \cdot a \longrightarrow \begin{cases} > 0 \\ = 0 \\ < 0 \end{cases}.$$

It is necessary in such non-Euclidean spaces to distinguish between **covariant** and a **contravariant** decompositions of vectors using corresponding sets of different basis vectors. The basis of the contravariant decomposition in an n-dimensional space is denoted by

$$e_1, e_2, \ldots, e_n.$$

It is characterised by a symmetric matrix,

$$e_\nu \cdot e_\mu = g_{\nu\mu} = g_{\mu\nu}.$$

The decomposition of a vector a in this space has the form

$$a = \sum_{\mu=1}^{n} a^\mu e_\mu, \qquad (C.5)$$

with the contravariant coordinates a^μ.

The basis vectors

$$e^1, e^2, \ldots, e^n$$

with the metric matrix

$$e^v \cdot e^\mu = g^{v\mu} = g^{\mu v}$$

span the covariant decomposition of the vector a

$$a = \sum_{\mu=1}^{n} a_\mu e^\mu . \tag{C.6}$$

The difference of the names of two decompositions are due to the fact, that for the covariant case the coordinates transform in the same way as the basis (thus co-). For a linear transformation with the transformation matrix (Λ), one has to deal with the basis transformation

$$(e')_\mu = \sum_{v=1}^{n} \Lambda^\mu_{\ v} e^v$$

and the (inverse) transformation of the coordinates

$$(a')_\mu = \sum_{v=1}^{n} a_v \Lambda^v_{\ \mu} .$$

The vector a can be reproduced with the transformed decomposition

$$a = \sum_{\mu=1}^{n} (a')_\mu (e')^\mu = \sum_{\mu,v,v'} a_v \Lambda^v_{\ \mu} \Lambda^\mu_{\ v'} (e)^{v'}$$

$$= \sum_{v,v'} a_v \delta^v_{\ v'} (e)^{v'} = \sum_{v=1}^{n} a_v e^v .$$

The different forms of the Kronecker symbol δ, used here, have the standard meaning, independent of the position of the indices

$$\delta^\mu_\kappa = \delta^\kappa_\mu = \delta^{\mu\kappa} \equiv \delta_{\mu\kappa} .$$

In the case of the contravariant decomposition one has a similar chain of relations

$$a = \sum_{\mu=1}^{n} (a')^\mu (e')_\mu = \sum_{\mu,v,v'} a^v \Lambda^\mu_{\ v} \Lambda^v_{\ \mu'} (e)_{\mu'}$$

$$= \sum_{v,v'} a^v \delta^{v'}_{\ v} e_{v'} = \sum_{v=1}^{n} a^v e_v .$$

The sets of basis vectors of the two decompositions are related by the equations

$$e_\nu = \sum_{\mu=1}^n g_{\nu\mu} e^\mu \qquad (C.7)$$

respectively

$$e^\mu = \sum_{\nu=1}^n g^{\mu\nu} e_\nu . \qquad (C.8)$$

The matrix $[g^{\mu\nu}]$ is the inverse matrix with respect to $[g_{\mu\nu}]$, if the statement

$$e^\mu = \sum_{\nu=1}^n g^{\mu\nu} e_\nu = \sum_{\nu,\kappa=1}^n g^{\mu\nu} g_{\nu\kappa} e^\kappa = \sum_{\kappa=1}^n \delta^\mu_\kappa e^\kappa = e^\mu$$

holds.

The consistency of the transformation between the two sets basis vectors is confirmed by the following argument

$$e^{\mu'} \cdot e^\mu = \sum_{\nu,\nu'=1}^n g^{\mu'\nu'} g^{\mu\nu} e_{\nu'} \cdot e_\nu = \sum_{\nu,\nu'=1}^n g^{\mu'\nu'} g^{\mu\nu} g_{\nu'\nu}$$

$$= \sum_\nu g^{\mu\nu} \delta^{\mu'}_\nu = g^{\mu'\mu} .$$

The scalar product of a covariant basis vector with a contravariant basis vector yields

$$e^\mu \cdot e_\nu = \sum_{\kappa=1}^n g^{\mu\kappa} e_\kappa \cdot e_\nu = \sum_{\kappa=1}^n g^{\mu\kappa} g_{\kappa\nu} = \delta^\mu_\nu . \qquad (C.9)$$

The two sets of coefficients representing the same vector are related by the transformation

$$a = \sum_{\mu,\nu=1}^n a^\mu g_{\mu\nu} e^\nu = \sum_{\nu=1}^n a_\nu e^\nu \quad \text{with} \quad a_\nu = \sum_{\mu=1}^n a^\mu g_{\mu\nu} .$$

The scalar product of two vectors can be evaluated either in the contra- or the covariant basis

$$a \cdot b = \sum_{\mu\nu} a^\mu b^\nu e_\mu \cdot e_\nu = \sum_{\mu\nu} a^\mu b^\nu g_{\mu\nu} = \sum_\mu a^\mu b_\mu = \sum_\nu a_\nu b^\nu$$

$$= \sum_{\mu\nu} a_\mu b_\nu e^\mu \cdot e^\nu = \sum_{\mu\nu} a_\mu b_\nu g^{\mu\nu} = \sum_\mu a_\mu b^\mu = \sum_\nu a^\nu b_\nu \, .$$

In either way, one finds (naturally) the same result in the form of a specific **contraction**, which constitutes a sum over all products of covariant and contravariant components with the same index.

A direct example of a space with a metric, which is also called pseudo-Euclidean, is the four-dimensional **Minkowski space**. This space is defined by the specification of the metric matrix

$$g_{\mu\nu} = 0 \quad \text{for} \quad \mu \neq \nu \quad \text{and} \quad g_{00} = -1, \quad g_{11} = g_{22} = g_{33} = 1 \, ,$$

which can, in view of its original definition as a product of two basis vectors also called a metric tensor (of rank 2).[1]

---

[1] The association of the Minkowski coordinates with numbers or the specification of the metric is not handled consistently in the literature. Variants are e.g. the use of the numbers 1 to 4, where the time coordinate is indexed with 4 or the diagonal of the metric tensor with a sequence $1, -1, -1, -1$ instead of $-1, 1, 1, 1$.

Index

A
Aberration, 200, 261, 263
AC generator, 66
Addition theorem
    spherical harmonics, 318
    of velocities, 195, 205
        relativistic, 206
Aether, 198
Ampère
    definition, 299
    extended law, 280
    law, 10, 14
Antenna
    linear, 115
        multipole expansion, 119
        radiation pattern, 117
    thin, centrally fed, 168

B
Babinet's principle, 100
Bessel function
    integral representation, 99
    spherical, 113, 165
Bremsstrahlung, 127
    radiation pattern, 128, 187
Brewster angle, 77

C
Causality, 48, 123
Cerenkov
    cone, 132
    radiation, 130
CGS-system, 15, 297
Coefficient
    mutual induction, 66
    reflection, 78
    self-induction, 66
    transmission, 78
Cone
    hyper light, 219
    light, 219
Confluent hypergeometric functions, 320
Constant
    dielectric, 300
Continuity conditions
    electromagnetic field, 73
Coordinates
    complex, 229
    contravariant, 223
    covariant, 217
    Minkowski, 216, 222, 229
Coulomb
    gauge, 43
Coulomb's law, 14, 252, 280
    Gauss form, 10
Covariance, 245
Crystal optics, 69
Current density
    displacement, 12

D
d'Alembert
    equations, 45
    operator, 45, 246
Dielectric
    constant, 300
Differential equation
    Bessel function
        spherical, 113
    d'Alembert, 45
    Helmholtz, 102
    wave, 16, 45

© The Author(s), under exclusive license to Springer-Verlag GmbH, DE, part of Springer Nature 2025
R. M. Dreizler, C. S. Lüdde, *Electrodynamics and Special Theory of Relativity*,
https://doi.org/10.1007/978-3-662-69942-3

Diffraction, 91
  on a circular opening, 96
  Fraunhofer, 96
  Fraunhofer zone, 92
  Fresnel, 96
  Fresnel zone, 92
  Kirchhoff's integral representation, 95
Dipole
  Hertz, 104
  radiation
    Hertz, 105
Dipole moment
  electric, 105
Dispersion, 32
  relation, 81
  wave packets, 54
Displacement current density, 12

E
Eigentime, 234
Eigenzeit, 234
Einstein's
  sum convention, 229
Electric field
  wave equation, 26
Electromagnetic field
  conservation of momentum, 42
  continuity conditions, 73
  energy conservation, 37
  energy density, 36, 255
  energy-momentum tensor, 255
  field tensor, 248
  Liénard-Wiechert, 124
  Lorentz invariant, 253
  momentum density, 40
Electromagnetic wave, 13, 15
  energy transport, 35
  properties, 52
  wave equation
    one dimensional, 16
    three dimensional, 23
    two dimensional, 21
Elliptic integrals, 320
Energy
  conservation law
    differential form, 37
    integral form, 38
  conservation of, 287
  inner, 239
  kinetic (relativistic), 238
  at rest, 239
Energy density
  electromagnetic, 36

Energy momentum relation (relativistic), 241
Euclidean space, 216, 322
Event, 218
  causally, 219
  vector, 222
Experiment
  Michelson-Morley, 198
  relativistic addition theorem of velocities, 208
  time dilatation, 214

F
Faraday's
  law, 3, 14
Flux
  energy, 77
Form invariant, 196
Formular
  Larmor, 127, 187
Four
  acceleration, 242
  current, 254
  current density, 245
  dimensional space, 216
  divergence, 254
  force, 254
  gradient, 227
  momentum, 236
    electromagnetic, 259
  potential, 246
  velocity, 234
Fraunhofer
  diffraction, 96
  diffraction zone, 92
Fresnel
  diffraction zone, 92
Fresnel's relation, 75
Function
  Bessel
    integral representation, 99
    spherical, 113, 165
  confluent hypergeometric, 320
  Hamiltonian (relativistic), 259
  Hankel
    spherical, 113, 165
  hypergeometric, 318
  Kummer, 321
  Lagrange (relativistic), 256
  Legendre
    associated, 315
  Neumann
    spherical, 165
  spherical harmonics, 316

Index

G
Galilei transformation, 195
Gauge
 Coulomb, 43
 hyperbola, 222
 Lorentz, 44, 246
 transverse, 44
Generalised momentum, 258
Green's
 theorem
  time dependent, 92
Green's function
 retarded, 48
Group velocity, 34

H
Hamiltonian function (relativistic), 259
Hankel
 function
  spherical, 113, 165
Helmholtz equation, 102
Hertz
 dipole, 104
 dipole radiation, 105
Huygens's principle, 91
Hypergeometric function, 318

I
Index of refraction, 32
Induction
 law of, 2, 11
 mutual, 6, 66
 self-, 6
Initial value problem, 20
Integrals
 elliptic, 320
Interferometer, 198

J
Joule's heat term, 36

K
Kirchhoff
 integral representation of diffraction, 95

L
Lagrange equations (relativistic), 256
Lagrangian (relativistic), 256
Larmor formula, 127, 187

Law
 Ampère, 10, 14
  extended, 280
 conservation of energy and momentum, 287
 conservation of momentum, 42
 Coulomb, 14, 252, 280
 energy conservation
  differential form, 37
  integral form, 38
 Faraday, 3, 14
 induction, 2, 11
 of reflection, 71
 of refraction, 71
Legendre
 associated functions, 315
 polynomials, 312
 series, 314
Lenz
 rule of, 3
Levi-Civita symbol
 four-dimensional extension of, 253
Liénard-Wiechert potentials, 121
Light
 cone, 219
  hyper, 219
 line, 218
 velocity, 26, 197
  in a medium, 121
Line
 light, 218
 world, 218
Long wave length approximation, 104
Lorentz contraction, 210
Lorentz gauge, 44, 246
Lorentz transformation, 197, 204
 general, 226
 graphical representation, 219, 230

M
Magnetic field
 wave equation, 26
Mass
 relativistic, 237
 rest, 237
Maxwell's
 equations, 14
  covariant, 249
 relation, 32
 stress tensor, 41, 62, 255
Metal optics, 68
Metric
 Euclidean, 216

pseudo-Euclidean, 217
tensor, 217
Michelson-Morley condition, 203, 223
Michelson-Morley experiment, 198
Minkowski
   coordinates, 216, 222, 229
   diagram, 218
   force, 242
   indices, 229
   space, 217, 325
      complex coordinates, 229
   world, 219
Momentum
   conservation of, 287
   generalised, 258
Multipole expansion
   linear antenna, 119
   radiation, 112
Mutual induction, 6

N
Neumann
   function
      spherical, 165

O
Operator
   d'Alembert, 45, 246

P
Phase velocity, 34
Poincaré transformation, 224
Potential
   electromagnetic, 42
      differential equation, 43
   Liénard-Wiechert, 121
Power (emitted), 107
Poynting vector, 37
Principle
   Babinet's, 100
   Huygens, 91
   of relativity, 197
   superposition, 19
Problem
   initial value, 20
Pseudo-Euclidean space, 217, 325

R
Radiation
   Čerenkov, 130

Radiation pattern
   bremsstrahlung, 128, 187
   Hertz dipole, 105, 108
   linear antenna, 117
   magnetic dipole, 110
   multipole expansion, 112
   quadrupole, 111
   quadrupole spheroidal, 111
Radiation zone, 103
Reflection
   coefficient, 78
   law of, 71
Refraction
   index of, 32
   law of, 71
Relation
   energy momentum (relativistic), 241
   Fresnel's, 75
   Maxwell's, 32
Relativity
   principle of, 197
Rest energy, 239
Rest mass, 237
Retardation, 48

S
Self-induction, 6
SI-system, 15, 297
Space-like, 219
Spherical harmonics, 316
   addition theorem, 318
Sum convention
   of Einstein, 229
Superposition principle, 19
System of units
   CGS, 15, 297
   SI, 15, 297

T
Telegraph equations, 80
Tensor
   Maxwell's stress, 41
   metric, 217
   quadrupole, 109
Theorem
   equivalence of energy and mass, 239
   of relativity, 195
Time dilatation, 213
Time-like, 219
Transformation
   Galilei, 195
   Lorentz, 197, 204

general, 226
graphical representation, 219, 230
Poincaré, 224
Transformer, 66
equation, 68
Transmission coefficient, 78
Twin paradox, 215

V

Velocity
light, 26, 197
in a medium, 121

W

Wave
electromagnetic, 13, 15
energy transport, 35
properties, 27
equation
inhomogeneous, 45
Kirchhoff representation, 50
one dimensional, 16
function, 33
harmonic, 18
length, 17
monochromatic, 18
number, 17
packet, 19, 32
plane
circularly polarised, 30
elliptically polarised, 31
three dimensional, 23
two dimensional, 22
propagation, 68
spherical, 103
TE, 89
TEM, 85
TM, 88
vector, 21
wire, 90
Wave equation
electric field, 26
inhomogeneous, 45
magnetic field, 26
Wave guides
TE-waves, 89
TEM-waves, 85
TM-waves, 88
Wave packet
dispersion, 54
group velocity, 34
phase velocity, 34
Wire waves, 90
World line, 218

The manufacturer's authorised representative in the EU is Springer Nature Customer Service Centre GmbH, Europaplatz 3, 69115 Heidelberg, Germany. If you have any concerns regarding our products, please contact ProductSafety@springernature.com

Printed and bound by CPI Group (UK) Ltd, Croydon, CR0 4YY

25/03/2026

02078191-0016